M. NIELSEN

Event History Analysis

Statistical Theory and Application in the Social Sciences

Hans-Peter Blossfeld
Alfred Hamerle
Karl Ulrich Mayer

Event History Analysis

Statistical Theory and Application in the Social Sciences

LEA LAWRENCE ERLBAUM ASSOCIATES, PUBLISHERS
1989 Hillsdale, New Jersey Hove and London

Lawrence Erlbaum Associates, Inc., Publishers
365 Broadway
Hillsdale, New Jersey 07642

Library of Congress Cataloging-in Publication Data

Blossfeld, Hans-Peter.
Event history analysis/ Hans-Peter Blossfeld, Alfred Hamerle, Karl Ulrich Mayer.
p. cm.
Includes index.
ISBN 0-8058-0126-X
1. Event history analysis. I. Hamerle, Alfred, 1947– . II. Mayer, Karl Ulrich. III. Title.
H61.B49 1989
001.4'3—dc19

88-7073
CIP

Printed in the United States of America
10 9 8 7 6 5 4

Foreword

In the social sciences, especially in economics and sociology, there is an increasing interest in the analysis of event histories. Compared to traditional panel or time-series data, event histories are often better suited to the dynamic nature of empirical phenomena. For each unit of analysis event histories provide information about the exact duration until a state transition as well as the occurrence and sequence of events. Examples of event histories include the survival rates of patients in medical studies; periods of unemployment in economic studies; the "lifetime" of political systems in the field of political science; the time span in which a technical apparatus works without defect in engineering science; required learning time in psychological research; periods of stability in migration and mobility analyses; recidivism in criminological studies; the length of time in which children remain living in their parent's household in youth and family sociological studies, and so on.

The statistical theory and practical examples of event history analysis presented in this book are thus of interest to readers in a large circle of disciplines. However, the examples presented in this book are especially designed for the needs of modern economic and social science research.

The book is written for students and scientists who want to learn how to analyze event history data. It also may be used as a handbook and reference guide for users in practical research. We have tried to present the statistical foundations of event history analysis and we have especially attempted to illustrate the entire research path required in applications of event history analysis: (1) the problems of recording event oriented data; (2) specific questions of data organization; (3) the application of statistical programs; and (4) interpretation of the obtained results.

Compared with other textbooks in this field of applied statistics, it was our special intention in writing this book to provide many examples of studies in which covariates are included in semiparametric and parametric regression models. We have also sought to complement practical examples with concise explanations of the underlying statistical theory. Parameter-free methods of analysis of event history data and the possibilities for their graphical presentation are also discussed in detail. Much space is devoted to the specific problems of multistate and multiepisode models, the introduction of time-depending covariates, and the question of unobserved population heterogeneity. Detailed examples demonstrate how to check the assumptions of the models, how to test hypotheses, and how to choose the right model.

The data used in examples throughout the book are drawn from the German Life History Study (GLHS) conducted by Karl Ulrich Mayer as principal investigator at the Max Planck Institute for Human Development and Education in Berlin. The original data collection was funded by the Deutsche Forschungsgemeinschaft (DFG) within its Sonderforschungsbereich 3 "Micro-analytical Foundations of Social Policy" (Mikroanalytische Grundlagen der Gesellschaftspolitik).

Work on this textbook has been part of the research project "Life Courses and Social Change" (Lebensverläufe und gesellschaftlicher Wandel) at the Max Planck Institute for Human Development and Education in Berlin in close collaboration with the Statistics Department of the University of Constance, FRG. We wish to express a note of thanks to the Max Planck Institute for Human Development and Education for its support in the publication of both the German and English version of this book.

Our special thanks go to Doris Gampig, who in a highly professional manner and with great care and precision prepared the manuscript. Also, we wish to express our gratitude to Ulrich Kuhnert and Dieter Schmidt for the layout and preparing the figures, tables, and program examples. Our appreciation also goes to Gottfried Pfeffer, Peter Wittek, and the copy editors at Lawrence Erlbaum Associates for their editorial assistance. Furthermore, we wish to thank Gerhard Tutz from the University of Regensburg, FRG, for commenting on parts of the manuscript. Bettina Althainz, Peter Baumann, Holger Hainke, and Rolf Hackenbroch provided research assistance in the preparation of the practical examples. Finally, we wish to express our thanks to Trond Petersen (Harvard University) for his kind permission to document his BMDP subprogram P3RFUN.

The English version of this book has been translated by Michael B. Gilroy of the Department of Economics (University of Constance, FRG) and revised by Constance A. Witte (Berlin) and Jacqui Smith (Max Planck Institute for Human Development and Education, Berlin). We are grateful to Frank Schwoerer of Campus Publishing House for granting permission for the English language publication. We are especially indebted to Aage B. Sørensen (Harvard University), David L. Featherman (University of Wisconsin at Madison), and John Nesselroade (Pennsylvania State University), who took the initiative for the English publication. Last but not least we would like to express our appreciation to Aage B. Sørensen (Harvard University) and Michael T. Hannan (Cornell University) for their comments on the statistical parts of this book. Although these friends and colleagues eliminated some of our mistakes, only the authors bear the responsibility for those that remain.

Berlin and Constance, April 1988

Hans-Peter Blossfeld
Alfred Hamerle
Karl Ulrich Mayer

Contents

Chapter 1: Aim and Structure of the Book

This book tries to give a comprehensive overview of the most important methods of event history analysis. By "event history analysis" we mean statistical methods used to analyze time intervals between successive state transitions or events. The number of states occupied by the analyzed units are finite, but the events may occur at any point in time. Consequently, in event history analyses statistical methods for analyzing stochastic processes with discrete states and continuous time are used.

A wide range of statistical tools are available today to analyze event history data as exemplified in a variety of models, approaches, and methods. These statistical methods have, however, not yet found their place in standard statistics textbooks. There are several reasons for this. First, event history analysis applies stochastic models that are not often found in normal applications. Second, incomplete or censored data frequently occur only in very specific research designs. And third, due to the development and application of these methods in various disciplines such as medicine, demography, technology, economics, and the social sciences, the terminology is not uniform and thus the methods are not easily accessible to the user.

Consequently, the aim of this book is to show how modern statistical methods can be used to analyze event history data as well as to give some examples of event history analysis in practical research. To complement a comprehensive presentation of the statistical background we will use examples taken from current sociological research to illustrate the applications of event history analysis.

Following this overview (Chapter 1), Chapter 2 first illustrates three different ways in which in event history analysis problems are conceptualized and solved. Section 2.1 discusses the wide palette of subject areas in which event history analysis may be applied. Section 2.2 reviews the design of the German Life History Study (GLHS) which forms the empirical base of the examples used in Chapters 4 to 6. Finally, the theoretical and methodological advantages of collecting and analyzing event history data as compared to cross-sectional and traditional panel data are discussed in Section 2.3.

The statistical foundations of event history analysis are presented in Chapter 3. In addition to the classification of event history analysis within the structure of stochastic processes, Section 3.1 presents the basic concepts of event history analysis such as the hazard rate, the survivor function, cumulative hazard rates, and so on, as well as nonparametric methods of estimation including the life table method and the Kaplan-Meier estimator (Section 3.2). Of special importance for the textbook is Section 3.3 in which the inclusion of explanatory variables in semiparametric Cox models and parametric models such as the exponential, Weibull, Gompertz-(Makeham), and the log-logistic model are presented. The general theory of multiple state and multiple event cases are then given in Sections 3.4 and 3.5. Section 3.6 follows with the maximum likelihood estimation of unknown model parameters and Section 3.7 discusses methods of constructing hypotheses tests and how to choose models. The inclusion of time-dependent covariates is dealt with in Section 3.8, and models with unobserved population heterogeneity are presented in Section 3.9. Finally, the theoretically oriented Chapter 3 closes with a brief presentation of hazard rate models with discrete time.

Chapters 4 to 6 are specifically designed for the potential users of event history analysis in research. One may, however, also use the material in these chapters as a type of workbook to introduce the empirical analysis of occupational and job trajectories within labor market research. Based on the GLHS, the strategies required for preparing and evaluating event history data are discussed in a stepwise fashion. Using concrete examples, it is then shown how the formulation of a research question may be realized at the methodological and statistical level, which available computer program packages are adequate (SPSS, BMDP, GLIM, RATE, SAS) for specific analytical aims, how the control cards must be structured, and how the results of the analyses are to be interpreted and evaluated.

Chapter 4 looks at aspects of the technical implementation of event history data structures (Section 4.1) and the various ways to present them graphically (Section 4.2). The application of the life table method and the Kaplan-Meier estimator are also discussed (Section 4.3).

Chapter 5 focuses on the application of Cox models and the partial likelihood estimation. After an examination of the proportionality assumption in Section 5.1, the interpretation of the Cox model is discussed in detail in Section 5.2. Model choice with the aid of stepwise regression is demonstrated in Section 5.3. Especially important for the application of event history analysis in economics and the social sciences are those instances in which time-dependent covariates are introduced into a Cox model (Section 5.4) and the practical application of the multiple state cases (Section 5.5).

Chapter 6 is devoted to the application of parametric models. After the graphical examination of the distribution assumptions in Section 6.1, Section 6.2 discusses in detail the exponential model, its interpretation and residual analysis. This is followed by examples of introducing time-dependent covariates with the aid of episode splitting (Section 6.3) and examples of models with periodical durations (Section 6.4). Special duration models are presented in Section 6.5, whereby extensive interpretative examples and residual tests of the Gompertz-(Makeham) (Section 6.5.1), the Weibull (Section 6.5.2), the log-logistic (Section 6.5.3), and the lognormal distributions (Section 6.5.4) are given. Section 6.6 concludes with applications and examples of unobserved population heterogeneity for parametric models.

Chapter 2:
Domains and Rationale for the Application of Event History Analysis

In the fields of economics and the social sciences there are many good reasons for studying the processes and course of development. First of all, an adequate description of reality necessitates the systematic characterization of processes, change, and transitions. Naturally, this proposal is not new. However, interest in characterizing change has increased in a time that is seen as the turning point for many middle and long-term economic and social developments. Recently, it has been recognized that explanations based upon cross-sectional data are appropriate only in the relatively rare cases where there is no change in causal variables (Tuma and Hannan, 1984; Petersen, 1988). In other situations processes of change are best comprehended with the aid of longitudinal data. Furthermore, only those models of processes that capture the right causal mechanisms, and so do more than just account for certain outcomes, should be used as the basis of rational political intervention.

In the past, in the field of economics and the social sciences, the possibility to measure and formalize processes using mathematical models was rather limited. This was due, not only to the lack of available data, but also to the lack of mathematical and statistical methods. The application of differential equations requires continuously measured metric variables over time (Hannan, Blossfeld, and Schömann, 1988). These variables are sometimes available in economics as monetary units, but rarely in other social sciences. Two- and multiple-wave panel studies collect—as we show in Section 2.3—processes over time incompletely, and as a rule are distorted by externally set time points of data collection. On the other hand, time series analysis and the numerous types of econometric models require a large number of points of measurement with constant intervals.

Nowadays event histories are increasingly being collected or made available in which the exact time of transition between states of the unit analyzed are registered. Such data offers information about the exact duration until events and their sequence occur. In addition to these durations or waiting times, variables that individually or in combination influence the timing of an event are of interest. These may be time stable characteristics or attributes that vary over time.

2.1 Application Examples

In the following, some examples are presented that illustrate the specific way in which in event history analysis problems are conceptualized. It should then become quite clear, that event history analysis is amenable to a wide range of questions.

Example 1: Unemployment Studies

In labor market research, event history analysis has been applied to the study of unemployment (Heckman and Borjas, 1980; Flinn and Heckman, 1983; Heckman and Singer, 1982, 1984a; Hamerle, 1988; Sørensen, 1988; Hujer and Schneider, 1988). These studies start from the idea that in analyzing unemployment, cross sections of unemployed or the number of entrants into unemployment in a given period are only partially informative and may even be misleading. Such indicators do not permit differentiation between short and long-term unemployment, and time-dependent covariates may not be included in the analysis.

In unemployment studies, the successive phases of unemployment a worker experiences represent the "duration" variable that is included in event history analyses. Periods of unemployment might be terminated due to various reasons, for example, by beginning a new occupation, through governmental job programs, re-education or re-training, retirement or the recognition of an employment disability. Such different end states may be formulated and examined as "competing risks" or multiple state models.

Example 2: Consumer Behavior Studies

A wide range of applications for event history analysis are to be found in the area of consumer research. Various product brands are offered for sale in a market. Consumers choose and purchase one of the brand names and, at a later point in time, they may either purchase the same brand again or switch to another brand. In this example, an episode or duration is equivalent to the time a consumer sticks to a given product. The states are initiated by the various brand names.

According to the methods presented in this book, the durations of brand loyalty may be related to exogeneous influences, some of which may change over time. Such influence factors include demographic variables (e.g., age, sex, family status, household size), socio-economic characteristics (e.g., income, education, occupation, social status), geographical aspects (metropolitan area, countryside location), or psychological conditions (e.g., personal attitudes, preferences, price awareness, quality awareness, buying habits). Furthermore, the duration a product is purchased, may also depend upon previous experiences with the commodity. Data from the prior history of the consumer process can be included in models and analyses of consumer behavior with the aid of the methods presented in this book.

Example 3: Medical Studies on the Course of Illness

In recent years the methods discussed in this book have been used in analyses of the healing process and survival time in medical and epidemiological studies (see, e.g., Kalbfleisch and Prentice, 1980 and the examples discussed there). Most of these studies deal with one or more absorbing end states. Here, "absorbing" means that once a respective end state has been obtained it is no longer possible to leave it, as, for example, is the case in the death of a patient.

There are few medical multiepisode models in the empirical literature although they are often appropriate. For example, the course of an illness is usually a succession of various stages characterized by events such as remission or death. Hamerle (1985b), for example, studied in female patients, the periods of nonillness following a breast cancer operation. One of the interesting points with regard to this example is the finding that the time period of nonillness appeared to be an especially good predictor of the final survival rate. In this example, separate examinations of the respective phases with single episode models were not adequate descriptors of the problem because they did not take into consideration the inherent dependence of the events and their temporal occurrence. In this book, we suggest methods to deal with special cases like these.

Example 4: Learning Experiments in Psychology and Instruction Research

In the psychology of learning, event history analysis may be used to obtain information regarding the temporal process of learning. The durations being modeled and analyzed here are simply the time spells required for learning some specific fact or task. Here it is possible to observe the speed of learning in relationship to personal and environmental factors.

In practical research on instruction, for example, event history analysis based upon video recordings has been applied to evaluate the concentration spells of pupils. One would ask, for example, whether instructional groups within classes or level of achievement influenced the concentration levels of pupils (Felmlee and Eder, 1983).

Example 5: Insurance and Accident Studies

Application of event history analysis methods may be used to research accidents, especially the conditional risk of an accident occurring based on time dependency as well as numerous other risk factors. For example, the duration might simply be characterized by the length of time a driver has driven without having an accident. Possible explanatory variables could include age, number of miles driven, traffic context, type of automobile, and so on.

Example 6: Studies of Migration

When analyzing residential mobility and migration between regions, event history analysis also proves to be especially useful. The states are represented here by living in a given apartment or house, city or region and the episodes by the respective durations of residence. The rate of migration could be related to various factors and motives such as earning opportunities, the housing market, access to services, tenure status, or the recreational value of a place. Important in this regard, is the question whether migration is influenced by varying resources of persons or through other life events such as job change, marriage or childbirth (Courgeau, 1984, 1985; Mayer and Wagner, 1986; Sandefur and Scott, 1981; Wagner, 1987a, 1987b).

Example 7: Analysis of Family Formation and Fertility

Event history analysis is especially suited to the study of marriage, fertility, and divorce behavior. Although work on population research up till now has mostly used the simple life table method and has mostly not moved beyond aggregate data to individual life histories, a growing number of applications of event history analysis have been published in recent years (Michael and Tuma, 1985; Diekmann, 1987; Huinink, 1987; Papastefanou, 1987; Sørensen and Sørensen, 1986; Wu, 1988). Of interest is how an individual's age or length of prior marriage is related to fertility or divorce, for example, or how the "risks" for marriage, divorce, or pregnancy are distributed over time. For example, the risk of divorce is minimal directly after marriage, increases, and then decreases monotonically after several years of marriage. In all of these examples, the problem arises as to the choice of an appropriate functional relationship path. How such parametric models may be chosen and whether it is better to apply methods that leave the temporal development of risks unspecified is discussed in detail.

The divorce example is also illustrative of the problem of "heterogeneous populations." Consider the case in which, due to religious convictions, certain groups within a population may not face the risk of getting divorced at all (Diekmann and Mitter, 1984).

Example 8: Criminology Studies and Legal Research

In criminology, event history analysis may be used to study the inclination of a criminal recidivism amongst ex-prisoners. Duration here is defined as a spell of time between prison release and commitment of new criminal acts and may be related to certain resocialization and rehabilitation measures or to income and living conditions. Also, the number and duration of previous prison sentences may be included in an explanation (Diekmann, 1980). Another area of application could be legal research. Here, the length of time that passes until the conclusion of civil or criminal court cases may be analyzed dependent on court procedures, characteristics of the judge, or in relation to changes in legal statutes.

Example 9: Organization and Management Research

The survival time of political regimes, of firms, working groups, and similar institutions may also be effectively analyzed using event history analysis. Especially interesting in this new area of research is the so-called "organization ecology" (Carroll and Huo, 1985, 1986; Hannan and Freeman, 1977; Freeman, Carroll, and Hannan, 1983; Carroll, 1984; Carroll and Delacroix, 1982). So far, event history analysis has been used to examine the "births" and "deaths" of newspapers, restaurants, and local worker union organizations of the 19th century.

These illustrative examples of the application of event history analysis in various empirical research areas naturally do not exhaust the potential uses. Many other areas of application can be imagined, especially in technology. In industrial reliability studies (where simpler survival models are well established), the determination of the influence of time-dependent covariates on the life span of a technical apparatus, particularly tests under extreme conditions of stress or general "accelerated life tests" appear very promising. With regard to the study of employment trajectories and occupational careers, we present in Chapters 4 to 6 further examples of practical applications of the method of event history analysis.

2.2 Life Course Studies and the German Life History Study

The recent rapidly increasing demand for longitudinal studies in the field of economics and the social sciences is closely linked to the general rise in interest in the study of the life course.

By the term *life course research* we would like to designate an interdisciplinary paradigm that has been emerging over the last decade or so. Its main objective is the representation of societal processes and the explanation of individual life events and life trajectories within a common formal, conceptual, and empirical frame of reference. The unit of analysis is the individual life course as an institutionalized sequence of activities and events in various life domains.

The observation plan involves mapping the flow of successive cohorts through institutionally defined events (such as leaving home, marriage, birth of children, job entry and exit, or retirement) and states or role incumbencies (such as class membership, marital or employment status, household membership, or schooling).

The life course paradigm is innovative not least in the sense that it is breaking down century-old barriers between scientific disciplines and schools of theory and is transcending long-held distinctions between micro- and macroanalysis. What in the past was separated into the fields of microeconomics, aggregate demography, migration theory, sociology of the family, social mobility, and status attainment research is being brought into a common and therefore—in regard to explanatory claims—competitive framework.

This development is documented by numerous research projects that have led to or are leading to many comprehensive event oriented data sets.

In the Federal Republic of Germany these include:

- the German Life History Study (GLHS) conducted at the Max Planck Institute for Human Development and Education in Berlin (see Mayer et al., 1988);
- the socio-economic panel, carried out within the framework of the DFG-Sonderforschungsbereich 3 and based at the German Institute for Economic Research (DIW), Berlin (see Hanefeld, 1987; Krupp and Hanefeld, 1987);
- the follow-up survey of former Gymnasium (upper secondary school) pupils (Gymnasiasten-Wiederholungsbefragung), conducted by Meulemann and Wiese at the Central Survey Archive in Cologne (see Meulemann et al., 1984);
- and the project "Generative behavior in Nordrhein-Westfalen" (see Strohmeier, Schultz, and Kaufmann, 1985) as well as the project "Labor market dynamics, family development, and generative behavior" conducted at the Institute for Population Research and Social Policy of the University of Bielefeld (see Birg et al., 1985).

Whereas most of these projects are at a relatively early stage (see Mayer and Tuma, 1987, 1988), a number of other similar studies have already been concluded in other countries, especially in connection with American, Norwegian, French, and Israeli life history studies (Featherman and Sørensen, 1983; Matras, 1983; Courgeau, 1984; Michael and Tuma, 1985).

A common approach in all these studies is the examination of educational and occupational histories from a dynamic perspective based on certain historical time periods; the changes observed are not restricted to the field of education and work, but also include other spheres of life (such as social origin, family, spatial mobility, etc.); finally the educational and occupational histories are recorded retrospectively, prospectively, or on the basis of process-induced data.

Similarly, it is the aim of the GLHS (Mayer et al., 1988) to demonstrate and reconstruct the German social history since the end of the Second World War using quantitative life histories. This study also examines the effects of social institutions, especially the educational system, the employment system, and the family on the individual life course. The study attempts to answer the following questions: What do the processes of family formation look like and to what extent have the life courses of women changed? Is the social system in the Federal Republic of Germany characterized by age norm processes and what effect do they have on individual life courses? How has the relationship between the educational and the employment system evolved and what transitional effects can be observed in the choice of careers in different birth cohorts? What quantitative importance do certain migration paths play and what is the degree of spatial mobility within life courses?

Because, however, the main goal of this book is to demonstrate, in a stepwise fashion, how to apply and use event history analysis given an event history database, the examples presented concentrate primarily on life events and trajectories—rather than their macrosocial implications—and specifically on selected aspects of labor market processes.

The GLHS is useful for this purpose since it provides detailed data about the life histories of 2,171 women and men from the birth cohorts 1929–31, 1939–41, and 1949–51, collected in the years 1981–1983. These birth cohorts were chosen so that the respondents' phase of transitions from school to work fell in particularly significant periods in history: for the 1930 cohort, this transition phase lies in the immediate postwar period; those born around 1940 left school in a time of large-scale economic growth, and the cohort 1949–51 entered the labor market during a phase marked by the expansion of the welfare state. The underlying hypothesis is that these specific historic conditions at the point of transition had a substantial impact on the respondents' subsequent careers.

The educational and occupational histories of the GLHS were recorded retrospectively in accordance with the event oriented observation plan. This method is demonstrated by an abstract from the questionnaire where respondents were interviewed about their work careers (Figure 2.1). It is characteristic that apart from collecting theoretically interesting information about the area of employment, number of working hours, income, and so on, the exact beginning and end of each job were recorded on a monthly basis. When this information about the sequence of job episodes is combined with records of periods of training and interruption, the educational and occupational history of an individual can be completely reconstructed. Such an event oriented observation plan provides detailed information on the states of a given respondent's career at any point in the period of observation.

A study conducted prior to the actual drawing of the GLHS sample demonstrated that the reliability of retrospectively recorded data about objective life histories is not systematically affected by a lack of ability to answer questions or deficient memory capacity (Papastefanou, 1980; Tölke, 1980). This study indicated that while the possibility of recall errors was minimal, the form and precision of the survey instrument proved to be of key importance with respect to the quality of the responses. In particular, it was important to divide the life history interview into different spheres of life (education/training, employment, residence, etc.). Lengthy and extensive data editing, data checks, and cross-comparisons also vouched for the quality of the collected information (Brückner et al., 1984). Finally, an examination of the representative quality of the life history data on the basis of census and microcensus surveys shows that the GLHS data provide a reliable picture of sociostructural cross-sections of the past (Blossfeld, 1987a).

Because the data cover not only educational and occupational histories, but also provide information on the whole spectrum of the various spheres of life (i.e., information on social background, family history, the spouse's history, residence history, etc.), it is possible to study the effects of events in

Figure 2.1: Example of an Event Oriented Observation Plan to Record Work Careers

400 Now I want to ask you about your occupation and employment. I shall proceed as I did for the other questions and go through all occupational activities, e.g., including part-time employment or temporary jobs you may have had. Any changes should be recorded as exactly as possible.
INT: If respondent was never employed—go on to Q414, p. 32.

401 Let's begin with your first job.
What occupation did you hold in your first job?
INT: Note exact job title in column 1, continue with Q402

401a What about your next job?
What was your occupation then?
INT: Continue with Q402

402 What was your exact activity at the beginning of this job?
INT: Note below and go on to Q403

403 How did your activity change during this job?—I'm also referring to e.g. changes between full-time and part-time jobs
INT: Let respondent describe the activities and note them down. For each activity go to the next box below. When all activities per page are filled in, go on to Q404

404 In what month and year did the job begin and in what month and year did the job end?

405 *INT: 1st job: Q405a, all subsequent jobs: Q405b*

405a Was this job in the firm in which you did your apprenticeship/vocational training?
INT: Only ask for 1st job

405b Was this the same firm/place of employment as your previous job?

Occupation	Activity at the beginning and changes of activity		M.	Y.	Training establishment
(KA 3)		fr.			yes......... 1
		to			no 2
(KA 4)		fr.			same firm 1
		to			other firm 2
(KA 5)		fr.			same firm 1
		to			other firm 2
(KA 6)		fr.			same firm 1
		to			other firm 2
(KA 7)		fr.			same firm 1
		to			other firm 2
(KA 8)		fr.			same firm 1
		to			other firm 2
(KA 9)		fr.			same firm 1
		to			other firm 2
(KA 10)		fr.			same firm 1
		to			other firm 2

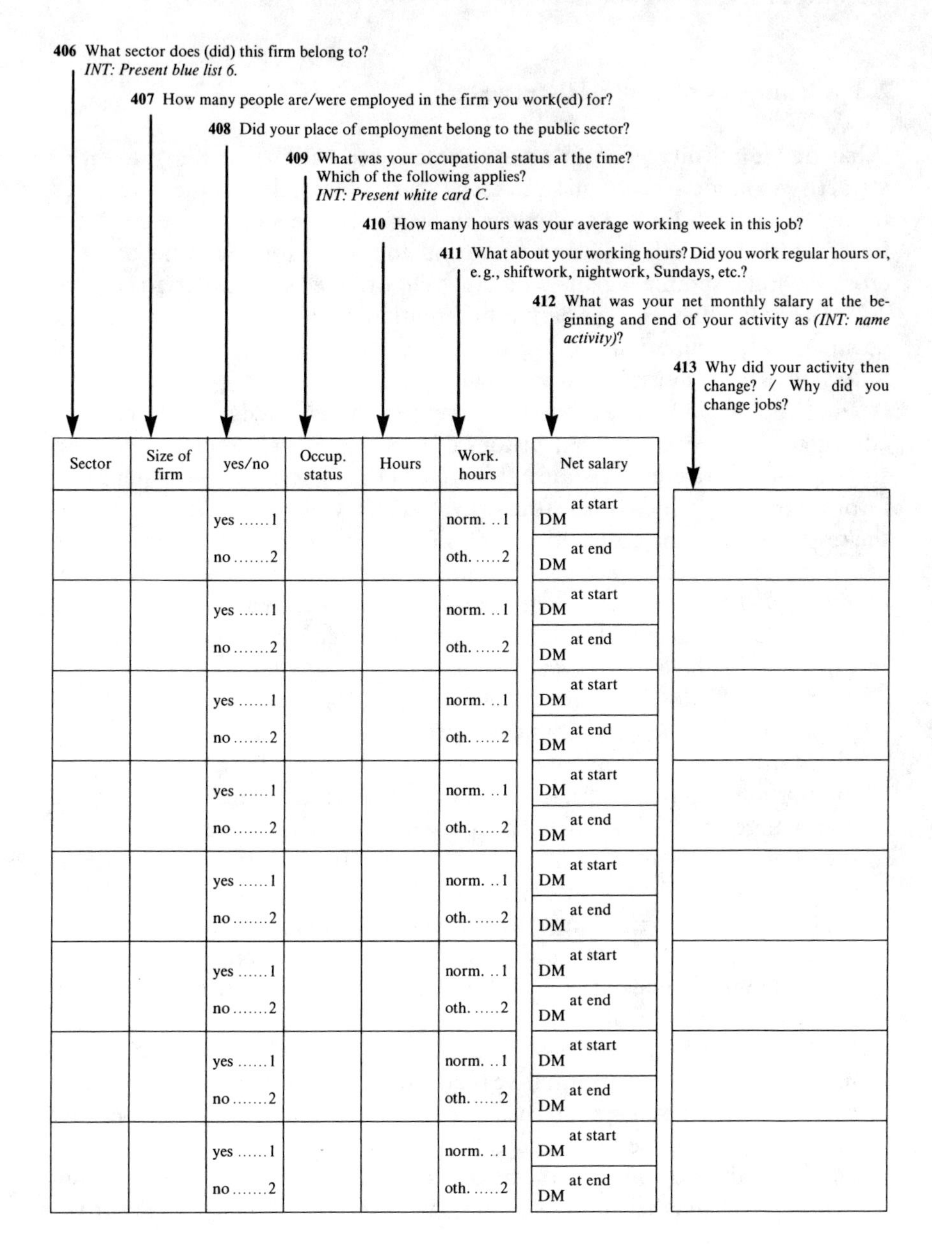

406 What sector does (did) this firm belong to?
INT: Present blue list 6.

407 How many people are/were employed in the firm you work(ed) for?

408 Did your place of employment belong to the public sector?

409 What was your occupational status at the time? Which of the following applies?
INT: Present white card C.

410 How many hours was your average working week in this job?

411 What about your working hours? Did you work regular hours or, e.g., shiftwork, nightwork, Sundays, etc.?

412 What was your net monthly salary at the beginning and end of your activity as *(INT: name activity)*?

413 Why did your activity then change? / Why did you change jobs?

Sector	Size of firm	yes/no	Occup. status	Hours	Work. hours	Net salary	
		yes 1			norm. ..1	DM at start	
		no 2			oth. 2	DM at end	
		yes 1			norm. ..1	DM at start	
		no 2			oth. 2	DM at end	
		yes 1			norm. ..1	DM at start	
		no 2			oth. 2	DM at end	
		yes 1			norm. ..1	DM at start	
		no 2			oth. 2	DM at end	
		yes 1			norm. ..1	DM at start	
		no 2			oth. 2	DM at end	
		yes 1			norm. ..1	DM at start	
		no 2			oth. 2	DM at end	
		yes 1			norm. ..1	DM at start	
		no 2			oth. 2	DM at end	
		yes 1			norm. ..1	DM at start	
		no 2			oth. 2	DM at end	

other parallel processes (e.g., in the case of family history, the event "marriage") have on occupational careers (e.g., "stability" of occupational trajectories). Similarly, prior history can be analyzed to examine the extent to which the subsequent career has been predetermined and channelled in certain directions. This database thus can be used as a good example for illustrating the various stages and potentials of event history analysis.

2.3 Advantages of Event History Data

What are the advantages of the event oriented observation plan that make it so attractive to modern economic and social science research? In order to answer this question let us look at a simple example in which data has been collected for an individual with regard to education and occupation with the aid of a cross-sectional sample, a panel, and an event oriented sample design (Figure 2.2). The individual's career path is differentiated into seven states (training, occupation 1, occupation 2, occupation 3, occupation 4, unemployment, and illness) which the individual may occupy.

First, looking at Figure 2.2 one observes that in a cross-sectional survey the educational and occupational history of a person is only represented by a single point, that being the state at the time of the interview. Somewhat more information is obtained by the four-wave panel in which the circumstances of the respondent can be observed at four different points in time. However, the career between the four waves of the panel remains unclear. It is only the *event oriented collection design* in which changes in states and their precise times are explored. Such a design allows the educational and occupational career to be reconstructed in detail in its various phases and at any point in time.

This example illustrates the following:

- As a rule, cross-sectional analysis presupposes a steady state (i.e., the distribution at any given point in time is only informative if the underlying process remains relatively stable over time). In cases of major fluctuations and changes, the "snapshot" of a cross-section will not be a good picture of the situation because the analysis will depend upon the specific conditions prevailing at the time of survey. In contrast, panel and event oriented data explicitly take into account change and the dynamics of empirical phenomena. Accordingly, any research survey related to economic and social policy should be backed up by information based on longitudinal data on the level of the units of analysis.
- Even if empirical conditions are predominantly stable, panel and event history data are more informative than cross-sections. Cross-sectional data can be regarded as a special case of panel and event history data because cross-sections can be reconstructed from the latter. Moreover, in cases of empirical application, only the recording of panel or event history data can demonstrate whether stability really exists over time. Finally, unlike cross-

Figure 2.2: Recording of a Person's Educational and Occupational Career on the Basis of a Cross-Sectional Sample, a Panel and an Event Oriented Collection Design

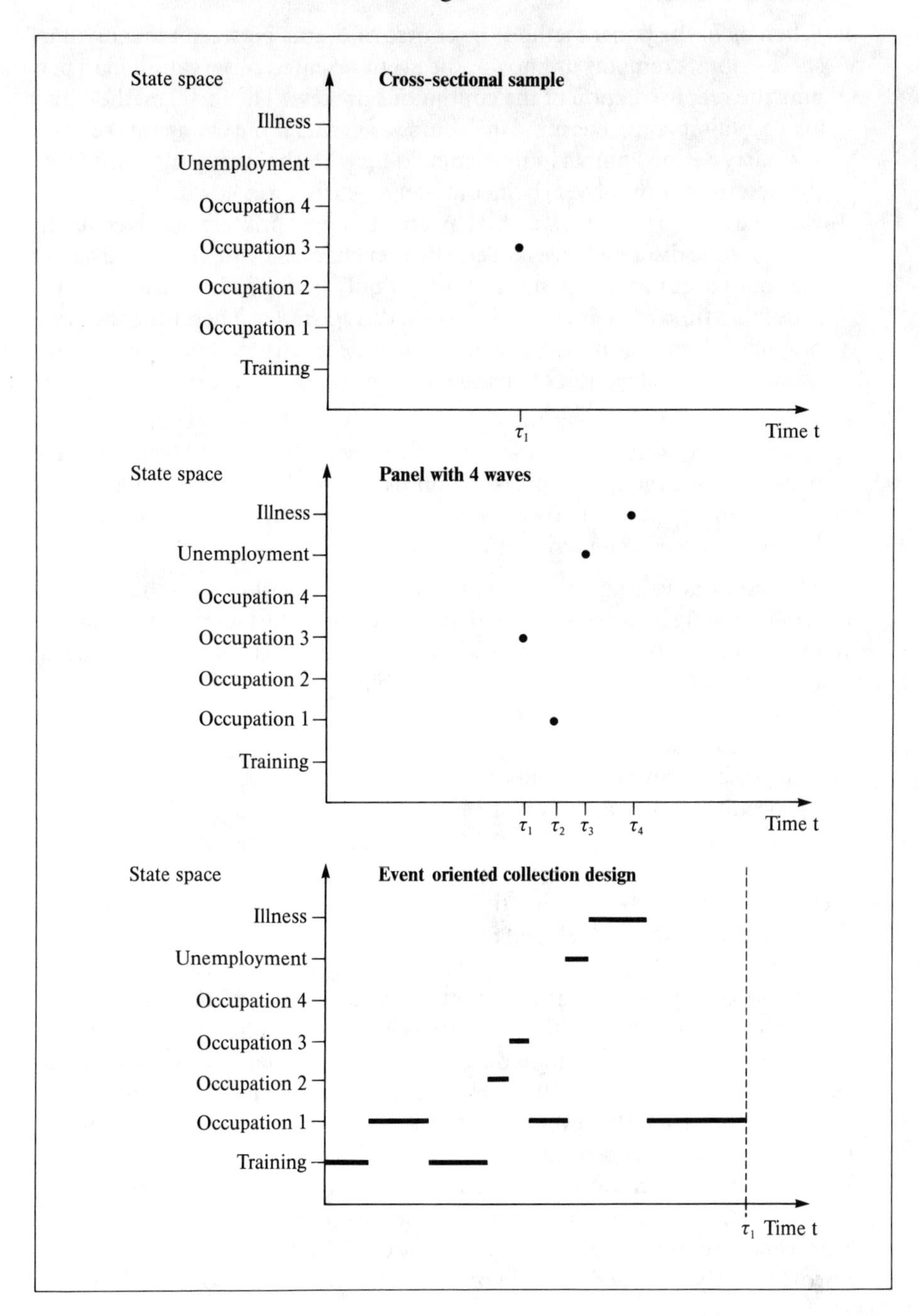

sections, panel and event history data provide information on prior history which can help to improve the explanatory and prognostic capacity of statistical models.

- Whereas in the panel method the course of events between the individual survey points remains unknown, the event oriented observation plan permits the reconstruction of the continuous process. The panel method may also be suitable to determine the course of events if the changes take place at clearly defined points in time coinciding with the survey intervals (e. g., the determination of yearly income on a yearly basis) or if a continuous variable (e. g., a person's weight) can only be appropriately observed on the basis of time discrete surveys. Yet, all other changes in qualitative variables that may occur at any point in time can only be fully reconstructed if the states and time of their changes are exactly registered. Therefore, the event oriented observation plan proves to be a necessary precondition for the adequate reconstruction of change in many fields of research.
- Finally, if one considers the dynamic analysis of *complex feedback processes*, the continuous survey of qualitative variables would seem to be the only adequate method to assess empirical change. This is particularly true if the events of parallel processes occur not only at arbitrary points in time, but also have an interactive effect at a later stage.

The major advantage of an event oriented observation plan that concurs with the growing interest in the analysis of change is the fact that it permits an adequate representation of changes in qualitative variables which may occur at any point in time. The question nevertheless remains: Why has the event oriented observation plan thus far only seldom been used in economic and social science research?

One explanation can certainly be found in the extensive and costly observation procedures necessary to record event histories. One way of doing this is to observe the process and follow the development of the characteristics of individuals with the survey instrument over a lengthy period of time. However, it then often takes a long time before the data is finally available and theory has sometimes developed in a different direction. Event history data is therefore often collected retrospectively. As was also the case in the GLHS, the history of events is thus reconstructed over a long period of time. This type of data collection is sometimes the only way of obtaining event oriented information (e. g., today, the educational and occupational careers of the 1929–31, 1939–41, and 1949–51 birth cohorts can only be recorded on a retrospective basis). However, such data is often criticized as being unreliable, in particular when the events to be recalled took place in the distant past. Retrospective recording of event history data therefore requires a greater degree of care and control and this can generally only be achieved by extensive data checking and time consuming data editing. Moreover, if the data is retrospectively recorded on only one occasion or for only one birth cohort, there is a considerable risk of the database becoming obsolete relatively quickly.

This is why in the case of the socio-economic panel (Hanefeld, 1987), the advantages of the traditional panel are combined with the retrospective recording of event history data. Thus each new panel wave provides not only up-to-date information for the time of survey, but by retrospective questions one also records the most important changes and their precise point of time between these waves (for comparison of panel and retrospective studies see Featherman, 1979-80).

Regardless which of the described procedures to record event history data is selected, it is always an *extensive and costly exercise.* However, this does not seem to be the main reason for the lack of usage of event oriented data. Another reason is certainly the fact that many economic and social science researchers simply do *not know how to use methods of dynamic analysis.* The structure of the data often is regarded as being too complex, the stochastic models that are part of event history analysis are also not well known, and the statistical programs required for samples with censored data are seldom used. This situation is changing rapidly as researchers recognize that in many cases it is unavoidable to base explanatory, causal, and dynamic inferences on event history data and corresponding stochastic models. Thus there exists a potentially strong demand for dynamic analysis of processes and courses in the fields of economics and the social sciences. This growing demand should lead to an increased supply of event history oriented data structures in these fields in the future.

Chapter 3: The Statistical Theory of Event History Analysis

In this chapter we present the statistical fundamentals of event history analysis. After discussing the classification of event history analysis within the framework of stochastic processes in Section 3.1, in Section 3.2 we thoroughly discuss the fundamental concepts of event history analysis. The basic concepts presented are the hazard function (frequently referred to in the literature as the failure rate, the instantaneous death rate, or the force of mortality), the survivor function, the cumulative hazard rate, as well as some important classes of distributions that are relevant for describing the episode (spell, duration, lifetime), which is the period of time between successive events. These concepts are then defined for the one-episode case, although many of these statistical concepts are applicable to more complex situations exhibiting recurrent episodes or competing risks. Finally, this section concludes with a presentation of nonparametric estimation methods such as the life table technique and the Kaplan-Meier estimate (commonly referred to as the product-limit estimate) of the survivor function as well as comparative tests of survivor functions.

Of central importance in this chapter is Section 3.3 in which the inclusion of covariates in the regression approaches is presented. Parametric models, such as the exponential, Weibull, Gompertz or the log-logistic regression models, and the semi-parametric Cox model are discussed in depth. The following sections of this chapter deal with the inclusion of covariates or prognostic factors.

In Sections 3.4 and 3.5 general multistate and multiepisode regression models are analyzed. A general theory for the presentation of the models is developed in which a central concept (element) is the episode and state specific hazard function.

The maximum likelihood estimation of unknown model parameters is the subject of Section 3.6. After a brief introduction on the general principles of the maximum likelihood estimation procedure, the censoring problem that may arise when analyzing event history data is dealt with. This is then followed by application of the maximum likelihood estimation method. For the Cox model, a modified approach is necessary.

In Section 3.7 we present methods of constructing tests of hypotheses and model choice, followed by the possibilities for the inclusion of time-dependent and stochastic covariates as exemplified in Section 3.8. Section 3.9 then discusses various methods, solutions, and potential problems of incorporating unobserved population heterogeneity in the analysis including individual specific disturbance terms.

Finally, we conclude this chapter with a brief presentation of regression methods for discrete hazard functions given "grouped" durations.

3.1 Event History Analysis: A Special Stochastic Process

The basic statistical model of an event history analysis examines the length of time intervals between consecutive changes of state defined by some qualitative variable within some observation period. Events are thus changes in the set of all distinct values that the chosen qualitative variable may take on within the state space. For the observation period, the points of time at which changes of state occur, or equivalently the occurrence of the sequence of events, is given. If the length of the time intervals or, the durations of the episodes can be measured exactly, we have a stochastic process with a continuous time parameter. Time is a continuous variable since changes of state may occur at any point in time. The state variable, on the other hand, possesses only a finite number of values. In the statistical model, points of time at which transitions occur are represented by a series of nonnegative random variables $0 = T_0 \leq T_1 \leq T_2 \leq \ldots$ and the state variable is characterized by the set of random variables with a finite state space $\{Y_k : k = 0, 1, 2, \ldots\}$. The corresponding stochastic process $(Y, T) = \{(Y_k, T_k) : k = 1, 2, \ldots\}$ may be described as

$$Z = \{Z(t) : t \geq 0\}$$

with $\quad Z(t) = Y_{k-1} \text{ for } T_{k-1} \leq t < T_k,\ k = 1, 2, \ldots$

which is a continuous time, discrete state stochastic process.

Although from a theoretical point of view there may be a countable number of states, practical application normally requires only observation of a finite number of states. For example, in examining aspects of unemployment, a division of the state space into "employed," "not in the labor force," or "unemployed" is meaningful (see Figure 3.1). Occasionally, it is a matter of observing the times of events that occur repeatedly. Examples of such occurrences are the intervals of time between successive childbirths in demographic studies or functionability of some technical apparatus until the appearance of some defect in industrial reliability studies. In such cases, the process Y_k is said to be a *degenerate* process, which simply means that the state space consists of only one element. The emphasis of the analysis is then upon the times T_k, k = 1, 2, ..., of the repeated occurrences of the event in question.

In any case, the important theoretical preliminary decision concerning the state space must be made in accordance with the nature of the substantive problem. The choice of the state space significantly affects the statistical model structure as well as the interpretation of the obtained results. With the exception of the above mentioned special case, the term "event" always corresponds to changes in Z(t), that is, with a transition from one state to another.

The term *episode* or *spell* designates the period of time between successive events. Of special interest are the duration intervals

$$V_k = T_k - T_{k-1} \quad , k = 1, 2, \ldots$$

which are commonly referred to as the *waiting times.*

Example:

In research on durations of repeated episodes of unemployment, the states "employed," "not in the labor force," and "unemployed" may be distinguished. The complete history of state occupancies and times of changes, the *sample path* of an individual, is presented in Figure 3.1 below. The horizontal axis in Figure 3.1 represents time and the vertical axis states the person's status at time point t with regard to the three possible states.

Figure 3.1: Hypothetical Employment History of a Person

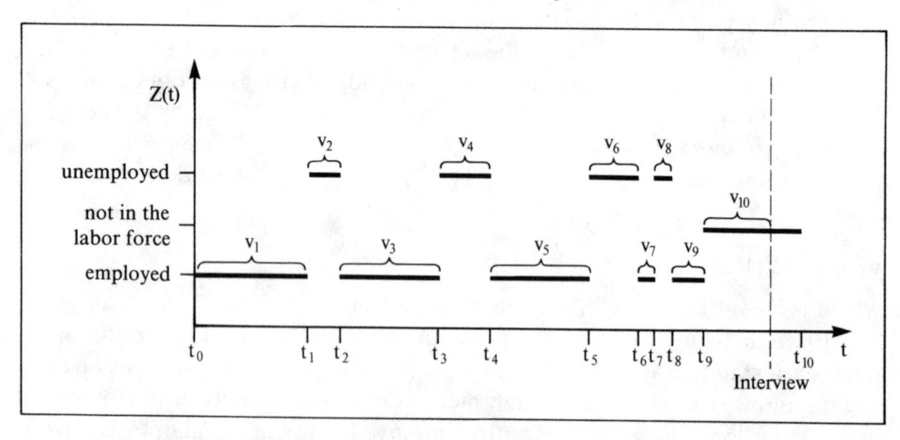

A very important special case that is often relevant in research studies is one in which processes exhibit only *one* episode, *one* initial state, and *one* destination state. Methods of dealing with this sort of data have been developed mainly in "survival analysis" in which research focuses on the distribution of lifetimes. If multiple outcomes on destination states exist, the method of analysis needed may be characterized as a *multistate model* as commonly found in biometric literature under the catchword: competing risk models. Finally, *multiepisode models* are characterized by repeated transitions from one state to another or when a specific event may occur repeatedly.

Since the termination of the entire observation period is, as a rule, exogenous (e. g., due simply to the time point of the retrospective collection of event history data), the endpoint of the last episode of an individual or subject may not be observed. In such situations, event history data are said to be *right censored.* For example, it is conceivable that event history data for individual employment histories might have been collected on the first day of an individual's initial act of unemployment registration and thereafter the event history may be followed over a specific time span until some target date. The length of time between successive episodes in which the observed persons are unemployed will thus be measured. Under such a scenario it is possible that the end of the last unemployment episode in the final target date is not observable. For a detailed description of possible censoring mechanisms, the reader is referred to Kalbfleisch and Prentice (1980, Chapter 5), or Lawless (1982). The statistical modeling of the main censoring mechanisms will be presented below in Section 3.6.2.

Alternatively, there is the possibility that event history data are *left censored,* such that the length of time that an individual or subject has already resided in state y_0 is unknown. Left censoring is more difficult to handle than right censoring, since it is usually not possible to calculate the effects of the unknown event history data upon future events. For sake of simplicity, the following always assumes that either the starting point in time and the original state are given (without loss of generality that $t_0 = 0$), or that the previous history of state occupancies is irrelevant for the future event history process. Since all of the survey samples of the cohorts 1929–31, 1939–41, and 1949–51 of life histories applied for illustrative purposes in Chapters 4–6 include the individual event histories in their entirety, the problem of left censored data does not arise. For the incorporation of left censored observation analyzing hazard rate models, see Hamerle (1988a).

Often, not only the waiting times but also various covariates or prognostic factors that may affect the waiting time singly or in combination are included in the analysis. One important goal of the statistical analysis of event histories is, therefore, the quantitative determination of the impact of such exogenous or endogenous variables with suitable regression models. In such analyses the covariates may be either quantitative or qualitative. Categorical covariates can be coded in the regression model by appropriate dummy variables. This is discussed further below in Section 3.3.1. Some of the covariates may also be time dependent and stochastic.

In the simplest case, a time dependent covariate is a fixed exogenous function of time, such as age. Some covariates may, however, be stochastic processes that are active parallel to the main process being observed. As such, one may be dealing with an external process whose time path is not influenced by the waiting time of an individual. Alternatively, it may also be dependent upon the waiting time of the individual in the observed state.

For an individual's *k*th episode, all information regarding the previous event course history of the process is collected in H_{k-1}. H_{k-1} contains the previous event course history till the time point t_{k-1}, that is,

$$H_{k-1} = \{t_0, y_0, t_1, y_1, \{\mathbf{x}_1(u) : u < t_1\}, \ldots, t_{k-1}, y_{k-1}, \{\mathbf{x}_{k-1}(u) : t_{k-2} \leq u < t_{k-1}\}\}.$$

In the above, individual subscripts have been omitted for the sake of simplicity. In the following sections, the covariates will be assumed to be time invariant. Section 3.8 offers a detailed discussion of potential problems that may arise when time dependent variables are included in the analysis.

A complete *event history* of an individual over some observed time span requires the following information:

1. y_{i0} — initial state;
2. n_i — number of episodes in the observation period;
3. $t_{i1} \leq t_{i2} \leq \ldots \leq t_{in_i}$ — the points in time, at which some state transition has occurred or a specific event has taken place;
4. $y_{i1}, y_{i2}, \ldots, y_{in_i}$ — state occupancies corresponding to the above points of time;
5. δ_i — an indicator that distinguishes whether or not the n_i-th episode is censored or not;
6. $\mathbf{x}_{i1}, \mathbf{x}_{i2}, \ldots, \mathbf{x}_{in_i}$ — covariate vector, to be measured at the beginning of each episode. For the time being, it is assumed that the covariates are time invariant.

3.2 Fundamental Statistical Concepts (Single Spell Model)

The simplest example of an event history analysis to be considered is characterized solely by measurement of the entrance into some initial state until attainment of some final state. Such situations often arise, for example, in research on life expectancy or survival studies in medical tests and also in analysis of the lifetimes of political or social organizations. Many of the concepts developed for the one-episode case may be applied to more complex situations with repeated episodes or competing risks. The duration of an episode is represented in the statistical model by a nonnegative stochastic variable T. If time is exactly measured, the variable T is a continuous stochastic variable. On the other hand, if only time intervals can be stated from which the terminal state may be obtained, T is a discrete stochastic variable. $T = t$ implies therefore that in the *t*th time interval a transition has occurred. In Sections 3.2–3.9, the variable T will be assumed to be continuous. The discrete model will be discussed briefly in Section 3.10.

3.2.1 Density, Distribution, and Survivor Functions, and the Hazard Rate

In this section, important statistical concepts of event history analysis for the one-episode case given initial and final states are introduced. We assume initially that the population under study is homogeneous, that there is no interindividual heterogeneity. Introduction of covariates and prognostic factors will be considered in some detail in Section 3.3.

The density and distribution function of the duration T $(T \geq 0)$ are denoted by $f(t)$ and $F(t)$, respectively. As usual the following relationship is valid, that is,

$$F(t) = P(T \leq t) = \int_0^t f(u)\, du, \tag{3.2.1}$$

and for all points for which $F(t)$ may be differentiated

$$f(t) = F'(t). \tag{3.2.2}$$

The *survivor function*

$$S(t) = P(T \geq t) \tag{3.2.3}$$

expresses the probability that an individual remains in the state ("survives") until time t, that is, that an event has not yet occurred and the episode is still continuing.

Measuring waiting time continuously, we have

$$S(t) = 1 - F(t). \tag{3.2.4}$$

Therefore, the survivor function is a non-increasing function of time (see Figure 3.2), approaching zero as time elapses.

The *hazard rate* (hazard function or failure rate) is defined as

$$\lambda(t) = \lim_{\substack{\Delta t \to 0 \\ \Delta t > 0}} \frac{1}{\Delta t} P(t \leq T < t + \Delta t \mid T \geq t). \tag{3.2.5}$$

The hazard function may be interpreted as the instantaneous probability that episodes in the interval $[t, t + \Delta t]$ are terminating provided that the event has not occurred before the beginning of this interval. Common terminology often found in application of the hazard function are *intensity* or *risk functions, transition* or *mortality rates.*

It is important to note that the values of the hazard function themselves are not (conditional) probabilities. Although they are always nonnegative, they may be greater than one. For a small Δt, $\lambda(t)\Delta t$ can be interpreted as an approximation of the conditional probability $P(t \leq T < t + \Delta t \mid T \geq t)$.

The *cumulative hazard function* is represented by the integral

$$\Lambda(t) = \int_0^t \lambda(u)\, du. \tag{3.2.6}$$

Figure 3.2: Typical Shape of a Survivor Function

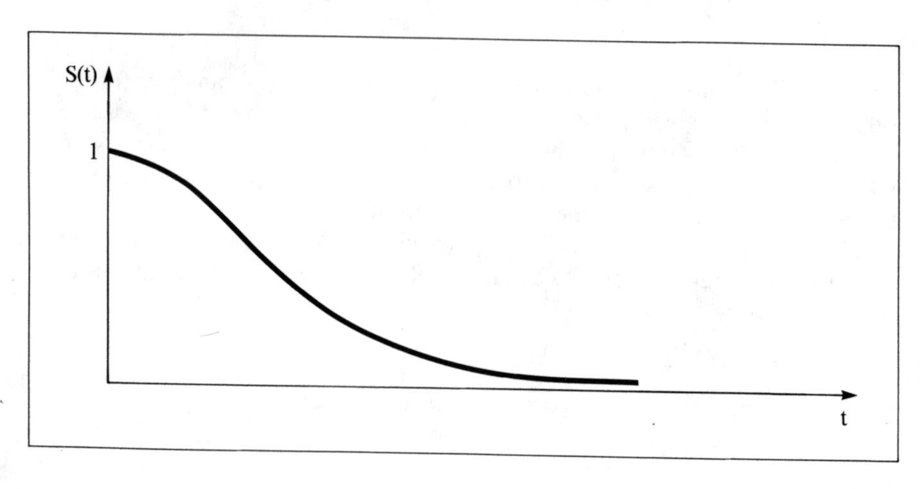

Figure 3.3: Hazard Function for Human Mortality

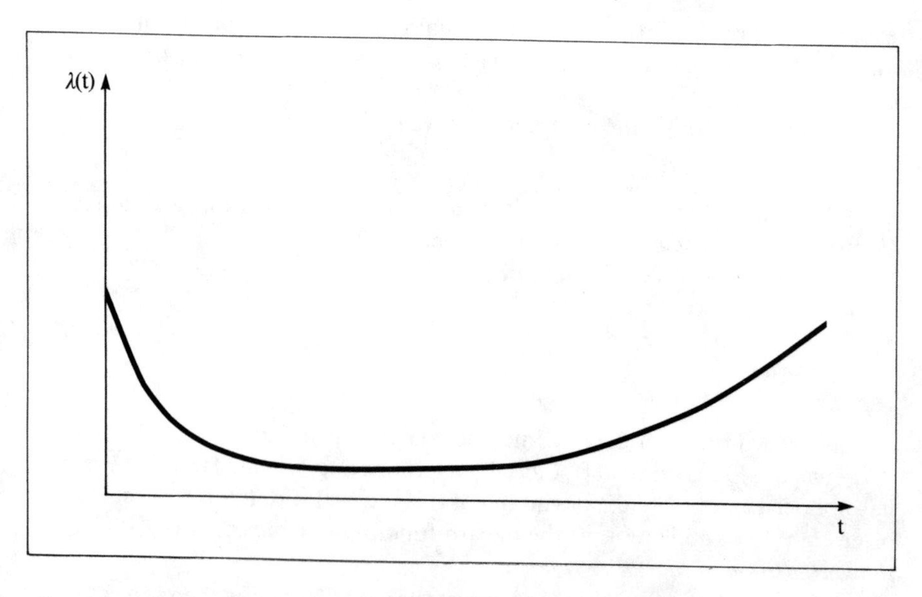

The hazard function is a key concept used in the analysis of event history data. If an individual "survives" to time point t, the hazard function contains information concerning the "future process (course)." Generally, the analyst possesses at least some degree of qualitative initial information concerning the form of the hazard function. This may be exemplified by the mortality rate of a population. The hazard function is characterized by the typical "bathtub-shaped" curve (see Figure 3.3).

At the beginning of the life process, the risk of dying, caused by infant mortality, is relatively high. Then the curve falls and remains constant for some time at a low level until aging causes it to rise gradually again. Similar time profiles of hazard functions for technical appliances have often been observed in industrial reliability studies. Due to "infant illnesses" and "first-trial defects" the failure rate is at first relatively high, then decreases to rise again as the aging process occurs and material endurance limits (fatigue) arise.

Naturally, alternative forms of the hazard function are conceivable, such as strictly increasing or decreasing hazard functions.

From definition (3.2.5), one immediately obtains the relationship between the hazard rate and the survivor function

$$\lambda(t) = \frac{f(t)}{S(t)}, \tag{3.2.7}$$

and since T has been assumed to be continuous we have

$$\lambda(t) = \frac{f(t)}{1 - F(t)}. \tag{3.2.8}$$

Inversely, one may derive the relationship between the survivor function and the hazard function by integration of $\lambda(t)$. From (3.2.7) and (3.2.8), we have

$$\begin{aligned} \int_0^t \lambda(u)\,du &= \int_0^t \frac{f(u)}{1 - F(u)}\,du = -\ln(1 - F(u)) \Big|_0^t \\ &= -\ln(1 - F(t)) = -\ln S(t). \end{aligned} \tag{3.2.9}$$

This leads to the important relationship

$$S(t) = \exp(-\int_0^t \lambda(u)\,du). \tag{3.2.10}$$

The density f(t) is obtained from (3.2.7) and (3.2.10) as a function of the hazard function

$$f(t) = \lambda(t) \cdot S(t) = \lambda(t) \cdot \exp(-\int_0^t \lambda(u)\,du). \tag{3.2.11}$$

Considering the relationships (3.2.1) to (3.2.11) it becomes evident that each of the three quantities f(t), S(t), and $\lambda(t)$ may be used to describe the duration of an episode. If one of these functions is known, the derivation of both the other functions is always possible. In particular, if one knows the hazard rate, the probability law of the process is completely characterized.

Occasionally, the application of a further relationship is helpful. From (3.2.3) and (3.2.11), the survivor function may be derived as

$$S(t) = \int_t^\infty f(u)\,du = \int_t^\infty \lambda(u)\,S(u)\,du. \tag{3.2.12}$$

In (3.2.10), S(t) is expressed in terms of past history until time point t, whereas in (3.2.12), S(t) is expressed in terms of the future.

Finally, one obtains as the (conditional) probability, that in the interval $[t_1, t_2]$ for $t_1 < t_2$ an event occurs given that an event has not yet occurred until t_1:

$$P(t_1 \leq T \leq t_2 | T \geq t_1) = \frac{S(t_1) - S(t_2)}{S(t_1)}. \tag{3.2.13}$$

For a set of successive points of time $t_0 = 0 < t_1 < t_2 < \ldots < t_k$ it follows that

$$\begin{aligned} S(t_k) &= \exp(-\int_0^{t_k} \lambda(u)\, du) = \exp(-\sum_{i=0}^{k-1} \int_{t_i}^{t_{i+1}} \lambda(u)\, du) \\ &= \prod_{i=0}^{k-1} \exp(-\int_{t_i}^{t_{i+1}} \lambda(u)\, du) = \prod_{i=0}^{k-1} \frac{S(t_{i+1})}{S(t_i)}. \end{aligned} \tag{3.2.14}$$

Thus,

$$S(t_k) = \prod_{i=0}^{k-1} P(T \geq t_{i+1} | T \geq t_i). \tag{3.2.15}$$

Relationship (3.2.15) is especially relevant in the construction of life tables, as will be presented below in Section 3.2.3.

3.2.2 Special Probability Distributions for Durations

Exponential Distribution—Time Invariant Hazard Rate

Here we present a short overview of some important distributions for durations. For a more detailed description, see, for example, Kalbfleisch and Prentice (1980, Section 2.2). One of the most commonly applied distributions for the waiting time and lifetime is the exponential distribution. It is characterized by a constant hazard rate

$$\lambda(t) = \lambda, \qquad t \geq 0, \quad \lambda > 0. \tag{3.2.16}$$

Thus, the respective survivor and density functions are (see Figure 3.4)

$$S(t) = \exp(-\lambda t), \tag{3.2.17}$$

$$f(t) = \lambda \exp(-\lambda t). \tag{3.2.18}$$

For the "mean duration" one obtains

$$E(T) = \frac{1}{\lambda}.$$

The greater the "risk" λ, the shorter the expected duration will be. The variance is calculated simply as

$$Var(T) = \frac{1}{\lambda^2}.$$

Figure 3.4: Density Function, Survivor Function, and Hazard Rate of the Exponential Distribution

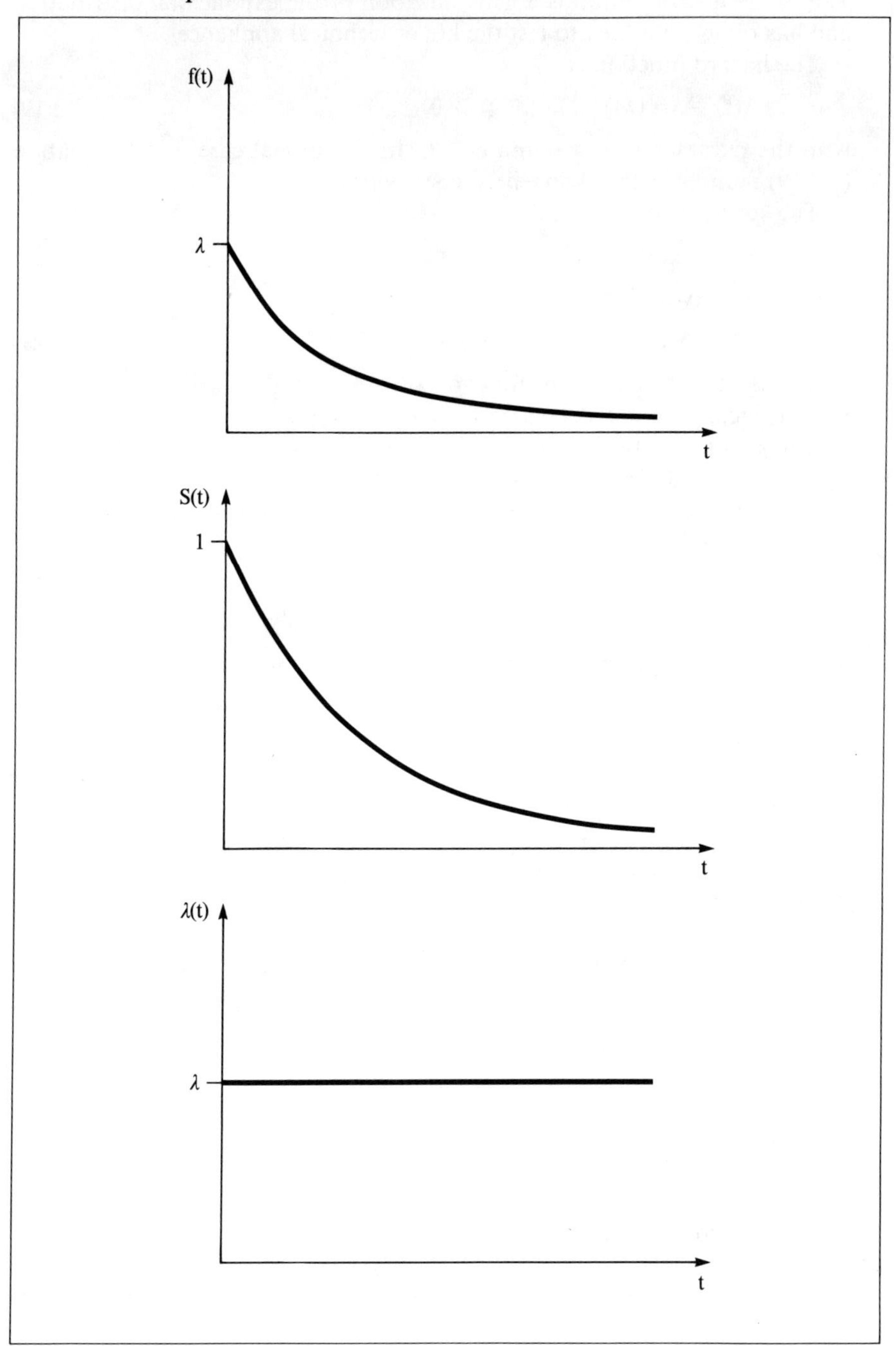

Weibull Distribution

The Weibull distribution is a generalization of the exponential distribution and has often been used to test the life of technical appliances.

The hazard function is

$$\lambda(t) = \lambda\alpha\,(\lambda t)^{\alpha-1} \qquad (t > 0) \tag{3.2.19}$$

with the parameters $\lambda > 0$ and $\alpha > 0$. In the special case $\alpha = 1$, equation (3.2.19) reduces to the exponential distribution.

The survivor function is

$$S(t) = \exp(-\,(\lambda t)^{\alpha}) \tag{3.2.20}$$

and the density of T is

$$f(t) = \lambda\alpha\,(\lambda t)^{\alpha-1}\exp(-\,(\lambda t)^{\alpha}). \tag{3.2.21}$$

The hazard or transition function of the Weibull distribution increases monotonically if $\alpha > 1$, decreases if $\alpha < 1$, and is constant if $\alpha = 1$. The Weibull model is quite flexible and therefore adaptable to a wide variety of models of durations and lifetimes (see Figure 3.5).

The expected value of the waiting time E(T) of the Weibull distribution is derived to be

$$E(T) = \Gamma\left(\frac{1+\alpha}{\alpha}\right) / \lambda, \tag{3.2.22}$$

where $\Gamma(\cdot)$ is the gamma function. The variance is obtained as

$$\mathrm{Var}(T) = \left(\Gamma\left(\frac{\alpha+2}{\alpha}\right) - \left(\Gamma\left(\frac{\alpha+1}{\alpha}\right)\right)^2\right) / \lambda^2.$$

Extreme Value Distribution

The Weibull distribution is closely related to the extreme value distribution. The extreme value distribution possesses the following survivor and density functions respectively,

$$S(y) = \exp\left[-\exp\left(\frac{y-\mu}{\sigma}\right)\right] \qquad -\infty < y < \infty \tag{3.2.23}$$

$$f(y) = \frac{1}{\sigma}\exp\left[\frac{y-\mu}{\sigma} - \exp\left(\frac{y-\mu}{\sigma}\right)\right] \qquad -\infty < y < \infty \tag{3.2.24}$$

and the hazard function is

$$\lambda(y) = \frac{1}{\sigma}\exp\left(\frac{y-\mu}{\sigma}\right) \qquad -\infty < y < \infty. \tag{3.2.25}$$

The relationship between Weibull and extreme value distribution is characterized as follows: If T possesses a Weibull distribution with a density function (3.2.21), then $Y = \ln T$ possesses an extreme value distribution with $\sigma = \alpha^{-1}$ and $\mu = -\ln\lambda$.

Figure 3.5: Density Function, Survivor Function, and Hazard Rate of the Weibull Distribution (given $\alpha = 0.5$, $\alpha = 1$, and $\alpha = 3$)

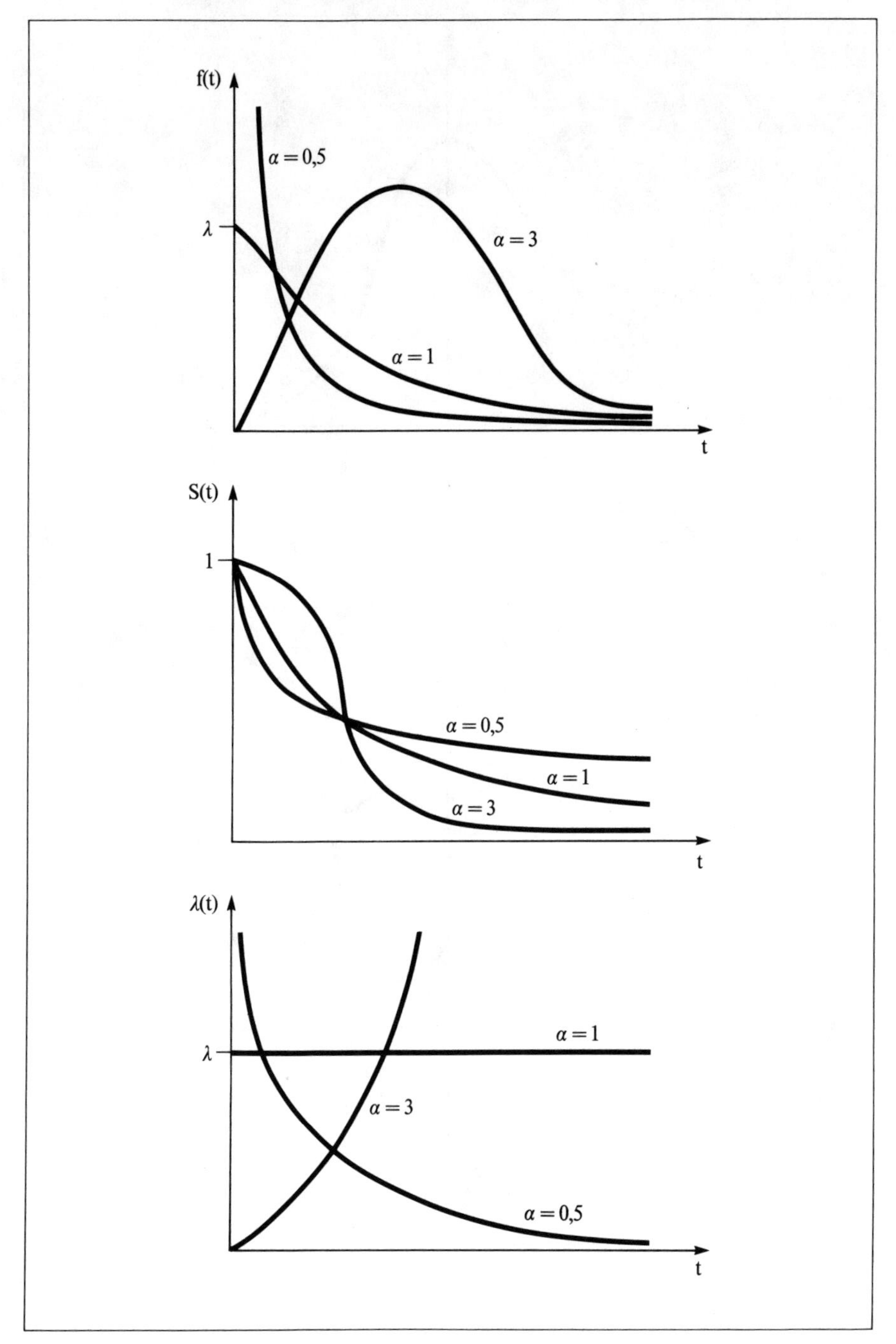

Figure 3.6: Density Function, Survivor Function, and Hazard Rate of the Standard Extreme Value Distribution

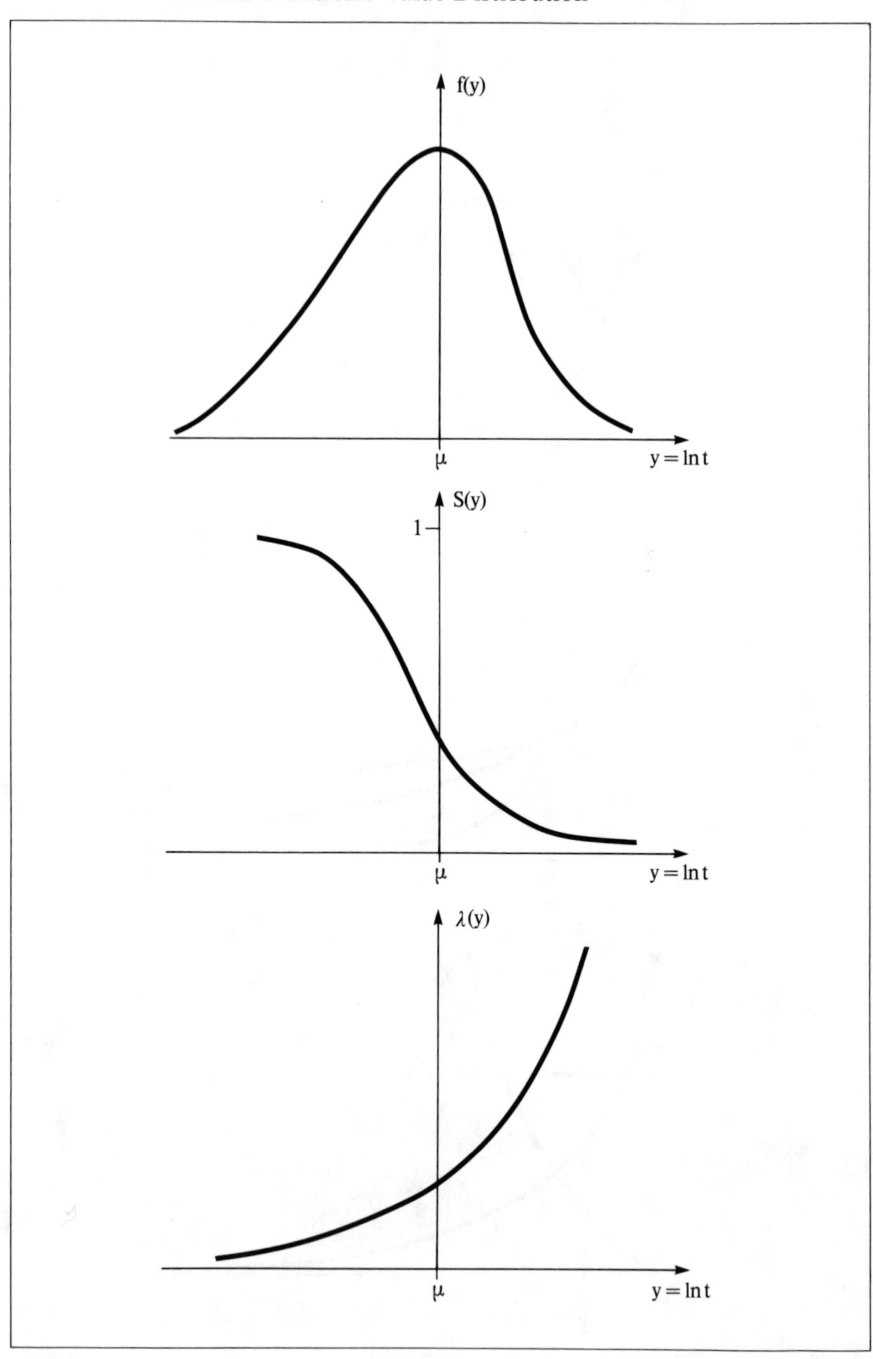

Since lifetime studies commonly work with logarithmic data, the extreme value distribution may be adequately applied to such data.

The special case $\mu = 0$ and $\sigma = 1$ is referred to as the "standard extreme value distribution" (see Figure 3.6).

Deriving the expected value and variance one obtains

$$E(Y) = \mu + \sigma \cdot \Gamma'(1)$$

and

$$Var(Y) = \frac{\sigma^2 \pi^2}{6}, \tag{3.2.26}$$

where $\Gamma'(1)$ represents the first derivative of the gamma function $\Gamma(n)$ evaluated at $n = 1$.

Log-Logistic Distribution

Setting

$$\ln T = \mu + \sigma \omega$$

and postulating a logistic distribution for ω with the density function

$$f(\omega) = \frac{\exp(\omega)}{[1 + \exp(\omega)]^2}, \tag{3.2.27}$$

one obtains for T itself the log-logistic distribution with the density function

$$f(t) = \lambda \alpha (\lambda t)^{\alpha-1} [1 + (\lambda t)^{\alpha}]^{-2}, \tag{3.2.28}$$

with $\lambda = e^{-\mu}$ and $\alpha = \sigma^{-1}$.

The survivor and hazard functions are given by

$$S(t) = \frac{1}{1 + (\lambda t)^{\alpha}} \tag{3.2.29}$$

and

$$\lambda(t) = \frac{\lambda\alpha (\lambda t)^{\alpha-1}}{1 + (\lambda t)^{\alpha}} \tag{3.2.30}$$

(see Figure 3.7).

In addition to the distributions applied in the event history analysis presented above, occasionally the gamma, lognormal, Pareto, or the generalized F-distribution are used. Illustrations of such applications may be found, for example, in Kalbfleisch and Prentice (1980, Chapter 2), Miller (1981, Chapter 1), or Lawless (1982, Section 1.3).

Demographic studies are often based upon the Gompertz distribution which has proven to be quite useful since Gompertz originally introduced it in 1825. We now turn to a short description of the Gompertz-(Makeham) distribution.

Figure 3.7: Density Function, Survivor Function, and Hazard Rate of the Log-Logistic Distribution

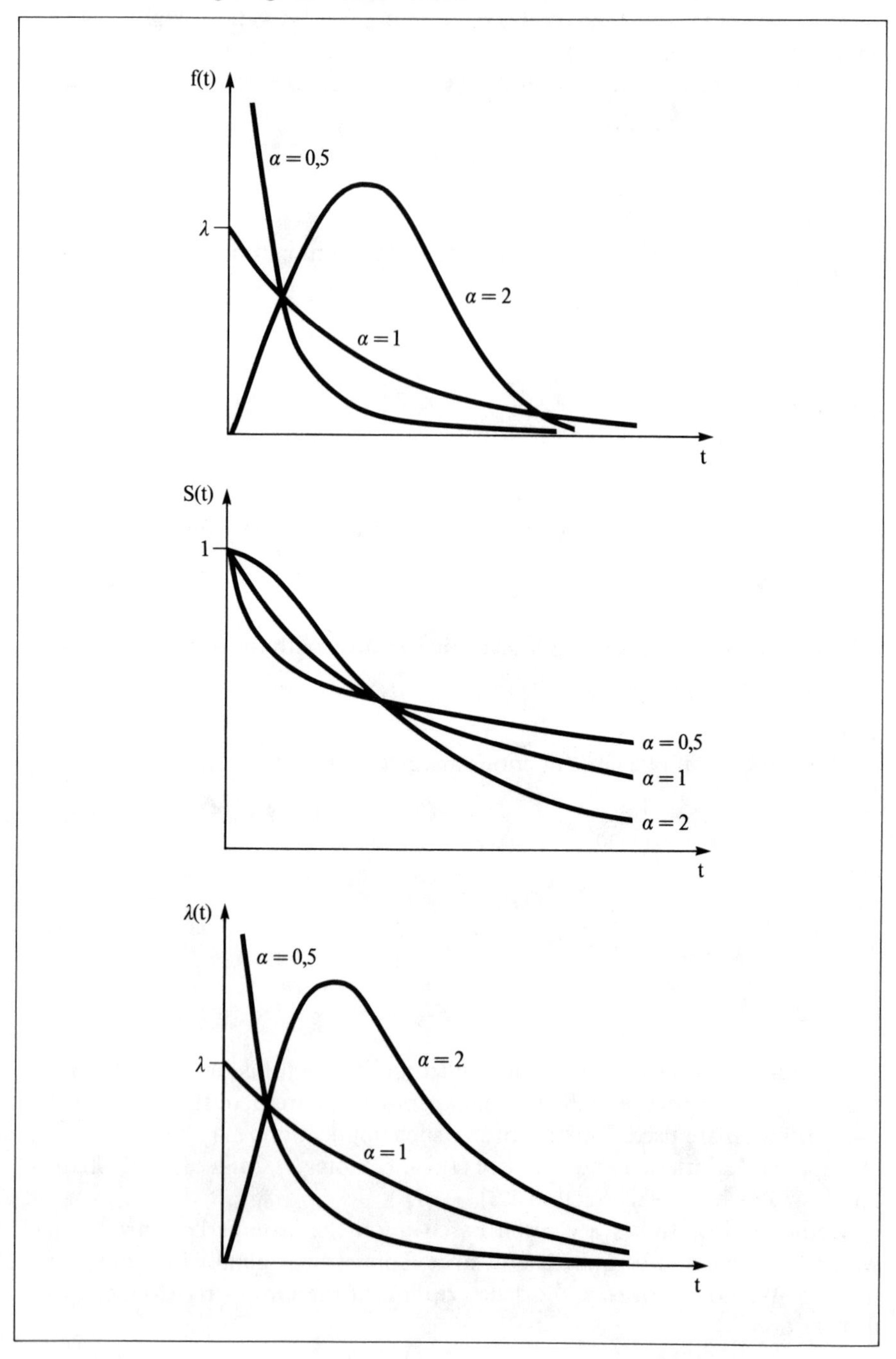

Figure 3.8: Density Function, Survivor Function, and Hazard Rate of the Gompertz Distribution

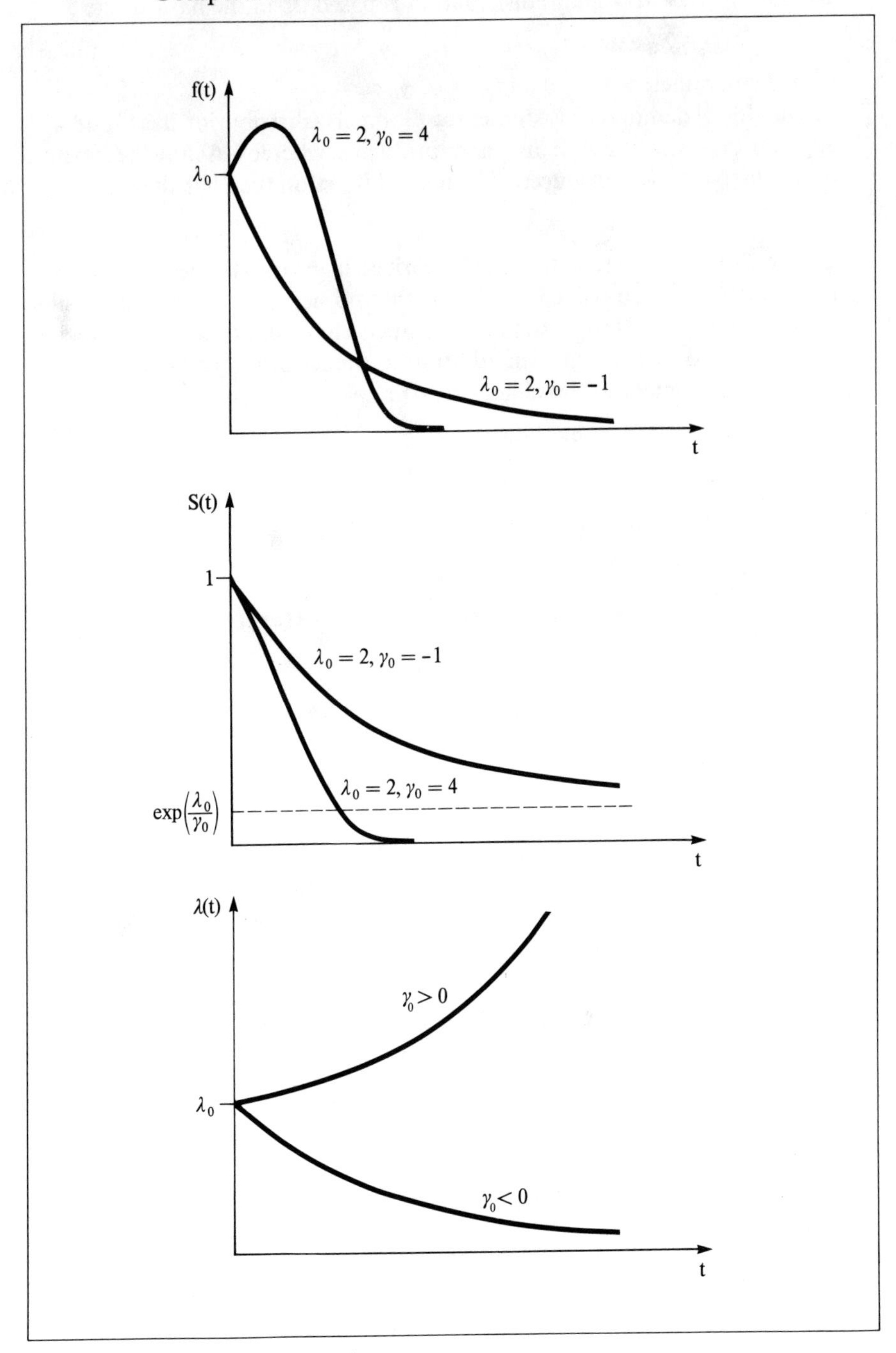

Gompertz-(Makeham) Distribution

The Gompertz-(Makeham) distribution is based upon the hazard rate

$$\lambda(t) = \lambda_0 \exp(\gamma_0 t) \quad (t \geq 0), \tag{3.2.31}$$

with the parameters $\lambda_0 > 0$ and $-\infty < \gamma_0 < \infty$.

Outside of demographic studies the Gompertz distribution has found wide application in dealing with insurance problems, whereby usually the substitution $\exp(\gamma_0) = c$ is introduced. The hazard function then has the form

$$\lambda(t) = \lambda_0 c^t. \tag{3.2.32}$$

The Gompertz distribution can be derived from an extreme value distribution which has been truncated at 0 so that no negative values are possible. Adding a constant α_0 ($\alpha_0 > 0$) to the Gompertz hazard rate, in order to include deaths caused by accidents (in addition to the natural mortality rates), one obtains the Gompertz-Makeham hazard rate

$$\lambda(t) = \alpha_0 + \lambda_0 \exp(\gamma_0 t). \tag{3.2.33}$$

With (3.2.10) and (3.2.11) the survivor function

$$S(t) = \exp(-\alpha_0 t - \frac{\lambda_0}{\gamma_0}(\exp(\gamma_0 t) - 1))$$

may be derived and for the density function one obtains

$$f(t) = (\alpha_0 + \lambda_0 \exp(\gamma_0 t)) \exp(-\alpha_0 t - \frac{\lambda_0}{\gamma_0}(\exp(\gamma_0 t) - 1))$$

(see Figure 3.8).

3.2.3 Life Table Method

The life table method is one of the most common methods applied to analyze waiting times and life expectancies. The method is basically nonparametric and has mainly been applied in demographic and insurance studies in the form of a population life table.

The time axis is divided into $q+1$ intervals $[a_{k-1}, a_k)$, $k = 1, \ldots, q+1$, where $a_0 = 0$ and $a_{q+1} = \infty$. Usually the last possible time point observable is chosen to represent a_q. The intervals need not be equidistant. The life table method is a technique for studying "grouped" time periods without explicit consideration of covariates. Models with covariates for grouped lifetimes are discussed below in Section 3.10.

The hazard rate of the *k*th interval

$$\lambda_k = P(T \epsilon [a_{k-1}, a_k) \mid T \geq a_{k-1}) \tag{3.2.34}$$

represents the conditional probability that in the *k*th time interval an event occurs given that the beginning of the interval has been reached.

Assuming

$$p_k = 1 - \lambda_k = P(T \geq a_k \mid T \geq a_{k-1})$$

and

$$P_k = P(T \geq a_k)$$

one obtains

$$P_k = P(T \geq a_k \mid T \geq a_{k-1}) \cdot \ldots \cdot P(T \geq a_1 \mid T \geq a_0)\, P(T \geq a_0)$$
$$= p_k \cdot \ldots \cdot p_1. \tag{3.2.35}$$

The data consists of:

- n total number of individuals or subjects to be studied at the outset of analysis,
- d_k number of events during the *k*th interval,
- w_k number of individuals who are censored in the *k*th interval, $k = 1, \ldots, q + 1$.

For the "risk set" R_k, that is, the number of individuals or study subjects who had no event until the beginning of the *k*th interval and have not been censored, one obtains

$$R_1 = n \quad \text{and}$$
$$R_k = R_{k-1} - d_{k-1} - w_{k-1} \quad \text{for } k = 2, \ldots, q + 1.$$

If during the *k*th interval no censoring has occurred, then the hazard rate λ_k may be estimated directly by the relative frequency d_k/R_k. Given the case $w_k > 0$, however, the relative frequency usually underestimates the actual hazard rate. The life table method thus corrects the risk set R_k through reducing it by $w_k/2$. The estimator obtained is then

$$\hat{\lambda}_k = \frac{d_k}{R_k - w_k/2}. \tag{3.2.36}$$

Applying $\hat{p}_k = 1 - \hat{\lambda}_k$ one obtains from equation (3.2.35) the estimates

$$\hat{P}_k = \hat{p}_k \cdot \ldots \cdot \hat{p}_1 \tag{3.2.37}$$

for the survivor function $S(a_k)$. The estimated event probability in the *k*th interval is thus directly obtained as

$$\hat{P}(T \in [a_{k-1}, a_k)) = \hat{P}_{k-1} - \hat{P}_k. \tag{3.2.38}$$

The corresponding "density" for the *k*th interval is

$$\hat{f}_k = \frac{\hat{P}_{k-1} - \hat{P}_k}{h_k} = \frac{\hat{P}_{k-1}\,\hat{\lambda}_k}{h_k}, \tag{3.2.39}$$

with $h_k = a_k - a_{k-1}$ characterizing the length of the *k*th interval. Normally, h_k is a chosen fixed time period, for example, a day, a month, or a year.

$\hat{f}_k$ from equation (3.2.39) may now be applied to estimate the hazard rate of the underlying continuous duration. An "average hazard rate" in the midpoint m_k of the kth interval can be estimated as

$$\hat{\lambda}(m_k) = \frac{\hat{f}_k}{\hat{P}(T \geq m_k)} = \frac{\hat{f}_k}{(\hat{P}_{k-1} + \hat{P}_k)/2} = \frac{2\,\hat{\lambda}_k}{h_k(\hat{P}_{k-1} + \hat{P}_k)/\hat{P}_{k-1}} = \frac{2\,\hat{\lambda}_k}{h_k\,(1 + \hat{p}_k)}. \tag{3.2.40}$$

3.2.4 Product-Limit Estimator (Kaplan-Meier Estimator) of the Survivor Function

One possible method of estimating the survivor function has already been presented in the last section with (3.2.37). This method requires a decomposition of the time axis, whereby the choice of the boundaries is arbitrary. The so-called product-limit estimator that has been derived by Kaplan and Meier (1958) is similar. The difference between the two methods is that now the observed event times are chosen to represent the interval boundaries.

Assume $t_{(1)} < t_{(2)} < \ldots < t_{(m)}$ represent the ordered event times ($m \leq n$). At first, we assume that no ties are present. Then the intervals

$$[0,\ t_{(1)}),\ [t_{(1)},\ t_{(2)}),\ \ldots,\ [t_{(m)},\ \infty)$$

are constructed. It follows that

$$\hat{p}_k = 1 - \frac{1}{R_k},$$

with R_k representing the risk set at time $t_{(k-1)}$. One obtains the estimator for the survivor function as

$$\hat{S}(t) = \begin{cases} 1 & \text{for } t \leq t_{(1)} \\ \prod\limits_{k|t_{(k)}<t} (1 - \frac{1}{R_k}) & \text{for } t > t_{(i)}. \end{cases} \tag{3.2.41}$$

In the case of ties, that is, when more than one event occur simultaneously, $1 - \frac{1}{R_k}$ in (3.2.41) must be replaced by $1 - \frac{d_k}{R_k}$ in which d_k represents the number of events at $t_{(k)}$.

When censored observations occur at the same time as events, the assumption is made that the events occur just before the censorings.

If the last observation is censored, then $\hat{S}(t) > 0$ for $t \to \infty$. In this case, $\hat{S}(t)$ is defined only over the length of time till the largest event time.

The product-limit estimator can also be derived as the maximum likelihood estimator. For the explicit derivation, the reader is referred to Kalbfleisch and Prentice (1980, pp. 10ff.), Lawless (1982, pp. 74ff.), or Johansen (1978). Furthermore, it may be shown that the product-limit estimator may be derived with the aid of the obtained life table estimation of the survivor function given that $q \to \infty$ and $\max|a_k - a_{k-1}| \to 0$.

An estimation of the (asymptotic) variance of $\hat{S}(t)$ is

$$\hat{Var}(\hat{S}(t)) = \hat{S}(t)^2 \sum_{t_{(k)} \leq t} \frac{d_k}{R_k (R_k - d_k)}. \tag{3.2.42}$$

An estimation of the expected average duration time or lifetime is

$$\hat{\mu} = \sum_{k=1}^{n} \hat{S}(t_{(k-1)}) (t_{(k)} - t_{(k-1)}), \tag{3.2.43}$$

where in (3.2.43) the observations have been ordered by time.

Given that the observation $t_{(n)}$ is censored, generally $\hat{\mu}$ will underestimate the true expected value. A formula for the estimated variance of $\hat{\mu}$ may also be derived (see Gross and Clark 1975).

It is also possible to obtain estimates of arbitrary quantities of the distributions of duration times or lifetimes. Without going into detail at this point, the interested reader is referred once again to Gross and Clark (1975).

3.2.5 Comparisons of Survivor Functions

Stratifying a population into s subsets makes it possible to estimate a survivor function for each subdivision of the population, for example, applying the product-limit method. Especially interesting is a comparison of the subgroup survivor functions. Are they similar or greatly different from one another? Due to $S(t) = 1 - F(t)$, such an examination is equivalent to see if the distributions of the duration times or lifetimes coincide. If no censoring occurs, familiar nonparametric methods for the multiple sample case may be applied. For a detailed analysis of such methods, see, for example, Schaich and Hamerle (1984, Chapter 5).

Usually, however, event analysis is characterized by censored data that requires modification of the procedure. In the literature on this subject numerous methods have been developed to deal with this problem. It would be beyond the scope of such an introductory text to give a detailed outline of these methods. Instead, we only present the principle procedures. For simplicity, the simple case of two groups ($s = 2$) will first be discussed, basing our discussion largely upon the presentation of Tarone and Ware (1977).

Assume that $u_1, \ldots, u_{n_1}$ and $v_1, \ldots, v_{n_2}$ represent the observations of the two samples, $n = n_1 + n_2$, and δ_i and ϵ_j with $i = 1, \ldots, n_1$ and $j = 1, \ldots, n_2$ are the respective censoring indicators. The survivor functions are $S_1(t)$ and $S_2(t)$, respectively. We now test the hypothesis

$$H_0 : S_1(t) = S_2(t) \quad \text{for all } t.$$

Pooling both samples, the values are arranged according to their magnitude.

The uncensored observations of the pooled sample ($m \leq n$) arranged according to magnitude are thus

$$T_{(1)} \leq T_{(2)} \leq \ldots \leq T_{(m)}.$$

The risk set R_i is once again the number of persons who have not experienced an event and are not censored till just before time $T_{(i)}$.

For each uncensored time point $T_{(i)}$, $i = 1, \ldots, m$, a 2×2 contingency table in the following form is constructed:

	event occurrence	no event	
U	a_i	b_i	r_{i1}
V	c_i	d_i	r_{i2}
	e_{i1}	e_{i2}	R_i

a_i represents the number of persons in the first group who experienced an event at time point $t_{(i)}$. If no ties are present, a_i is either 0 or 1 and e_{i1} is always 1. If ties are present at time point $t_{(i)}$, e_{i1} designates the number of ties and a_i and c_i represent how the events at time point $t_{(i)}$ are distributed over the two groups. r_{i1} is the risk set of members of the first group, and r_{i2} for the second group. The following simple example is illustrative of the above described procedure.

Example:

The $n_1 = 4$ values (U_i, δ_i) of the first group are

(5, 1), (4.5, 1), (12, 1), (6, 0)

and the $n_2 = 3$ values (V_j, ϵ_j) of the second group are

(7, 1), (5, 1), (8, 0).

Pooling both samples together, and ordering the uncensored values according to size, one obtains (m = 5)

4.5 5 5 7 12.

The following table includes the values for R_i, e_{i1}, r_{i1}, and a_i underneath the five stated observations.

$z_{(i)}$	4.5	5 5	7	12
R_i	7	6	3	1
e_{i1}	1	2	1	1
r_{i1}	4	3	1	1
a_i	1	1	0	1

In principle, it suffices for each 2×2 table to determine the value a_i, since the other values may be derived from the given marginal values.

With $E_0(a_i)$ and $Var_0(a_i)$ designating the mean and variance of a_i that may be calculated if the null hypothesis is true, the following test statistic is constructed:

$$S = \frac{\sum_{i=1}^{m} \omega_i (a_i - E_0(a_i))}{[\sum_{i=1}^{m} \omega_i^2 Var_0(a_i)]^{1/2}}, \tag{3.2.44}$$

where the ω_i's represent weights.

Chosing $\omega_i = 1$ for all i, one obtains the Mantel-Haenszel statistic. This test is also known under the name log-rank test or Mantel-Cox test. Chosing $\omega_i = R_i$, one obtains Gehan's test which may also be derived from the Mann-Whitney variant of the Wilcoxon test (see, e.g., Schaich and Hamerle, 1984, pp. 116ff.). In the Gehan test smaller observations are given more weight.

Tarone and Ware propose $\omega_i = \sqrt{R_i}$ and find that this choice of weights offers an especially high efficiency for the test with regard to a large class of alternative hypotheses.

For $s > 2$ samples the procedure must be generalized accordingly; however, the basic principle remains the same.

The Breslow test is an extension of the Gehan test. The Mantel-Haenszel test may also be generalized for samples larger than $s > 2$. For the exact formulas of the test statistics, see, for example, Breslow (1970), Gehan (1965), Lee and Desu (1972), or Tarone and Ware (1977).

If H_0 is true the test statistics have an asymptotic χ^2 distribution with $s - 1$ degrees of freedom. The above mentioned tests are all Omnibus tests, which merely check the global null hypothesis $H_0: S_1(t) = \ldots = S_s(t)$ for all t. It is, however, possible to construct tests for trends against the alternative hypothesis $S_1(t) > S_2(t) > \ldots > S_s(t)$ (see, e.g., Miller 1981, pp. 110ff.).

Especially when dealing with small samples one must be careful that the pattern of censoring in the subgroups does not vary too much, since different censoring mechanisms may influence the distribution of the test statistic.

3.3 Introducing Covariates: Regression Models

3.3.1 Quantitative and Qualitative Covariates

In addition to the duration or lifetime, generally, various covariates or prognostic factors are collected for each individual or subject, and an important aim of statistical analysis is to ascertain the quantitative influence of these exogenous or endogenous variables on the hazard or transition rate.

The covariates may be quantitative or qualitative. A quantitative variable x_j is handled as in a traditional multiple regression. It is weighted with a parameter β_j and introduced into the model as $x_j \beta_j$. Categorical variables are treated as in analysis of variance by coding categories using dummy variables.

One method of coding the categories of qualitative variables is the (0,1)-coding ("cornered effects"). If a variable A possesses I categories (features, classes, factor levels), they may be characterized through the I − 1 dummy variables of the form

$$x_i^A = \begin{cases} 1 & \text{if category i of variable A occurs} \\ 0 & \text{otherwise} \end{cases}$$

$$i = 1, \ldots, I-1. \tag{3.3.1}$$

The *i*th dummy variable x_i^A $(i = 1, \ldots, I-1)$ thereby codes only the presence or nonpresence of the *i*th category. The occurrence of the *I*th category is implicitly expressed by the coding $x_i^A = 0$ for $i = 1, \ldots, I-1$. The choice of the *I*th category as the reference category is arbitrary. With regard to the interpretation of the results, a reference category should, however, be chosen for which all other categories may be easily compared, since the parameter β_j measures the respective "distance" between the *j*th category and the reference category.

The special case of a dichotomous independent variable is especially simple. Then I = 2 and one obtains only one dummy variable

$$x^A = \begin{cases} 1 & \text{given category 1} \\ 0 & \text{given category 2.} \end{cases}$$

In the general case, all categories of the qualitative variables A may be coded by $x_1^A, x_2^A, \ldots, x_{I-1}^A$. The corresponding regression coefficients β_j are commonly called the *main effects* as in analysis of variance.

The (0,1)-coding is highly useful for approaches with mixed quantitative and qualitative covariates. Given only qualitative covariates, the "centered effects" coding is commonly applied, which is directly related to traditional ANOVA (see, e.g., Hamerle, Kemény, and Tutz, 1984, p. 214).

Within the framework of regression models for durations and lifetimes, especially with categorical independent variables, the incorporation of *interaction effects* is important. They measure the common influence of a certain combination of categories of two or more independent variables. Formally, they may be included in a simple manner in the regression analysis by forming the corresponding products of the dummy variables. For the two-factor interaction effects of factors A and B, one obtains the product $x_i^A x_j^B$, $i = 1, \ldots, I-1$, $j = 1, \ldots, J-1$, for the three-factor interaction, the product $x_i^A x_j^B x_k^C$, etc.

The values of the quantitative covariates of an individual or study subject i, as well as the codings of all main effects and the interaction effects of the qualitative covariates included in the model are then collected in a design vector $\mathbf{x}_i$. $\mathbf{x}_i$ has the dimension p.

We are interested in finding how the covariates influence the durations or lifetimes. It is common practice, much like that of the traditional regression approaches, to assume that the influence of the covariates or prognostic factors is linear in the parameters

$$\eta_i = \mathbf{x}_i'\boldsymbol{\beta}$$

with an unknown p-dimensional parameter vector $\boldsymbol{\beta}$. The parameters $\beta_1, \ldots, \beta_p$ represent the weight of the influence of the covariates. Contrary to classical multiple regression analysis, it is not assumed that the linear combination $\eta_i = \mathbf{x}_i'\boldsymbol{\beta}$ directly influences the durations or lifetimes T_i, rather it influences a function of T_i, for example, $\ln T_i$.

Another important difference as compared to cross-sectional regression is that in the duration models some covariates may be time dependent. Such is the case, for example, when a specific medical therapy is applied only during a certain time period. The aim of such a study might be to examine the influence of the therapy during the effective application period or to check for possible secondary effects (side-effects). To accomplish this, for example, two dummy variables may be defined, $x_1(t)$ and $x_2(t)$ with

$$x_1(t) = \begin{cases} 1 & \text{during the time period of active participation of an individual in a therapy or program,} \\ 0 & \text{otherwise} \end{cases}$$

$$x_2(t) = \begin{cases} 1 & \text{after the individual's termination of the "medical treatment" within a specific therapy or program,} \\ 0 & \text{otherwise.} \end{cases}$$

If the hazard rate depends on the independent variables and given that the corresponding regression coefficients are negatively (positively) significant, a therapy is then effective and lowers (increases) the instantaneous probability of a change of state. Furthermore, if the first coefficient is larger in absolute value than the second, then the effect of the therapy sinks (rises) after termination of the therapy.

One possible method of analyzing the influence of covariates upon the duration times or lifetimes is to formulate a regression model in which the distribution of the duration times or lifetimes are dependent upon the covariates. Designating $\mathbf{x}$ as the covariate vector, one has to specify a model for the duration times or lifetimes T under the given covariate vector $\mathbf{x}$.

Obviously, an analogous procedure to traditional regression analysis may be chosen and the above introduced distributions, such as the exponential or Weibull distribution, may be generalized such that one or more parameters are assumed to depend on the covariates $\mathbf{x}$. Usually the duration distribution will be determined by the regression coefficients corresponding to the covariates and a further parameter vector $\boldsymbol{\theta}$. An alternative approach has been proposed by Cox (1972). Cox's semiparametric approach requires fewer assumptions about the underlying duration time and lifetime distributions. The Cox model will be discussed after presentation of the full parametric models.

3.3.2 Parametric Regression Models

Exponential Model

Applying a simple model of an exponentially distributed duration time or lifetime, the following offers a demonstration of the procedure for constructing regression models. According to (3.2.18), the density of T without considering covariates is

$$f(t) = \lambda \exp(-\lambda t) \qquad , t \geq 0, \lambda > 0.$$

The exponential distribution is determined by the parameter λ. The mean duration is $1/\lambda$. Now let the mean duration depend on the covariates, for example, in the form $1/\lambda = g(\mathbf{x};\boldsymbol{\beta})$ with the unknown parameter vector $\boldsymbol{\beta}$. This corresponds to the procedure in traditional multiple regression analysis in which given a $N(\mu, \sigma^2)$ distributed dependent variable y, the mean is parameterized in the form $\mu = \mathbf{x}'\boldsymbol{\beta}$. Assuming that the covariates influence the duration time or lifetime T through the linear combination $\mathbf{x}'\boldsymbol{\beta}$, one obtains the parameterization

$$1/\lambda = g(\mathbf{x}'\boldsymbol{\beta}).$$

Now the function g must be specified. The choice of the identical function $g(z) = z$, that is,

$$1/\lambda = \mathbf{x}'\boldsymbol{\beta}$$

is not appropriate here since the parameter λ must be nonnegative. Applying the approach $1/\lambda = \mathbf{x}'\boldsymbol{\beta}$ given such a restriction could lead to unwanted and possibly uncontrollable restrictions for the parameter $\boldsymbol{\beta}$ (see, e. g., Mantel and Myers, 1971). It is thus better to chose a function g that may only take on positive values. One simple possibility is

$$g(\mathbf{x}'\boldsymbol{\beta}) = \exp(\mathbf{x}'\boldsymbol{\beta})$$

or

$$g(\mathbf{x}'\boldsymbol{\beta}) = \lambda_0 \exp(\mathbf{x}'\boldsymbol{\beta}) \qquad (\lambda_0 > 0). \tag{3.3.2}$$

Setting $\beta_0 = \ln \lambda_0$, one obtains for (3.3.2)

$$g(\mathbf{x}'\boldsymbol{\beta}) = \exp(\beta_0 + \mathbf{x}'\boldsymbol{\beta}). \tag{3.3.3}$$

λ_0, or respectively β_0, represents the constant term of the regression model. Given (3.3.3), T possesses an exponential distribution with the parameter $\lambda = \exp(-\beta_0 - \mathbf{x}'\boldsymbol{\beta})$, and the hazard rate is

$$\lambda(t|\mathbf{x}) = \exp(-\beta_0 - \mathbf{x}'\boldsymbol{\beta}). \tag{3.3.4}$$

Usually the model (3.3.4) is reparameterized with new parameters which are the negative of the β's in (3.3.4).

The heterogeneity within the population is expressed in the covariates and is constant over time. If two individuals are characterized by different covariates, the hazard rates of the individuals will also differ between them, as is demonstrated below in Figure 3.9.

Figure 3.9: Hazard Rate of Two Individuals Possessing an Exponentially Distributed Duration

The hazard rates are time invariant, and the rate of one individual is a multiple of the rate of another individual. The last feature of the "proportional hazards," which is fulfilled for the exponential model, is valid for some important classes of models with time dependent hazard rates.

The similarity with the traditional models of multiple regression becomes even more evident when one uses the logarithm of the duration as the dependent variable. The variable $y = \ln T$ possesses the density function

$$g(y) = \exp(y + \ln \lambda - \exp(y + \ln \lambda)) \qquad -\infty < y < \infty. \tag{3.3.5}$$

Writing the regression model in the usual form ($\ln \lambda = -\beta_0 - \mathbf{x}'\boldsymbol{\beta}$)

$$y = \beta_0 + \mathbf{x}'\boldsymbol{\beta} + \omega, \tag{3.3.6}$$

the disturbance term ω possesses a standard extreme value distribution with the density

$$f(\omega) = \exp(\omega - \exp(\omega)) \qquad -\infty < \omega < \infty. \tag{3.3.7}$$

A graphical presentation of the density function of the standard extreme value distribution has been presented above in Figure 3.6. It must be noted that the disturbance term ω in the regression model (3.3.6) possesses a distribution in which no parameter may be selected freely. This implies a strong limitation which is not the case in the traditional multiple regression approach. Assuming the simplest model, the disturbance term is distributed according to $N(0;\sigma^2)$, where σ^2 is an unknown parameter which must be estimated from the data. Therefore, we want to extend the model presented in (3.3.6) as in the multiple regression

$$y = \beta_0 + \mathbf{x}'\boldsymbol{\beta} + \sigma\ \omega,$$

where ω has the distribution (3.3.7). This is demonstrated in the next section and is closely linked with the Weibull distribution for the duration times or lifetimes.

Weibull Regression Model

Applying the regression approach ($y = \ln T$)

$$y = \beta_0 + \mathbf{x}'\boldsymbol{\beta} + \sigma\ \omega, \tag{3.3.8}$$

where ω once again has a standard extreme value distribution with the density (3.3.7), we may easily derive the distribution of the duration T. Given the covariate vector $\mathbf{x}$, T possesses the density function

$$f(t|\mathbf{x}) = \frac{\delta}{\exp(\beta_0 + \mathbf{x}'\boldsymbol{\beta})} \cdot \left(\frac{t}{\exp(\beta_0 + \mathbf{x}'\boldsymbol{\beta})}\right)^{\delta-1} \exp\left[-\left(\frac{t}{\exp(\beta_0 + \mathbf{x}'\boldsymbol{\beta})}\right)^{\delta}\right]$$

$$t \geq 0,\ \delta = 1/\sigma. \tag{3.3.9}$$

Note that (3.3.9) corresponds exactly to the density of the Weibull distribution in which the parameter λ has the form $\lambda = \exp(-\beta_0 - \mathbf{x}'\boldsymbol{\beta})$ or $1/\lambda = \exp(\beta_0 + \mathbf{x}'\boldsymbol{\beta})$. The other parameter $\delta = 1/\sigma$ is the same for all individuals.

The hazard rate of the Weibull regression model is

$$\lambda(t|\mathbf{x}) = \frac{\delta}{\exp(\beta_0 + \mathbf{x}'\boldsymbol{\beta})} \left(\frac{t}{\exp(\beta_0 + \mathbf{x}'\boldsymbol{\beta})}\right)^{\delta-1}. \tag{3.3.10}$$

The Weibull regression model belongs to the family of proportional hazards models, as may be easily verified. If two individuals possess the covariate vectors $\mathbf{x}_1$ and $\mathbf{x}_2$, the quotient of the respective hazard rates is

$$\frac{\lambda(t|\mathbf{x}_1)}{\lambda(t|\mathbf{x}_2)} = \exp((\mathbf{x}_2 - \mathbf{x}_1)'\boldsymbol{\beta}\ \delta)$$

which is time invariant.

The assumption regarding the identical values of the parameter $\delta = 1/\sigma$ for all individuals or subjects is equivalent to the supposition of variance homogeneity in traditional multiple regression.

For $\delta = \sigma = 1$, the exponential regression model is again obtained.

The analogy of (3.3.8) and the traditional multiple regression model is quite obvious. In the traditional multiple regression model it is postulated that ω is characterized by a standard normal distribution. The disturbance term ω thus has an expected value of zero. This is not the case for the standard extreme value distribution. The expected value in the standard extreme distribution case is $-0.5772\ldots$, where $0.5772\ldots = \Gamma'(1)$ is Euler's constant. The variance of the standard extreme value distribution is $\frac{\pi^2}{6} = 1.64493\ldots$. The variance of y is $\frac{\pi^2}{6}\sigma^2$ and the expected value of y is

$$E(y) = \beta_0 + \mathbf{x}'\boldsymbol{\beta} - 0.5772\ \sigma.$$

Note that E(y) depends on the parameter σ.

Lognormal Distribution Regression Model

One obvious possibility is to assume for ω in (3.3.8) a standard normal distribution. The duration time or lifetime T then possesses a lognormal distribution with the density function

$$f(t|\mathbf{x}) = \frac{1}{\sqrt{2\pi}\ \sigma t} \exp\left[-\frac{1}{2}\left(\frac{\ln t - \mu(\mathbf{x})}{\sigma}\right)^2\right] \qquad t>0, \tag{3.3.11}$$

where $\mu(\mathbf{x}) = \beta_0 + \mathbf{x}'\boldsymbol{\beta}$.
The survivor function and the hazard rate are

$$S(t|\mathbf{x}) = 1 - \Phi\left(\frac{\ln t - \mu(\mathbf{x})}{\sigma}\right)$$

and

$$\lambda(t|\mathbf{x}) = f(t|\mathbf{x}) \,/\, S(t|\mathbf{x}) \tag{3.3.12}$$

respectively, where $\Phi(z)$ is the distribution function of the standard normal distribution.
At first the hazard rate increases to a maximum value and then decreases to zero for $t \to \infty$. The lognormal regression models do not belong to the class of proportional hazards models.

Log-Logistic Regression Models

According to (3.2.30), the log-logistic hazard rate is

$$\lambda(t) = \frac{\lambda\ \alpha(\lambda t)^{\alpha-1}}{1+(\lambda t)^{\alpha}}.$$

A regression model is obtained by parameterization of the parameter λ with regard to the covariates, for example, in the form $\lambda(\mathbf{x}) = \exp(\mathbf{x}'\boldsymbol{\beta})$. Thus,

$$\lambda(t|\mathbf{x}) = \frac{\exp(\mathbf{x}'\boldsymbol{\beta})\ \alpha(\exp(\mathbf{x}'\boldsymbol{\beta})\ t)^{\alpha-1}}{1+(\exp(\mathbf{x}'\boldsymbol{\beta})t)^{\alpha}}. \tag{3.3.13}$$

General Log-Linear Regression Models

The regression approaches (3.3.6), (3.3.8), and (3.3.11) belong to the *log-linear regression models* where a linear relationship between the covariates and the logarithm of the duration or lifetime is assumed. A log-linear regression model is generally characterized in the form

$$y = \ln T = \beta_0 + \mathbf{x}'\boldsymbol{\beta} + \sigma\ \omega \tag{3.3.14}$$

where the distribution of ω does not depend on $\mathbf{x}$. The distribution of y may be written as

$$\frac{1}{\sigma} f(u)$$

with $u = (y - \beta_0 - \mathbf{x}'\boldsymbol{\beta})/\sigma$. The parameter σ is a scale parameter and β_0 represents the general localization of lnT.

Gompertz-(Makeham) Regression Model

The Gompertz-Makeham hazard rate without covariates (for a homogeneous population) is in (3.2.33) characterized by

$$\lambda(t) = \alpha_0 + \lambda_0 \exp(\gamma_0 t). \tag{3.3.15}$$

As in the approaches presented above the possibility exists to model one or more of the parameters in (3.3.15) as being dependent upon the covariate vector $\mathbf{x}$, for example in the form $\lambda_0(\mathbf{x}) = \exp(\mathbf{x}'\beta)$ und $\gamma_0(\mathbf{x}) = \mathbf{x}'\gamma$. Of course, we must be careful to check that no identification problems arise, that is, that no two parameter constellations are observationally equivalent and lead to the same probability distribution for the observed data.

Periodical Changes of Covariates or Parameters

Another possible method of introducing time dependent hazard rates as presented by Tuma and Hannan (1984, Section 7.2) is to divide the time axis into intervals and to model the hazard rate within the intervals. The covariates and/or parameters may vary from interval to interval.

Divide the time axis into q + 1 intervals

$$[a_0, a_1), [a_1, a_2), \ldots, [a_{q-1}, a_q), [a_q, \infty),$$

where $a_0 = 0$ and interpret a_q as the terminal time point of the observation period. The simplest hazard rate is

$$\lambda_p(t|\mathbf{x}_p) = \exp(\mathbf{x}_p'\boldsymbol{\beta}_p) \qquad \text{for } t \in [a_{p-1}, a_p) \tag{3.3.16}$$

which implies that within each interval an exponentially distributed duration is postulated.

In (3.3.16) it is assumed that the interval limits for all individuals are the same. The interval limits must be exogeneously determined. In addition, even given a medium-sized number of time intervals the model requires a relatively large number of parameters that have to be estimated and the required amount of data is accordingly large. Special cases of (3.3.16) are also possible. For example, the covariates might vary from interval to interval but the parameter $\boldsymbol{\beta}$ is constant over time.

In estimating the unknown parameter vector $\boldsymbol{\beta}_p$, the survivor function is important (see Section 3.6.3). Analogous to (3.2.10) one obtains

$$S(t|\mathbf{x}) = \exp(-\int_0^t \lambda (u|\mathbf{x})\, du).$$

Due to the linearity of the integral it follows that

$$S(t|\mathbf{x}) = \exp(-\sum_{i=1}^{p-1} \exp(\mathbf{x}_i'\boldsymbol{\beta}_i)\ (a_i - a_{i-1}) - \exp(\mathbf{x}_p'\boldsymbol{\beta}_p)\ (t - a_{p-1}))$$

$$\text{for } a_{p-1} \leq t < a_p, \quad k = 1, \ldots, q+1 \quad (a_{q+1} = \infty). \quad (3.3.17)$$

The covariate vector **x** in (3.3.16) and (3.3.17) thereby contains all measured covariates of an individual up to the time point t.

In conclusion, we note that in a concrete practical application, instead of applying (3.3.16), one may also consider an application of the discrete regression models described in Section 3.10.

3.3.3 Cox's Proportional Hazards Regression Model

The proportional hazards model (PH model) was proposed by Cox (1972). Whereas in the approaches we have examined above the hazard rate and thus the duration distribution was known up to a finite number of unknown parameters, the Cox model is a semi-parametric approach with an unspecified "baseline" hazard rate.

The covariates are once again collected in a p-dimensional vector **x**. $\boldsymbol{\beta}$ represents the corresponding parameter vector and T designates the duration time or lifetime. The hazard rate of Cox's PH model is

$$\lambda(t|\mathbf{x}) = \lambda_0(t)\ \exp(\mathbf{x}'\boldsymbol{\beta}). \quad (3.3.18)$$

$\lambda_0(t)$ is the arbitrary, unspecified baseline hazard rate. While such modeling achieves more flexibility, different methods must be used to estimate the parameters.

The proportionality of the hazard rates means, as above, that the quotient

$$\frac{\lambda(t|\mathbf{x}_1)}{\lambda(t|\mathbf{x}_2)}$$

for two individuals or subjects with the covariates $\mathbf{x}_1$ and $\mathbf{x}_2$ does not depend on t.

From (3.3.18) it follows that

$$\frac{\lambda(t|\mathbf{x}_1)}{\lambda(t|\mathbf{x}_2)} = \exp((\mathbf{x}_1 - \mathbf{x}_2)'\boldsymbol{\beta}).$$

This quotient is not time dependent (see Figure 3.10). The proportionality of the hazard rate is due basically to the factorization of the hazard rate into one term dependent upon time and another dependent only upon the covariates. Generally,

$$\lambda(t|\mathbf{x}) = \lambda_0(t)\ g(\mathbf{x};\boldsymbol{\beta}) \quad , g(\cdot) > 0. \quad (3.3.19)$$

The Cox model sets $g(\mathbf{x};\boldsymbol{\beta}) = \exp(\mathbf{x}'\boldsymbol{\beta})$.

The assumption of proportionality naturally limits the possible applications of the model. For example, given the introduction of the covariate "sex," the quotient of the hazard rate of males and females should not vary over time. The prerequisite of a proportional hazard rate may be somewhat slackened by the introduction of class specific hazard rates. If one or more (categorical) covariates does not have a multiplicative influence upon the hazard rate, and the different categories or levels produce hazard rates where the proportionality assumption is violated, we can extend the model (3.3.18). We define the hazard rate for an individual in the *j*th stratum (or category, or level) as

$$\lambda_j(t|\mathbf{x}) = \lambda_{0j}(t)\ \exp(\mathbf{x}'\boldsymbol{\beta}) \qquad j = 1, \ldots, J \tag{3.3.20}$$

where J is the number of strata.
The covariate vector **x** now contains only the remaining covariates.

With the aid of specially constructed time dependent covariates, it is possible to construct a statistical test of proportionality (see Section 3.7 for a detailed discussion). According to (3.2.10), the survivor function of the Cox model can be written as

$$\begin{aligned} S(t|\mathbf{x}) &= \exp(-\int_0^t \lambda(u|\mathbf{x})\, du) \\ &= \exp(-\int_0^t \lambda_0(u) \exp(\mathbf{x}'\boldsymbol{\beta})\, du) \\ &= \exp(-\int_0^t \lambda_0(u)\, du)^{\exp(\mathbf{x}'\boldsymbol{\beta})} \\ &= S_0(t)^{\exp(\mathbf{x}'\boldsymbol{\beta})} \end{aligned} \tag{3.3.21}$$

and the duration time or lifetime density function is

$$f(t|\mathbf{x}) = \lambda(t|\mathbf{x}) \cdot S(t|\mathbf{x}) = \lambda_0(t)\ \exp(\mathbf{x}'\boldsymbol{\beta})\ S_0(t)^{\exp(\mathbf{x}'\boldsymbol{\beta})}. \tag{3.3.22}$$

The proportional hazards model was proposed by Cox (1972), who also proposed an estimation method for $\boldsymbol{\beta}$ and $\lambda_0(t)$ without requiring special assumptions (with the exception $\lambda_0(t) \geq 0$). In the recent past, numerous studies of the Cox model have appeared and it is the most widely applied model. A detailed discussion of the PH model is given in Kalbfleisch and Prentice (1980, Chapters 4 and 5) or Lawless (1982, Chapter 7). Assuming a special parametric form of the baseline hazard rate, the full parametric proportional hazards model is obtained. Special cases here are the Weibull and exponential regression models, which have already been discussed. In the Weibull regression model,

$$\lambda_0(t) = \delta\ \lambda_0\ (\lambda_0 t)^{\delta-1}, \tag{3.3.23}$$

and the exponential model is obtained from (3.3.23) by setting $\delta = 1$.

Figure 3.10: Illustration of Two Proportional Hazard Rates

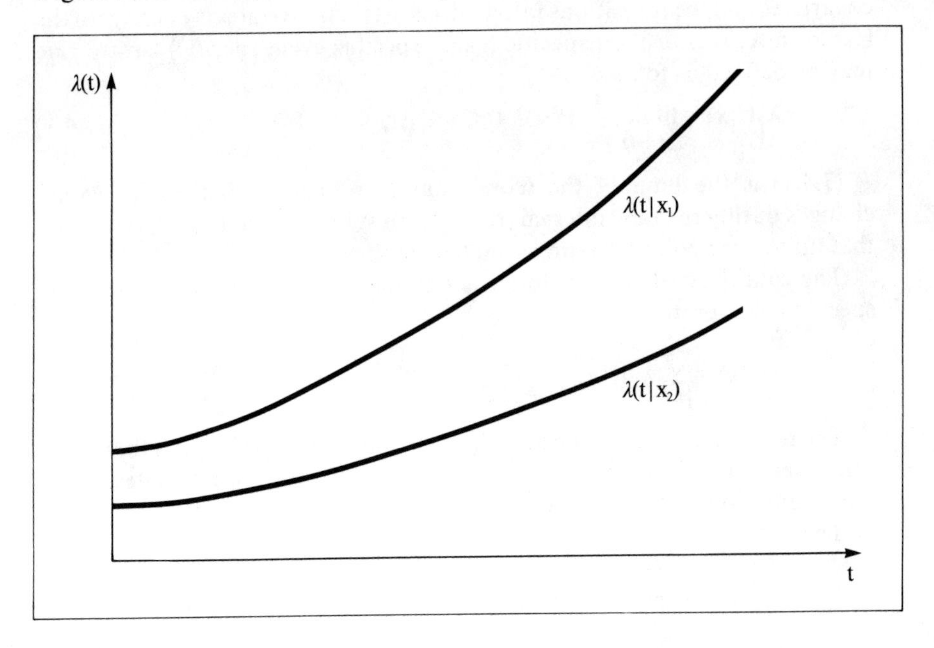

3.4 Multistate Models—Competing Risks

The previous section assumed a one-episode model that results simply in a transition to an absorbing final state. The next obvious extension of the model is to consider transitions into multiple end states. In survival analysis, such "competing risks" are commonly represented as different causes of death (or event types). In industrial reliability studies, various possible defects are included which may cause a technical appliance or machine to stop functioning. Extensive literature exists concerning competing risks models without the inclusion of covariates. For a survey of the representative literature, see Gail (1975). The one-episode case with covariates is dealt with in David and Moeschberger (1978), Seal (1977), and in the "survival" literature in Holt (1978), Prentice and Breslow (1978), as well as Prentice et al. (1978).

For each individual or subject, a state variable Y is now observed in addition to the duration time or lifetime T. The state variable may take on values out of the set of destination states $\{1, \ldots, m\}$. Furthermore, each individual is then characterized by various covariates or prognostic factors which, together with possible interaction effects, are collected in the p-dimensional covariate vector $\mathbf{x}$. For the sake of simplicity, at present the covariates or prognostic factors are presumed to be time invariant.

An appropriate starting point for analyzing the relationships between covariates and the transitions into various destination states is once again the hazard rate. A transition specific (cause specific; event specific) hazard rate may be defined as follows

$$\lambda_j(t|\mathbf{x}) = \lim_{\Delta t \to 0} \frac{1}{\Delta t} P(t \leq T < t + \Delta t, Y = j | T \geq t, \mathbf{x}). \tag{3.4.1}$$

(3.4.1) is the limit of the (conditional) probability that an individual changes during the time interval $[t, t + \Delta t]$ to state j, given the covariates and that up to time point t no transition has occurred.

The total hazard rate at time t is obtained as the sum of all transition specific hazard rates

$$\lambda(t|\mathbf{x}) = \sum_{j=1}^{m} \lambda_j(t|\mathbf{x}). \tag{3.4.2}$$

The total hazard rate is the limit of the (conditional) probability that in the time interval $[t, t + \Delta t]$ a transition has taken place, given the covariates, under the assumption that up to time point t no transition has occurred.

The survivor function is

$$S(t|\mathbf{x}) = \exp(-\int_0^t \lambda(u|\mathbf{x})\, du). \tag{3.4.3}$$

If one substitutes the transition specific hazard rates into (3.4.3) one obtains

$$S(t|\mathbf{x}) = \exp(-\int_0^t \sum_{j=1}^{m} \lambda_j(u|\mathbf{x})\, du) = $$

$$= \prod_{j=1}^{m} \exp(-\int_0^t \lambda_j(u|\mathbf{x})\, du). \tag{3.4.4}$$

Sometimes it is helpful to also introduce the "transition specific density" $f_j(t|\mathbf{x})$ defined as

$$f_j(t|\mathbf{x}) = \lim_{\Delta t \to 0} \frac{1}{\Delta t} P(t \leq T < t + \Delta t,\ Y = j|\mathbf{x})$$

$$= \lambda_j(t|\mathbf{x}) \cdot S(t|\mathbf{x}), \qquad j = 1, \ldots, m. \tag{3.4.5}$$

It must be noted, however, that $f_j(t|\mathbf{x})$ is not the density function of the duration time or lifetime. In particular

$$\int_0^\infty f_j(t|\mathbf{x})\, dt = P(Y = j|\mathbf{x}) = \pi_j(\mathbf{x}). \tag{3.4.6}$$

$\pi_j(\mathbf{x})$ is the probability of transition into the *j*th state ($j = 1, \ldots, m$) given the covariate vector $\mathbf{x}$, with the relationship

$$\sum_{j=1}^{m} \pi_j(\mathbf{x}) = 1.$$

Specification of the Hazard Rate

In principle, all of the previously discussed regression models may be adapted to the multistate case. Thereby, the respective parameters of the covariates as well as the other parameters of the hazard rate may depend on the destination state. One must, however, insure that the entire model does not contain too many parameters, since this reduces the precision of the estimates and requires very large samples.

A main goal of model building is to describe the empirical phenomena relatively simply, which implies a model with the fewest necessary parameters. Of course, such a simple model should adequately represent the empirical relationships.

Postulating a Cox model, the transition specific hazard rates are

$$\lambda_j(t|\mathbf{x}) = \lambda_{0j}(t) \exp(\mathbf{x}'\boldsymbol{\beta}_j) \qquad j = 1, \ldots, m. \tag{3.4.7}$$

The baseline hazard functions $\lambda_{0j}(t)$ as well as the regression coefficients $\boldsymbol{\beta}_j$ depend upon the destination state. An important special case is obtained if one postulates baseline hazard rates that are proportional to another through the proportionality factor $\exp(\beta_{0j})$, such that

$$\lambda_{0j}(t) = \lambda_0(t) \exp(\beta_{0j}). \tag{3.4.8}$$

In order to assume that the relationship (3.4.8) is identifiable, one can, for example, set $\beta_{01} = 0$. Substituting (3.4.8) into (3.4.7), we have

$$\lambda_j(t|\mathbf{x}) = \lambda_0(t) \exp(\beta_{0j} + \mathbf{x}'\boldsymbol{\beta}_j). \tag{3.4.9}$$

One consequence of model (3.4.9) is the simple calculation of $P(Y = j|\mathbf{x}) = \int_0^\infty f_j(t|\mathbf{x})dt$, which is the probability that an individual with a given covariate vector $\mathbf{x}$ reaches the *j*th final state. It may be shown (Kalbfleisch and Prentice, 1980, p. 171) that

$$P(Y = j|\mathbf{x}) = \frac{\exp(\beta_{0j} + \mathbf{x}'\boldsymbol{\beta}_j)}{\sum_{k=1}^{m} \exp(\beta_{0k} + \mathbf{x}'\boldsymbol{\beta}_j)} \qquad j = 1, \ldots, m. \tag{3.4.10}$$

(3.4.10) is a logistic model for the transition probabilities.

Another important specification is the Weibull regression approach for competing risks. If one lets the regression coefficients and the parameters of the baseline hazard rate vary from destination state to destination state, after a reparameterization one obtains from (3.3.23) and (3.3.10)

$$\lambda_j(t|\mathbf{x}) = \delta_j \, \lambda_{0j} \, (\lambda_{0j}t)^{\delta_j - 1} \exp(\mathbf{x}'\boldsymbol{\beta}_j) \qquad j = 1, \ldots, m \tag{3.4.11}$$

for the transition specific hazard rates.

The unknown parameters δ_j, λ_{0j} of the baseline hazard rate and the corresponding parameter vectors $\boldsymbol{\beta}_1, \ldots, \boldsymbol{\beta}_m$ of the covariates must then be estimated from the data.

3.5 Regression Models for the Multiepisode Case

The single-episode models of the previous sections have largely been applied in analyzing event histories in which the destination states are absorbing. If one is dealing with durations representing time intervals for which not all states are absorbing, it is possible that each individual or subject has more than one transition (e.g., the durations of various occupations in an examination of professional careers, the length of spells of unemployment, the time periods between moving to another region in migration and mobility studies, studies of the lifetimes of durable consumer commodities, etc.).

In the following, we consider first of all only the special case in which a certain event repeatedly occurs. In the next section, the more general case of multiepisode and multistate models is dealt with.

In the statistical models, times at which events occur are represented by nonnegative random variables $0 = T_0 < T_1 < T_2 < \ldots$. The durations, that is, length of an episode are given by

$$V_k = T_k - T_{k-1} \qquad , k = 1, 2, \ldots . \tag{3.5.1}$$

For each individual or subject, a vector of covariates $\mathbf{x}_k$ is measured for each episode. The covariates may also be time dependent. For simplicity, it is assumed at the present that the covariates are time invariant. The number of selected covariates may, however, vary from episode to episode.

The sample path of an individual is generated in the following manner:

1. The path of an individual begins at the initial time point $T_0 = 0$. If the beginning time point of the first episode is not known, this is designated as a *left censoring*. This problem is more difficult to deal with than right censoring. In the following we always presume that either the starting time point is given (without loss of generality, $t_0 = 0$) or that the prehistory of the process before the observation period has no influence upon the future course of the process. This is the case only when the durations are exponentially distributed. For a discussion of the problems which arise if left-censored observations are present, see Hamerle (1988a).

The length of the first episode is governed by the hazard rate

$$\lambda^1(t|\mathbf{x}_1) = \lim_{\Delta t \to 0} \frac{1}{\Delta t} P(t \le T_1 < t + \Delta t \mid T_1 \ge t, \mathbf{x}_1) \tag{3.5.2}$$

or the survivor function

$$S^1(t|\mathbf{x}_1) = \exp(-\int_0^t \lambda^1(u|\mathbf{x}_1)\, du). \tag{3.5.3}$$

For the density function of T_1, one obtains

$$f^1(t|\mathbf{x}_1) = \lambda^1(t|\mathbf{x}_1)\, S^1(t|\mathbf{x}_1) \tag{3.5.4}$$

as in the single-episode case.

2. The first event occurs at time point $T_1 = t_1$ for an individual and the second episode begins. The hazard rate of the second episode is characterized by

$$\lambda^2(t|\mathbf{x}_2, H_1) \qquad t \geq t_1. \tag{3.5.5}$$

Note that $\lambda^2(t|\mathbf{x}_2, H_1)$ is zero for $t < t_1$. H_1 contains the prehistory of the process (with the covariates). Thus $H_1 = \{t_1, \mathbf{x}_1\}$. The hazard rate $\lambda^2(t|\mathbf{x}_2, H_1)$ may also depend upon H_1. For example, the duration of a second period of unemployment is to a great extent dependent upon the duration of previous unemployment episodes. The concept of the survivor function may be applied here without difficulties. One obtains

$$S^2(t|\mathbf{x}_2, H_1) = P(T_2 \geq t|\mathbf{x}_2, H_1) = \exp(-\int_{t_1}^{t} \lambda^2(u|\mathbf{x}_2, H_1)\, du), \qquad t \geq t_1. \tag{3.5.6}$$

The duration of the second episode is

$$V_2 = T_2 - T_1. \tag{3.5.7}$$

The survivor function may also be expressed in terms of the duration V_2.

3. One proceeds in this manner, and the consecutive episodes for an individual are obtained. The points of time in which an event occurs are $t_1 < t_2 < t_3 < \ldots$ In each episode it is possible to select a new vector of covariates. For the *k*th episode, this vector is designated as $\mathbf{x}_k$. The prehistory is

$$H_{k-1} = \{t_1, \mathbf{x}_1, \ldots, t_{k-1}, \mathbf{x}_{k-1}\}.$$

Occasionally, the relevant part for the *k*th episode of the prehistory, such as the duration of the k-1 previous episodes or certain previously chosen covariates, for example, are introduced into the current covariate vector $\mathbf{x}_k$. In this case, the dependence of the hazard rate or survivor function on H_{k-1} may be omitted.

The hazard rate of the *k*th episode (k = 1,2, ...) is

$$\lambda^k(t|\mathbf{x}_k, H_{k-1}) = \lim_{\Delta t \to 0} \frac{1}{\Delta t} P(t \leq T_k < t + \Delta t \mid T_k \geq t, \mathbf{x}_k, H_{k-1}), \qquad t \geq t_{k-1}. \tag{3.5.8}$$

For $t < t_{k-1}$, it follows that $\lambda^k(t|\mathbf{x}_k, H_{k-1}) = 0$.

The survivor function is derived as

$$S^k(t|\mathbf{x}_k, H_{k-1}) = P(T_k > t|\mathbf{x}_k, H_{k-1}), \qquad t \geq t_{k-1}, \tag{3.5.9}$$

and dependent upon the hazard rate as

$$S^k(t|\mathbf{x}_k, H_{k-1}) = \exp(-\int_{t_{k-1}}^{t} \lambda^k(u|\mathbf{x}_k, H_{k-1})\, du). \tag{3.5.10}$$

Finally, given $\mathbf{x}_k$ and the prehistory H_{k-1}, the density of T_k is

$$f^k(t|\mathbf{x}_k, H_{k-1}) = \lambda^k(t|\mathbf{x}_k, H_{k-1}) \exp(-\int_{t_{k-1}}^{t} \lambda^k(u|\mathbf{x}_k, H_{k-1})\, du), \quad t \geq t_{k-1}. \tag{3.5.11}$$

Each of the functions (3.5.8), (3.5.10), and (3.5.11) may be used to analyze multiple spell regression models. If one of these functions is known, the other two may be derived. For regression analysis, it is most appropriate to model the hazard rates.

In principle, all of the regression approaches presented in Section 3.3 may be adapted to the multiepisode case. Assuming, for example, a Weibull regression model, the hazard rate of the *k*th episode is

$$\lambda^k(t|\mathbf{x}_k) = \delta_k \, \lambda_{0k} \, (\lambda_{0k} \, t)^{\delta_k - 1} \exp(\mathbf{x}_k' \boldsymbol{\beta}_k), \tag{3.5.12}$$

assuming the relevant part of the H_{k-1} has already been included in $\mathbf{x}_k$. For the general proportional hazards model

$$\lambda^k(t|\mathbf{x}_k) = \lambda_{0k}(t) \exp(\mathbf{x}_k' \boldsymbol{\beta}_k) \tag{3.5.13}$$

with the unspecified baseline hazard rate $\lambda_{0k}(t)$. It is also possible to choose different classes or models for successive episodes. For example, one may choose a Weibull regression model for the first episode and a Cox model for the second, etc. Of course, the choice of an appropriate approach always depends on the context of the application. An assortment of variants can be used in order to model the dependency of the hazard rate of the *k*th episode upon the durations of previous episodes. One possibility, for example, applied in examining the duration of unemployment, is to include as a covariate for the hazard rate of the *k*th episode only the average length of previous spells of unemployment postulating

$$\lambda^k(t|\mathbf{x}_k, H_{k-1}) = \lambda^k(t|\mathbf{x}_k, \frac{1}{k-1} \sum_{j=1}^{k-1} (t_j - t_{j-1})).$$

Braun and Hoem (1978) have chosen this type of specification without consideration of covariates.

Heckman and Borjas (1980) have studied the duration of unemployment using a hazard rate which depends on a time invariant covariate vector and upon the number of previous unemployment episodes a person experienced.

Multiepisode and Multistate Regression Models

The presentation of the general multiepisode and multistate model follows Hamerle (1988b). The stochastic transition times are once again represented by nonnegative stochastic variables $0 = T_0 < T_1 < T_2 < \ldots$, and the durations of the episodes are

$$V_k = T_k - T_{k-1} \quad , k = 1,2, \ldots .$$

In addition, a state variable $\{Y_k : k = 0,1,2, \ldots\}$ is also defined as a sequence of stochastic variables in a finite state space. We consider the *k*th episode, k = 1,2, … . The (k–1)th episode terminates in state y_{k-1}. The set of attainable states given state y_{k-1} is designated as $M(y_{k-1})$. This set may generally vary from state to state and from episode to episode. For simplicity, we always base the analysis here on the entire state space with all m states. If a specific transition is impossible this is expressed by setting the corresponding transition specific hazard rate or respective transition probability equal to zero.

Given $Y_k = j$, the transition specific hazard rate (transition rate) of the *k*th episode is defined as

$$\lambda_j^k(t|\mathbf{x}_k, H_{k-1}) = \lim_{\Delta t \to 0} \frac{1}{\Delta t} P(t \leq T_k < t + \Delta t, Y_k = j | T_k \geq t, H_{k-1}, \mathbf{x}_k), \tag{3.5.14}$$

where the prehistory of the process until time t_{k-1} is collected in H_{k-1}, thus

$$H_{k-1} = \{t_0, y_0, t_1, y_1, \mathbf{x}_1, \ldots, t_{k-1}, y_{k-1}, \mathbf{x}_{k-1}\}.$$

Once again it should be noted that $\lambda_j^k(t|\mathbf{x}_k, H_{k-1})$ is identical to zero for $t < t_{k-1}$.

The probability law of the transition process and the duration process can be completely expressed in terms of transition specific hazard rates. Therefore, transition specific hazard rates are the basic concept for the construction of regression models for duration data. If we specify the transition specific hazard rates, all relevant duration distributions and transition probabilities can be calculated.

For a complete specification of the hazard rate we must fix (1) the time dependence of the hazard rate, (2) how much of the previous history is included, and (3) the dependence on the incorporated covariates. Completely or partially parameterized models may be applied, as presented in Sections 3.3 and 3.4. The parameter vectors and/or the baseline hazard rate may depend on the number k of the current episode, the origin state y_{k-1}, and the destination state y_k, or may even depend on the previous history in a more general manner. But the number of unknown parameters increases rapidly in this case. An important special case is presented at the end of this section.

The total hazard rate $\lambda^k(t|\mathbf{x}_k, H_{k-1})$ expressing "the risk" of a transition from state y_{k-1} in the *k*th episode is

$$\lambda^k(t|\mathbf{x}_k, H_{k-1}) = \sum_{j=1}^{m} \lambda_j^k(t|\mathbf{x}_k, H_{k-1}). \tag{3.5.15}$$

The survivor function can also be extended to cover the multiepisode and multistate case. It expresses the probability that starting from state y_{k-1} one "survives" time point t. In other words, up to this point in time a *k*th transition has not occurred

$$S^k(t|\mathbf{x}_k, H_{k-1}) = P(T_k > t|\mathbf{x}_k, H_{k-1}) \quad \text{for } t \geq t_{k-1}. \tag{3.5.16}$$

Defining the duration

$$V_k = T_k - T_{k-1},$$

the survivor function is

$$S^k(v|\mathbf{x}_k, H_{k-1}) = P(V_k > v|\mathbf{x}_k, H_{k-1}), \quad v \geq 0. \tag{3.5.17}$$

The relationship between the survivor function and the hazard rate is characterized by

$$S^k(t|\mathbf{x}_k, H_{k-1}) = \exp[-\int_{t_{k-1}}^{t} \lambda^k(u|\mathbf{x}_k, H_{k-1})\, du], \quad t \geq t_{k-1} \tag{3.5.18}$$

and with (3.5.15)

$$S^k(t|\mathbf{x}_k, H_{k-1}) = \exp[-\int_{t_{k-1}}^{t} \sum_{j=1}^{m} \lambda_j^k(u|\mathbf{x}_k, H_{k-1})\, du] =$$

$$= \prod_{j=1}^{m} \exp[-\int_{t_{k-1}}^{t} \lambda_j^k(u|\mathbf{x}_k, H_{k-1})\, du]. \tag{3.5.19}$$

Finally, for the "density function" one obtains

$$f_j^k(t|\mathbf{x}_k, H_{k-1}) = \lim_{\Delta t \to 0} \frac{1}{\Delta t} P(t \leq T_k < t + \Delta t, \ Y_k = j|\mathbf{x}_k, H_{k-1})$$

dependent upon the hazard rate and survivor function, thus

$$f_j^k(t|\mathbf{x}_k, H_{k-1}) = \lambda_j^k(t|\mathbf{x}_k, H_{k-1})\, S^k(t|\mathbf{x}_k, H_{k-1}), \quad t \geq t_{k-1}. \tag{3.5.20}$$

An important special case of a multiepisode model, upon which Chapters 5 and 6 have been based, is obtained when the episode specific hazard rates $\lambda^k(t|\mathbf{x}_k, H_{k-1})$ depend only upon the duration $v = t - t_{k-1}$, such that

$$\lambda^k(t|\mathbf{x}_k, H_{k-1}) = \lambda^{*k}(t - t_{k-1}|\mathbf{x}_k, H_{k-1}). \tag{3.5.21}$$

If one presumes a Cox model as well as the assumption that the baseline hazard rate and the regression coefficients remain constant from episode to episode, then

$$\lambda^k(t|\mathbf{x}_k) = \lambda_0(v) \exp(\mathbf{x}_k'\boldsymbol{\beta}) \quad, v = t - t_{k-1}. \tag{3.5.22}$$

For model (3.5.22), the parameterization is thus simplified greatly as will be discussed below in Section 3.6.6. For the discussion of more general Cox models see Blossfeld and Hamerle (1988a).

3.6 Estimation

After the construction of a statistical model for the event history under discussion, the unknown parameters have to be estimated from the data. In this section, we first discuss the maximum likelihood method, which is the most suited to the requirements of event history analysis given right censored data. First, a brief presentation of the general theory of the maximum likelihood estimation (ML estimation) is given. Then the censoring problem and the necessary modifications are discussed. The following subsections explain the ML estimation procedure for various important parametric regression models (see Section 3.3) and the proportional hazards model partial likelihood estimation procedure developed by Cox. The ML estimator for competing risks models and for general multiepisode and multistate regression models are discussed in the next section. The last section deals with least squares estimation.

3.6.1 The General Theory of Maximum Likelihood Estimation

In the following, the fundamental principles of the maximum likelihood estimator are briefly sketched. For detailed presentation, which, however, occasionally presumes some knowledge of measure theory, the reader may refer to, for example, Cox and Hinkley (1974), Rao (1973), or Witting and Nölle (1970).

The starting point of maximum likelihood estimation are the observed data. In the simplest case one is dealing with independent realizations $x_1, \ldots, x_n$ of some random variable X. The density function (for a continuous variable) or the probability mass function (given a discrete variable) are specified entirely by an unknown parameter vector $\boldsymbol{\theta}$. This may be written as $f(\mathbf{x}; \boldsymbol{\theta})$. The parameter vector $\boldsymbol{\theta} = (\theta_1, \ldots, \theta_p)'$ is to be estimated. The likelihood function of the sample is

$$L(\boldsymbol{\theta}; x_1, \ldots, x_n) = \prod_{i=1}^{n} f_i(x_i | \boldsymbol{\theta}). \tag{3.6.1}$$

The right-hand side of (3.6.1) represents the common density or respective probability of $x_1, \ldots, x_n$ for the true, however unknown $\boldsymbol{\theta}$. The likelihood function $L(\boldsymbol{\theta}; x_1, \ldots, x_n)$ is considered to be a function of $\boldsymbol{\theta}$ given the sample values $x_1, \ldots, x_n$. $\boldsymbol{\theta}$ may vary within the parameter space Θ.

The *maximum likelihood principle* consists of choosing an estimated parameter value $\hat{\boldsymbol{\theta}} = \hat{\boldsymbol{\theta}}(x_1, \ldots, x_n) \in \Theta$ given the data $x_1, \ldots, x_n$ from which a maximal probability density (in the discrete case, a maximal probability) is obtained for the data.

$\hat{\boldsymbol{\theta}} = \hat{\boldsymbol{\theta}}(x_1, \ldots, x_n)$ is thus called the maximum likelihood (ML) estimate for $\boldsymbol{\theta}$, if

$$L(\hat{\boldsymbol{\theta}}; x_1, \ldots, x_n) \geq L(\boldsymbol{\theta}; x_1, \ldots, x_n) \quad \text{for all } \boldsymbol{\theta} \in \Theta. \tag{3.6.2}$$

Usually, in order to simplify the necessary technical calculations one maximizes the logarithm of the likelihood function, since the logarithm is a strictly monotone transformation and the maximum point remains unchanged.

The log likelihood function is

$$l(\boldsymbol{\theta}; x_1, \ldots, x_n) = \ln L(\boldsymbol{\theta}; x_1, \ldots, x_n) = \sum_{i=1}^{n} \ln f_i(x_i; \boldsymbol{\theta}). \tag{3.6.3}$$

Important for the concrete calculation of the ML estimate are the derivatives

$$\frac{\partial}{\partial \theta_j} l(\boldsymbol{\theta}; x_1, \ldots, x_n).$$

The column vector of these derivatives

$$s(\boldsymbol{\theta}; x_1, \ldots, x_n) = \left(\frac{\partial}{\partial \theta_1} l(\boldsymbol{\theta}; x_1, \ldots, x_n), \ldots, \frac{\partial}{\partial \theta_p} l(\boldsymbol{\theta}; x_1, \ldots, x_n)\right)'$$

is referred to as the *score function*.

Setting the score function equal to zero yields the ML equations

$$\frac{\partial}{\partial \theta_j} l(\boldsymbol{\theta}; x_1, \ldots, x_n) = 0 \qquad j = 1, \ldots, p. \tag{3.6.4}$$

Only in a few cases, such as for the normal or exponential distribution, an explicit solving of the ML equations is possible. Usually, one must calculate the ML estimate by iterative methods.

Newton-Raphson Technique

An important iterative method of solving maximization problems or the solving of ML equations is the Newton method. Although today modified or quasi-Newton methods are usually applied, the technique of the Newton method is still of fundamental importance. In the following, we only sketch the basic idea of the method. For a detailed introduction, see, for example, Luenberger (1973) or Stoer (1976, Chapter 5) and Fahrmeir and Hamerle (1984, Section 3.2.4).

The Newton-Raphson technique of maximization of the log likelihood function $\ln L(\boldsymbol{\theta}; x_1, \ldots, x_n)$ is based upon the Taylor expansion of the score function $s(\boldsymbol{\theta}) = \partial \ln L / \partial \boldsymbol{\theta}$. The method is iterative. Beginning from some initial value $\boldsymbol{\theta}_0$, successive approximations $\boldsymbol{\theta}_1, \boldsymbol{\theta}_2, \ldots, \boldsymbol{\theta}_k, \ldots$ are constructed. Expanding $s(\boldsymbol{\theta})$ around $\boldsymbol{\theta}_k$, one obtains in matrix notation

$$s(\boldsymbol{\theta}) \approx s(\boldsymbol{\theta}_k) + \frac{\partial^2 \ln L}{\partial \boldsymbol{\theta}\, \partial \boldsymbol{\theta}'} (\boldsymbol{\theta}_k)\, (\boldsymbol{\theta} - \boldsymbol{\theta}_k). \tag{3.6.5}$$

If this approximation were correct, one would obtain $s(\boldsymbol{\theta}) = 0$ for

$$\boldsymbol{\theta}_{k+1} = \boldsymbol{\theta}_k - \left(\frac{\partial^2 \ln L}{\partial \boldsymbol{\theta}\, \partial \boldsymbol{\theta}'} (\boldsymbol{\theta}_k)\right)^{-1} s(\boldsymbol{\theta}_k). \tag{3.6.6}$$

Since the approximation (3.6.5) is, however, usually not exact, (3.6.6) is applied iteratively until $\boldsymbol{\theta}_{k+1} \approx \boldsymbol{\theta}_k$ with sufficient exactness.

We now demonstrate the application of the Newton-Raphson technique (3.6.6) for the simple case of a single parameter θ. We have

$$\theta_{k+1} = \theta_k - \frac{s(\theta_k)}{\frac{d^2}{d\theta^2} \ln L(\theta_k)} = \theta_k - \frac{s(\theta_k)}{s'(\theta_k)}. \tag{3.6.7}$$

In Figure 3.11 the score function $s(\theta)$, whose zero positions $\hat{\theta}$ are to be found, is graphed.

Figure 3.11: Graphical Illustration of the Newton-Raphson Technique

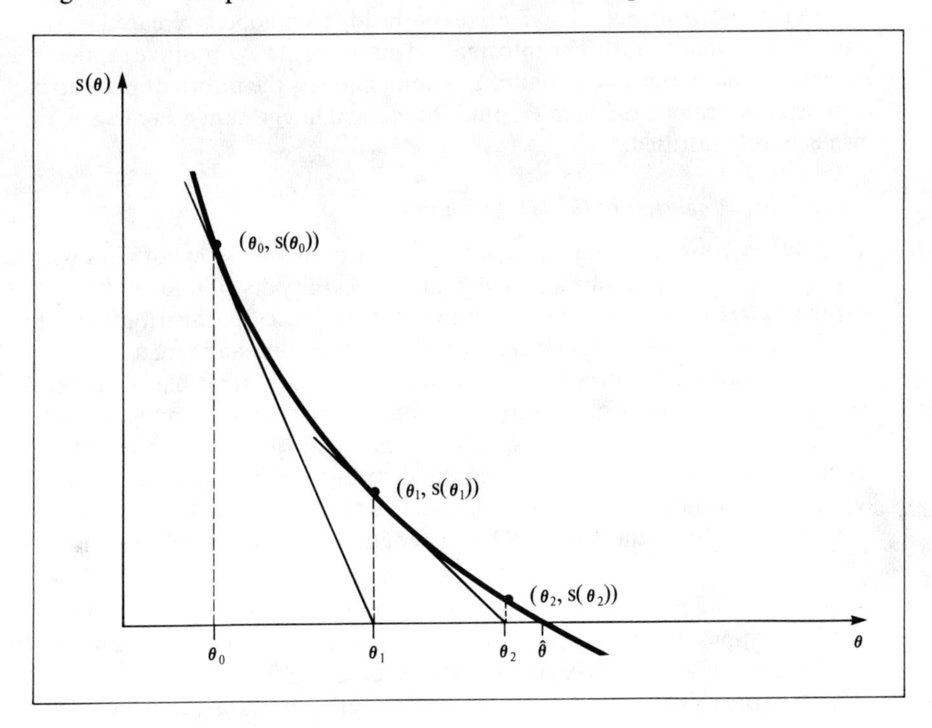

The initial value is θ_0. If one draws a tangent to the point $(\theta_0, s(\theta_0))$, the tangential equation is

$$y = s'(\theta_0)\,(\theta - \theta_0) + s(\theta_0).$$

Calculating the intersection point of the tangent line with the θ axis designated as θ_1, one obtains

$$\theta_1 = \theta_0 - \frac{s(\theta_0)}{s'(\theta_0)}.$$

This is exactly the iterative rule (3.6.7) for $k = 0$. This procedure is then successively applied until $\theta_{k+1} \approx \theta_k$.

Since a maximum of the log likelihood function is desired, the matrix of the second derivatives of $\ln L(\boldsymbol{\theta}; x_1, \ldots, x_n)$ must be negative definite. Let

$$\mathbf{H} = \frac{\partial^2 l}{\partial \boldsymbol{\theta}\, \partial \boldsymbol{\theta}'} = \left(\frac{\partial^2 l}{\partial \theta_i\, \partial \theta_j} \right)$$

be the $(p \times p)$ matrix of the second derivatives of the log likelihood function, then

$$\mathbf{I}(\boldsymbol{\theta}) = E\left(- \frac{\partial^2 l(\boldsymbol{\theta}; x_1, \ldots, x_n)}{\partial \boldsymbol{\theta}\, \partial \boldsymbol{\theta}'} \right)$$

represents the (Fisher) information matrix of the sample $x_1, \ldots, x_n$.

A procedure closely related to the Newton method is the *scoring algorithm* where the matrix of second derivatives of the log likelihood is replaced by the information matrix $\mathbf{I}(\boldsymbol{\theta})$. The information matrix is always positive semidefinite and usually positive definite. For some models the information matrix can be easier calculated than $\mathbf{H}$, but if numerical integration is necessary, its use is hardly justified.

Asymptotic Properties of the ML Estimator

In order to construct confidence intervals for $\hat{\theta}_j$ or to test hypotheses with regard to certain θ_j or subvectors of $\boldsymbol{\theta}$, the probability distribution of the ML estimator $\hat{\boldsymbol{\theta}}$ is needed. With the exception of special cases, the distribution of $\hat{\boldsymbol{\theta}}$ for finite samples of size n may not be stated. For certain standard situations, however, asymptotic properties of $\hat{\boldsymbol{\theta}}$ for $n \to \infty$ are derivable. One such standard situation is, for example, the case of independent and identically distributed observations $x_1, \ldots, x_n$. The maximum likelihood estimator is then consistent and asymptotically normally distributed with expectation $\boldsymbol{\theta}$ and variance-covariance matrix $(\mathbf{I}(\boldsymbol{\theta}))^{-1}$, which is the inverse of the information matrix, given that the log likelihood function satisfies certain regularity conditions.

This has the implication for practical applications that for large sample sizes, $\hat{\boldsymbol{\theta}}$ is approximately normally distributed, with an expected value $\boldsymbol{\theta}$ and a variance-covariance matrix $(\mathbf{I}(\boldsymbol{\theta}))^{-1}$. However, the information matrix contains the true value of the parameter $\boldsymbol{\theta}$. It is therefore necessary to estimate the information matrix. One possible approach is to evaluate the matrix of the second derivatives of the log likelihood function at point $\hat{\boldsymbol{\theta}}$ and to apply

$$-\hat{\mathbf{H}} = \left(- \frac{\partial^2 l}{\partial \boldsymbol{\theta}\, \partial \boldsymbol{\theta}'} (\hat{\boldsymbol{\theta}})\right) \tag{3.6.8}$$

as an estimation of the information matrix. By the application of the Newton method this does not present a problem, since this matrix is calculated during the last iteration.

The main diagonal elements of the matrix $(-\hat{\mathbf{H}})^{-1}$ contain the estimated (asymptotic) variances of the ML estimators $\hat{\theta}_j$, $j = 1, \ldots, p$ which are especially relevant for the construction of confidence intervals or tests.

It must be stressed that the models dealt with in this book do *not* generally belong to the above mentioned standard situations, especially if censored data is at hand. The asymptotic properties of the maximum likelihood estimates must then first be proven, which in some cases has yet to be fully solved. We return to this issue in Section 3.6.4.

If asymptotic properties are given or assumed to be valid, then particular parameters or certain model parts may be tested as to their significance. Such hypotheses may always be formulated as

$$H_0 : \mathbf{C}\, \boldsymbol{\theta} = 0 \qquad , m = \text{rank}\ (\mathbf{C}) \tag{3.6.9}$$

with a suitable matrix $\mathbf{C}$. The test of such a linear hypothesis may, for example, be carried out with the aid of the likelihood ratio test. The test statistic is

$$2(\ln L(\hat{\boldsymbol{\theta}}) - \ln L(\tilde{\boldsymbol{\theta}})), \tag{3.6.10}$$

where $\tilde{\boldsymbol{\theta}}$ represents the ML estimator under the restriction $\mathbf{C}\boldsymbol{\theta} = \mathbf{0}$ and $\hat{\boldsymbol{\theta}}$ signifies the ML estimator without restrictions. The test statistic has an asymptotic (central) χ^2-distribution with m degrees of freedom, if H_0 is valid.

In order to test the linear hypothesis $\mathbf{C}\boldsymbol{\theta} = \mathbf{0}$ one may also apply the score statistic or the Wald statistic. The asymptotic distributions are the same as for the likelihood ratio statistic. With regard to this aspect, compare further, for example, Rao (1973, pp. 351ff.).

3.6.2 Censoring

In applying the maximum likelihood estimator—as in other estimation methods—each sample element must be characterized as a realization of the stochastic variable in question. Since in event history analysis the termination of the entire observation time period is given, a spell may not be closed. In such a situation we have right censored data. The sample realization t_i of an individual simply states then that the duration of an episode is of at least t_i time units. Usually the sample consists of some t_i values which are complete durations, whereas the remaining t_i values are right censored. This is expressed with the aid of a *censoring indicator* δ_i as

$$\delta_i = \begin{cases} 1 & \text{if } t_i \text{ is not censored} \\ 0 & \text{if } t_i \text{ is censored,} \qquad i = 1, \ldots, n. \end{cases}$$

The possibility of simply ignoring the censored data and thereby reducing the size of the sample is not recommendable since this generally leads to biased estimates.

The maximum likelihood method offers the possibility of considering explicitly right censored data within the estimation procedure. Accordingly, one must analyze in depth the underlying censoring mechanism of the data and formulate a statistical model. Given the occurrence of censored data in a specific application, various statistical concepts are applicable. At the moment, only two models are discussed which are of special interest to the applications in this book.

Censoring Model I

In Model I, for each individual i, $i = 1, \ldots, n$ a fixed observation time period L_i is given. Of interest in a life course study is the time span between the individual's birth and the interview. The L_i time periods may vary from individual to individual. The durations T_i, with density function $f_i(t)$ and survivor function $S_i(t)$, may only be observed exactly given $T_i \leq L_i$. The censoring indicator is

$$\delta_i = \begin{cases} 1 & \text{if } T_i \leq L_i \\ 0 & \text{if } T_i > L_i \end{cases}$$

and the observations t_i in the sample are

$$t_i = \min(T_i, L_i) \quad , i = 1, ..., n.$$

The common probability distribution of (T_i, δ_i) may be calculated as follows:

Given $\delta_i = 0$, it is then valid that $T_i > L_i$ and thus $t_i = L_i$ and the probability is $S_i(L_i)$. The density of $(T_i, \delta_i = 1)$ is

$$\begin{aligned} f_i(t_i, \delta_i = 1) &= f_i(t_i \mid \delta_i = 1) \cdot P(\delta_i = 1) \\ &= f_i(t_i \mid T_i \leq L_i) \cdot P(T_i \leq L_i) \\ &= \frac{f_i(t_i)}{1 - S_i(L_i)} (1 - S_i(L_i)) \\ &= f_i(t_i) \end{aligned}$$

and the likelihood contribution of individual i is given by

$$L_i = f_i(t_i)^{\delta_i} S_i(L_i)^{1-\delta_i}. \quad (3.6.11)$$

The likelihood function is

$$L = \prod_{i=1}^{n} f_i(t_i)^{\delta_i} S_i(L_i)^{1-\delta_i}. \quad (3.6.12)$$

Censoring Model II (Random Censoring)

Another variation is to consider the time of censoring likewise as realizations of random variables and not to assume that they are fixed as in Model I. The durations T_i and the censoring times C_i are postulated to be independent random variables characterized by the respective densities $f_i(t)$, $g_i(t)$, and the corresponding survivor functions $S_i(t)$ and $G_i(t)$.

For a noncensored observation $(t_i, \delta_i = 1)$, that is $T_i = t_i$ and $T_i \leq C_i$, due to the independency of T_i and C_i (without considering the covariates) the joint distribution is

$$f_i(t_i)\ G_i(t_i),$$

for a censored observation $(t_i, \delta_i = 0)$, that is $T_i > C_i$, $C_i = t_i$ the joint distribution is

$$g_i(t_i)\ S_i(t_i).$$

For (t_i, δ_i) of an individual i, this may be summarized in the likelihood contribution

$$L_i = [f_i(t_i)\, G_i(t_i)]^{\delta_i}\, [g_i(t_i)\, S_i(t_i)]^{1-\delta_i} = f_i(t_i)^{\delta_i}\, S_i(t_i)^{1-\delta_i}\, g_i(t_i)^{1-\delta_i}\, G_i(t_i)^{\delta_i}. \tag{3.6.13}$$

If one assumes that the distributions of the censoring times are not dependent upon the relevant parameters for f_i and S_i, especially if they are not dependent upon the regression coefficients, both of the last terms in (3.6.13) may be collected in a constant term c (with regard to the relevant parameters) which has no effect upon the maximization of the likelihood function or the log likelihood function. Consequently, the total likelihood function is

$$L = c \cdot \prod_{i=1}^{n} f_i(t_i)^{\delta_i}\, S_i(t_i)^{1-\delta_i}. \tag{3.6.14}$$

Note that the likelihood function (3.6.14) is of the form (3.6.12). This form of likelihood function is typical for the ML estimation given right censored data. It is also valid for more general censoring mechanisms. With regard to this issue, see Kalbfleisch and Prentice (1980, Chapter 5) or Lagakos (1979). Principally, one must assume that the censoring mechanism is "noninformative" (see Kalbfleisch and Prentice, 1980, Chapter 5).

3.6.3 Maximum Likelihood Estimator for Parametric Regression Models

In this section the ML estimation procedure is demonstrated for regression models which possess a completely specified hazard rate, which means that the distribution of the duration of the episode is fully specified. We do not present each of the models of Section 3.3.2 here in detail since the calculations are analogous and practical application always implies the aid of a computer.

For each individual or subject i, i = 1, ..., n the following data is given:

t_i : duration of the episode (which may eventually be censored)
δ_i : censoring indicator
$\mathbf{x}_i$: covariate vector

According to (3.6.12) or (3.6.14), the likelihood function is thus (possibly up to a factor of proportionality),

$$L = \prod_{i=1}^{n} f_i(t|\mathbf{x}_i)^{\delta_i}\, S_i(t_i|\mathbf{x}_i)^{1-\delta_i}.$$

Considering the relationship existing between the hazard rate and the survivor function (see (3.2.10)), one obtains

$$\begin{aligned} L &= \prod_{i=1}^{n} \lambda_i(t_i|\mathbf{x}_i)^{\delta_i}\, S_i(t_i|\mathbf{x}_i) \\ &= \prod_{i=1}^{n} \lambda_i(t_i|\mathbf{x}_i)^{\delta_i}\, \exp\left(-\int_0^{t_i} \lambda_i(u|\mathbf{x}_i)\, du\right). \end{aligned} \tag{3.6.15}$$

The likelihood function (3.6.15) is dependent only upon the hazard rate. Especially interesting for the ML estimation is the log likelihood function which simplifies the necessary calculations. If the hazard rate is determined by an unknown parameter vector $\boldsymbol{\theta}$, which also contains the regression coefficients β_j, the log likelihood function is obtained as

$$l(\boldsymbol{\theta}; t, \mathbf{x}) = \ln L(\boldsymbol{\theta}; t, \mathbf{x}) = \sum_{i=1}^{n} (\delta_i \ln \lambda_i(t_i|\mathbf{x}_i, \boldsymbol{\theta}) - \int_o^{t_i} \lambda_i (u|\mathbf{x}_i, \boldsymbol{\theta})\, du). \tag{3.6.16}$$

The log likelihood function is to be maximized with respect to $\boldsymbol{\theta}$. Usually this is done by applying an iterative method such as the Newton method or a modified Newton method. As has already been described in Section 3.6.1, the inverse information matrix evaluated at $\hat{\boldsymbol{\theta}}$ is used as an estimation of the asymptotic covariance matrix $\mathbf{Cov}(\hat{\boldsymbol{\theta}})$.

Now we describe the ML estimation for the most simple parametric hazard rate regression model, the exponential model.

According to (3.6.2) or (3.6.3) the hazard rate is dependent upon the covariate vector $\mathbf{x} = (x_1, \ldots, x_p)'$, where the first component is equal to one. The hazard rate is thus

$$\lambda(t|\mathbf{x}) = \exp(\mathbf{x}'\boldsymbol{\beta})$$

resulting in the likelihood function ($\boldsymbol{\theta} = \boldsymbol{\beta}$)

$$L(\boldsymbol{\beta}; t, \mathbf{x}) = \prod_{i=1}^{n} \exp(\delta_i\, \mathbf{x}_i'\boldsymbol{\beta}) \exp(-\exp(\mathbf{x}_i'\boldsymbol{\beta})\, t_i) \tag{3.6.17}$$

and for the log likelihood function

$$l(\boldsymbol{\beta}; t, \mathbf{x}) = \sum_{i=1}^{n} [\delta_i\, \mathbf{x}_i'\boldsymbol{\beta} - \exp(\mathbf{x}_i'\boldsymbol{\beta})\, t_i]. \tag{3.6.18}$$

The score function possesses the components

$$s_j(\boldsymbol{\beta}; t, \mathbf{x}) = \frac{\partial l(\boldsymbol{\beta}; t, \mathbf{x})}{\partial \beta_j} = \sum_{i=1}^{n} x_{ij}\, (\delta_i - \exp(\mathbf{x}_i'\boldsymbol{\beta})\, t_i), \quad j = 1, \ldots, p,$$

and the ML equations are

$$\sum_{i=1}^{n} x_{ij}\, (\delta_i - \exp(\mathbf{x}_i'\boldsymbol{\beta})\, t_i) = 0, \quad j = 1, \ldots, p. \tag{3.6.19}$$

The matrix of the second derivatives of the log likelihood function is characterized in the *j*th row and *k*th column by the element

$$h_{jk}(\boldsymbol{\beta}) = \frac{\partial^2 l(\boldsymbol{\beta}; t, \mathbf{x})}{\partial \beta_j\, \partial \beta_k} = \sum_{i=1}^{n} x_{ij}\, x_{ik} \exp(\mathbf{x}_i'\boldsymbol{\beta})\, t_i.$$

The solution of the ML equations yields the ML estimators $\hat{\boldsymbol{\beta}}$, given that the matrix $\mathbf{H}(\hat{\boldsymbol{\beta}}) = (h_{jk}(\hat{\boldsymbol{\beta}}))$ is negative definite. $\mathbf{H}(\hat{\boldsymbol{\beta}})^{-1}$ is an estimate of the asymptotic covariance matrix.

Since $E(t_i \exp(\mathbf{x}_i'\boldsymbol{\beta})) = 1$, the information matrix is given by $\mathbf{I}(\boldsymbol{\beta}) = -\Sigma\, \mathbf{x}_i\mathbf{x}_i'$, and one can also use the scoring algorithm.

If one chooses another model for the hazard rate, such as the Weibull model or the Gompertz-Makeham rate, the procedure remains completely analogous, outside of the fact that one has to do a few more calculations since further unknown parameters must be considered along with the regression coefficients.

3.6.4 Cox Proportional Hazards Model: Partial Likelihood

In the proportional hazards regression model (PH model) first formulated by Cox (1972), the hazard rate is

$$\lambda(t|\mathbf{x}) = \lambda_0(t)\ \exp(\mathbf{x}'\boldsymbol{\beta}).$$

The second factor $\exp(\mathbf{x}'\boldsymbol{\beta})$ may also be replaced by another positive function $g(\mathbf{x}; \boldsymbol{\beta})$.

The likelihood function for the PH model (Cox model) is

$$L(\boldsymbol{\beta}, \lambda_0(t); t, \mathbf{x}) = \prod_{i=1}^{n} [\lambda_0(t_i) \exp(\mathbf{x}_i'\boldsymbol{\beta})]^{\delta_i} \exp[-\int_0^{t_i} \lambda_0(u) \exp(\mathbf{x}_i'\boldsymbol{\beta})\, du]. \quad (3.6.20)$$

(3.6.20) contains the unknown baseline hazard rate $\lambda_0(t)$, that is not only the unknown parameter $\boldsymbol{\beta}$ but the "nuisance function" $\lambda_0(t)$ as well. Therefore, (3.6.20) can not be used to estimate $\boldsymbol{\beta}$.

Cox (1972, 1975) suggested the factorization of the likelihood function (3.6.20). The uncensored durations are $t_{(1)} < \ldots < t_{(k)}$ $(k \leq n)$ and the "risk set" $R(t)$ is introduced as the set of individuals whose episode just before t has not yet ended and is not censored. Thus, from (3.6.20) one obtains

$$L(\boldsymbol{\beta}, \lambda_0(t); t, \mathbf{x}) = \prod_{i=1}^{k} \frac{\exp(\mathbf{x}_{(i)}'\boldsymbol{\beta})}{\sum_{l \in R(t_{(i)})} \exp(\mathbf{x}_l'\boldsymbol{\beta})} \sum_{l \in R(t_{(i)})} \lambda_0(t_{(i)}) \exp(\mathbf{x}_l'\boldsymbol{\beta}) \prod_{i=1}^{n} S_0(t_i)^{\exp(\mathbf{x}_i'\boldsymbol{\beta})} \quad (3.6.21)$$

with $S_0(t) = \exp(-\int_0^t \lambda_0(u)\, du)$.

The first factor

$$PL(\boldsymbol{\beta}; t, \mathbf{x}) = \prod_{i=1}^{k} \frac{\exp(\mathbf{x}_{(i)}'\boldsymbol{\beta})}{\sum_{l \in R(t_{(i)})} \exp(\mathbf{x}_l'\boldsymbol{\beta})}, \quad (3.6.22)$$

which is only dependent upon $\boldsymbol{\beta}$, is termed the "partial likelihood" by Cox (1972, 1975). Cox proposed to treat (3.6.22) as a usual likelihood function, and to maximize it with respect to $\boldsymbol{\beta}$.

Since the second factor on the right-hand side of (3.6.21) also contains $\boldsymbol{\beta}$, some information is lost. This may influence the obtained estimation results, especially if the sample is of small size. Yet the desired properties such as consistency and asymptotic normality, may still be proven to be valid under certain conditions. Regarding this, refer to Tsiatis (1981), Bailey (1983), Naes (1982), Prentice and Self (1983), and especially Andersen and Gill (1982).

Occasionally, another interpretation of the partial likelihood is useful. The term

$$\frac{\lambda\,(t_{(i)}|\mathbf{x}_i)}{\sum\limits_{l \in R(t_{(i)})} \lambda\,(t_{(i)}|\mathbf{x}_l)} = \frac{\exp(\mathbf{x}'_{(i)}\boldsymbol{\beta})}{\sum\limits_{l \in R(t_{(i)})} \exp(\mathbf{x}'_l\boldsymbol{\beta})}$$

on the right-hand side of (3.6.22) may be interpreted as the (conditional) probability that in time point $t_{(i)}$ an event has just occurred for individual i, presuming that for the individuals of the risk set $R(t_{(i)})$ just before time point $t_{(i)}$ the event in question has not yet occurred and that at time point $t_{(i)}$ exactly one event occurs. The product over all k points of time in which an event occurs results in the partial likelihood (3.6.22).

Application of the partial likelihood (3.6.22) requires that the durations t_i can be exactly measured so that no ties occur. In practical application, however, it is often the case that ties do arise due to inaccurate measurement or because only time intervals in which events occur may be stated. In such a case, the partial likelihood must be modified. Breslow (1974) has proposed approximating (3.6.22) by

$$PL(\boldsymbol{\beta}; t, \mathbf{x}) = \prod_{i=1}^{k} \frac{\exp(\mathbf{s}_i'\boldsymbol{\beta})}{[\sum\limits_{l \in R(t_{(i)})} \exp(\mathbf{x}_l'\boldsymbol{\beta})]^{d_i}}. \tag{3.6.23}$$

d_i thereby represents the number of equal durations $t_{(i)}$, and $\mathbf{s}_i$ is the sum of the covariate vector of the corresponding d_i individuals.

If the number of ties is large, the application of a discrete model is advisable, as discussed in Section 3.10.

Estimating Baseline Hazard Rate and Survivor Function

The survivor function of the PH model is

$$S(t|\mathbf{x}) = \exp(-\exp(\mathbf{x}'\boldsymbol{\beta})\int_0^t \lambda_0(u)\,du). \tag{3.6.24}$$

In order to estimate $S(t|\mathbf{x})$, estimates for $\hat{\boldsymbol{\beta}}$ as well as for the baseline hazard rate $\lambda_0(t)$ are required. There are various means of doing this, of which only two will be presented here which are implemented in common computer program packages (e.g. BMDP)

Breslow (1974) proposes to set $\lambda_0(t)$ to be constant between the observed durations $t_{(1)} < t_{(2)} < \ldots < t_{(k)}$

$$\lambda_0(t) = \lambda_i \qquad \text{for } t_{(i-1)} < t \le t_{(i)},\ i = 1, \ldots, k$$

where $t_{(0)} = 0$. Consequently, the ML estimator is

$$\hat{\lambda}_i = \frac{1}{t_{(i)} - t_{(i-1)}} \frac{d_i}{\sum_{l \in R(t_{(i)})} \exp(\mathbf{x}_l' \boldsymbol{\beta})} \quad \text{for } i = 1, \ldots, k. \tag{3.6.25}$$

d_i represents the number of episodes just ending at time point $t_{(i)}$. Given sufficiently exact measurement $d_i = 1$.

Link (1984) applies the integral of the Breslow estimator (this corresponds to a linear interpolation) to estimate the cumulated hazard rate

$$\Lambda_0(t) = \int_0^t \lambda_0(u)\, du.$$

One obtains

$$\hat{\Lambda}_0(t) = \int_0^t \hat{\lambda}_0(u)\, du = \sum_{i=1}^{s} (t_{(i)} - t_{(i-1)})\, \hat{\lambda}_i + (t - t_{(s)})\, \hat{\lambda}_{s+1}, \tag{3.6.26}$$

where for a given t the value s must be chosen such that $t_{(s)} < t$ and $t_{(s+1)} \geq t$. The estimator of the survivor function is

$$\hat{S}(t|\mathbf{x}) = \exp(-\hat{\Lambda}(t|\mathbf{x})) = \exp(-\exp(\mathbf{x}'\hat{\boldsymbol{\beta}})\, \hat{\Lambda}_0(t)). \tag{3.6.27}$$

3.6.5 Maximum Likelihood Estimation for Competing Risks Models

The statistical presentation of multistate or competing risks models in event history analysis was given in Section 3.4. In this section, we deal with the maximum likelihood estimation of the parameters of these models.

We assume that m different types of events or risks are possible, designated as y, $y \in \{1, \ldots, m\}$. The contribution of an individual i to the likelihood function given the covariate vector $\mathbf{x}_i$ is

$$L_i = f(t_i, y_i|\mathbf{x}_i)^{\delta_i}\, S(t_i|\mathbf{x}_i)^{1-\delta_i} = \lambda_{y_i}(t_i|\mathbf{x}_i)^{\delta_i}\, S(t_i|\mathbf{x}_i). \tag{3.6.28}$$

Dependent upon the transition specific hazard rates, this results in

$$\begin{aligned} L_i &= \lambda_{y_i}(t|\mathbf{x}_i)^{\delta_i} \exp\left(-\int_0^{t_i} \lambda(u|\mathbf{x}_i)\, du\right) \\ &= \lambda_{y_i}(t|\mathbf{x}_i)^{\delta_i} \exp\left(-\int_0^{t_i} \sum_{j=1}^{m} \lambda_j(u|\mathbf{x}_i)\, du\right) \\ &= \lambda_{y_i}(t|\mathbf{x}_i)^{\delta_i} \prod_{j=1}^{m} \exp\left(-\int_0^{t_i} \lambda_j(u|\mathbf{x}_i)\, du\right). \end{aligned} \tag{3.6.29}$$

For the total likelihood function, one obtains

$$L = \prod_{i=1}^{n} \lambda_{y_i}(t|\mathbf{x}_i)^{\delta_i} \prod_{j=1}^{m} \exp\left(-\int_0^{t_i} \lambda_j(u|\mathbf{x}_i)\, du\right). \tag{3.6.30}$$

Once again, δ_i represents a censoring indicator. In the following, (3.6.30) will be rearranged.

It is postulated that $t_{j1} < t_{j2} < \ldots < t_{jn_j}$ represents the n_j noncensored durations till the transition into state j. Then the likelihood function may be rewritten as

$$L = \prod_{j=1}^{m} \prod_{k=1}^{n_j} \lambda_j (t_{jk}|\mathbf{x}_{jk}) \prod_{i=1}^{n} S_j (t_i|\mathbf{x}_i). \tag{3.6.31}$$

$\mathbf{x}_{jk}$ is the covariate of an individual with the observed noncensored duration t_{jk} and

$$S_j(t_i|\mathbf{x}_i) = \exp(-\int_0^{t_i} \lambda_j(u|\mathbf{x}_i)\, du).$$

From (3.6.31) it is seen that the likelihood function may be divided into the product

$$L = \prod_{j=1}^{m} L_j \quad \text{mit } L_j = \prod_{k=1}^{n_j} \lambda_j(t_{jk}|\mathbf{x}_{jk}) \prod_{i=1}^{n} S_j(t_i|\mathbf{x}_i).$$

The L_j factors may by further rearranged as

$$L_j = \prod_{i=1}^{n} [\lambda_j(t_i|\mathbf{x}_i)]^{\delta_{ij}} S_j(t_i|\mathbf{x}_i) \tag{3.6.32}$$

$$\text{with } \delta_{ij} = \begin{cases} 1 & \text{if for individual i a transition to state j occurs at time } t_i, \\ 0 & \text{otherwise.} \end{cases}$$

The log likelihood function $\ln L = \sum_{j=1}^{m} \ln L_j$ may be maximized separately for each j given that the transition specific hazard rates $\lambda_j (t|\mathbf{x})$ are dependent upon the parameter vector $\boldsymbol{\theta}_j$, j = 1, ..., m, where the $\boldsymbol{\theta}_j$'s have no common components.

In particular, one may apply the implemented programs in which all transitions to a state other than j are considered to be censored observations. This is evident from (3.6.32).

Basically, one can apply any of the approaches presented in Section 3.3 in order to model the transition specific hazard rate. It must, however, be observed that with an increasing number of risks or types of events the number of parameters in the model also rapidly increase which may decrease the exactness of estimation. In any given practical application, one must always reach a compromise between the necessary model adequacy and the statistical consequences, especially when dealing with small or medium-sized samples.

3.6.6 Maximum Likelihood Estimation of the Multiepisode Case

Finally, in this section we deal with the extension of the estimation procedure given the multiepisode case. For each study subject i (i = 1, ..., n), the entire event history in the observation period must be known. This requires knowledge of the following data:

y_{i0}	initial state;
n_i	number of episodes within the observation period;
$t_{i1} < t_{i2} < \ldots < t_{in_i}$	points of time in which a state transition or event occurs;
$y_{i1}, y_{i2}, \ldots, y_{in_i}$	states which are occupied in the above mentioned points in time;
δ_i	an indicator expressing whether the n_i-th episode is censored or noncensored;
$\mathbf{x}_{i1}, \mathbf{x}_{i2}, \ldots, \mathbf{x}_{in_i}$	covariate vectors which are measured at the beginning of each episode.

For the sake of simplicity, we omit in the following the index i when constructing the contribution of an individual i to the likelihood function. The total likelihood function is once again obtained as the product of the contributions of all individuals of the sample. If it is at first assumed that no censored observations of the last episode are at hand, then $\delta_i = 1$ for all i. The contribution of an individual to the likelihood function, given the initial state, is

$$L_i = f(t_{n_i}, y_{n_i}, \mathbf{x}_{n_i}, \ldots, t_1, y_1, \mathbf{x}_1 | y_0), \tag{3.6.33}$$

given that individual i is in state y_o at time $t_0 = 0$. Using elementary properties of conditional probabilities, (3.6.33) can be written as

$$\begin{aligned} L_i &= \prod_{k=1}^{n_i} f(t_k, y_k, \mathbf{x}_k | H_{k-1}) \\ &= \prod_{k=1}^{n_i} f(t_k, y_k | H_{k-1}, \mathbf{x}_k)\, g(\mathbf{x}_k | H_{k-1}) \end{aligned} \tag{3.6.34}$$

where $H_0 = \{y_0\}$.

In the following we only use the first factor on the right-hand side of (3.6.34). If the marginal distribution $g(\mathbf{x}_k | H_{k-1})$ of the covariates does not depend on the relevant parameters, in particular on the regression coefficients which determine the distributions $f(t_k, y_k | H_{k-1}, \mathbf{x}_k)$, inference concerning these parameters can be based without loss of information on the conditional distributions $f(t_k, y_k | H_{k-1}, \mathbf{x}_k)$. If this assumption does not hold, it becomes necessary to specify a parametric form for the marginal distribution of the covariates. If this is impossible, one can nevertheless use the first factor on the right-hand side of (3.6.34) considering it as a partial likelihood, but there is a loss of efficiency. If special sampling schemes are used, for example, sampling of spells in progress, or if there is a considerable amount of left censored spells, the marginal distribution of the explanatory variables in the sample becomes informative for the parameters of interest (see Ridder, 1984), but we do not consider this here.

The densities $f(t_k, y_k | H_{k-1}, x_k)$ can be expressed as

$$f(t_k, y_k | H_{k-1}, x_k) = \lim_{\Delta t \to 0} \frac{1}{\Delta t} P(t_k \leq T_k < t_k + \Delta t, y_k | H_{k-1}, x_k)$$

$$= \lim_{\Delta t \to 0} \frac{1}{\Delta t} P(t_k \leq T_k < t_k + \Delta t, y_k | T_k \geq t_k, H_{k-1}, x_k) \, P(T_k \geq t_k | H_{k-1}, x_k). \tag{3.6.35}$$

The first term on the right-hand side of (3.6.35) is just the transition rate into state y_k. The second term expresses the probability during the *k*th episode of not leaving the state y_{k-1} till time point t_k. This probability is given through the survivor function (3.5.16). With the aid of (3.5.18), thus

$$L_i = \prod_{k=1}^{n_i} \lambda_{y_k}^k(t_k | H_{k-1}, x_k) \exp(-\int_{t_{k-1}}^{t_k} \lambda^k(u | H_{k-1}, x_k) \, du).$$

It must now be remembered that the last episode of an individual may be censored. One obtains

$$L_i = \prod_{k=1}^{n_i} [\lambda_{y_k}^k (t_k | H_{k-1}, x_k)]^{\delta_k} \exp(-\int_{t_{k-1}}^{t_k} \lambda^k(u | H_{k-1}, x_k) \, du) \tag{3.6.36}$$

with $\delta_k = 1$ for $k = 1, \ldots, n_i - 1$ and $\delta_{n_i} = 0$ if the last episode of an individual is censored, otherwise $\delta_{n_i} = 1$.

The total likelihood function is then

$$L = \prod_{i=1}^{n} \prod_{k=1}^{n_i} [\lambda_{y_{ik}}^k (t_{ik} | H_{i,\,k-1}, x_{ik})]^{\delta_{ik}} \exp(-\int_{t_{i,\,k-1}}^{t_{ik}} \lambda^k(u | H_{i,\,k-1}, x_{ik}) \, du). \tag{3.6.37}$$

In addition to (3.6.32), the likelihood function (3.6.37) may be rearranged as

$$L = \prod_{k} \prod_{j=1}^{m} \prod_{i=1}^{n} [\lambda_j^k(t_{ik} | H_{i,\,k-1}, x_{ik})]^{\delta_{ikj}} [S_j^k(t_{ik} | H_{i,\,k-1}, x_{ik})]^{\epsilon_{ik}}, \tag{3.6.38}$$

where

$$\delta_{ikj} = \begin{cases} 1 & \text{if for individual i a transition to state j occurs at } t_{ik} \\ 0 & \text{otherwise,} \end{cases}$$

$$\epsilon_{ik} = \begin{cases} 1 & \text{if individual i experiences the } k\text{th episode} \\ 0 & \text{otherwise} \end{cases}$$

and

$$S_j^k(t_{ik} | H_{i,k-1}, x_{ik}) = \exp(-\int_{t_{i,\,k-1}}^{t_{ik}} \lambda_j^k(u | H_{i,k-1}, x_{ik}) \, du).$$

From (3.6.38) it is evident that the log likelihood function may be maximized separately for each k and j if the hazard rates $\lambda_j^k(\cdot)$ are dependent upon the parameter vector $\boldsymbol{\theta}_{jk}$ possessing no common components. For the *k*th episode, only those individuals who have experienced at least k episodes are to be included in the sample. With regard to the individual event types, the procedure is the same as previously described at the end of the last section. If the risk j is presently being analyzed, then all durations of the current episode with a transition to a state other than j are to be considered censored observations. In order to model the hazard rate $\lambda_j^k(t|H_{k-1}, x_k)$, basically all of the approaches presented in Section 3.3 may be used. For example, the Cox model is characterized by the hazard rate

$$\lambda_j^k(t|H_{k-1}, \mathbf{x}_k) = \lambda_{0j}^k(t) \exp(\mathbf{x}_k'\boldsymbol{\beta}_j^k), \tag{3.6.39}$$

where the relevant part of the prehistory H_{k-1} is included in the actual covariate vector $\mathbf{x}_k$. For a detailed discussion of the construction of the partial likelihood, see Hamerle (1988b).

Instead of presenting the detailed discussion here, we briefly consider the generally applied special model (3.5.22) of Chapters 5 and 6. Its hazard rate is

$$\lambda^k(t|\mathbf{x}_k) = \lambda_0(v) \exp(\mathbf{x}_k'\boldsymbol{\beta}) \tag{3.6.40}$$

with $v = t - t_{k-1}$. A partial likelihood for this model is

$$\prod_k \prod_{i=1}^{d_k} \frac{\exp(\mathbf{x}_{ik}'\boldsymbol{\beta})}{\sum_k \sum_{l \epsilon R_k(v_{ik})} \exp(\mathbf{x}_{lk}'\boldsymbol{\beta})}, \tag{3.6.41}$$

in which $v_{ik}, \ldots, v_{d_k k}$ represent the d_k (noncensored) *k*th episode durations and $R_k(v)$ is the risk set of the *k*th episode at time v. The same partial likelihood is obtained, however, given that all episodes are collected and a single-episode model is applied. If a person has experienced n_i episodes, these are included in the single-episode model as n_i independent episodes with the corresponding covariate vectors. As such, the possibility of applying program systems originally conceived for the single-episode case may be utilized for the multi-episode case under model (3.6.40). It must be noted, however, that this is usually not valid for the more general multiepisode models such as (3.6.39). For further details see Hamerle (1988b) or Blossfeld and Hamerle (1988a).

3.6.7 Least Squares Estimation

The regression approaches presented in Section 3.3.2 for lnT presupposed a linear relationship between the covariates and the logarithmic duration time or lifetime $y = \ln T$ of the form

$$y_i = \beta_0 + \mathbf{x}_i'\beta + \omega_i.$$

The ω_i are thereby independent and identically distributed with $E(\omega_i) = 0$ and $Var(\epsilon_i) = \sigma^2$. The distribution of ω_i is independent of $\mathbf{x}$.

Assuming certain distributions for ω_i, one obtains some of the models discussed in Section 3.3.2; for example, an extreme value distribution of ω results in the Weibull model. The situation is, however, quite different if one possesses no knowledge with regard to the distribution of ω_i. Under such a circumstance, the maximum likelihood estimation may not be applied and one is inclined as in the traditional multiple regression case to attempt a least squares estimation. The least squares estimation presents no further problems and may be undertaken using the common estimation formula $\hat{\beta} = (\mathbf{X}'\mathbf{X})^{-1}\mathbf{X}'\mathbf{y}$, given that no censored observations are present.

If, however, the distribution of ω_i is known, then in applying the least squares estimation method rather large efficiency losses may occur. Illustrative of such efficiency losses is the widely applied Weibull model in practical applications (see Lawless, 1982, Section 6.7.1). For other distributions the losses may be insignificant. Prentice and Shillington (1975) and Williams (1978) discuss the efficiency of least squares estimation for some special extreme value regression models, the latter proposing a method of obtaining more efficient parameter estimates. Pereira (1978) gives the asymptotic relative efficiency when maximum likelihood fitting of the wrong form of the distribution is used.

The most straightforward procedure is to adjust censored observations in some way and to then compute least squares estimates as though the data were uncensored. One possible method is to apply the cumulated hazard rate $\Lambda(t|\mathbf{x})$. Thus,

$$S(t|\mathbf{x}) = \exp(-\Lambda(t|\mathbf{x})).$$

As such, the stochastic variable $r = \Lambda(t|\mathbf{x})$ has an exponential distribution with parameter equal to 1, and r may be interpreted as a generalized residual. With regard to this aspect, compare the following discussion below in Section 3.7.1. For a censored observation t_i^*, one may define a respective censored residual

$$r_i^* = \Lambda(t_i^*|\mathbf{x}_i) = \int_0^{t_i^*} \lambda(u|\mathbf{x}_i)\, du.$$

Since r_i possesses a standard exponential distribution, it follows that

$$E(r_i|r_i \geq r_i^*) = r_i^* + E(r_i) = r_i^* + 1.$$

From this expression one may then calculate for a censored duration t_i^* an adjusted duration t_i' which fulfills

$$\Lambda(t_i'|\mathbf{x}_i) = \Lambda(t_i^*|\mathbf{x}_i) + 1.$$

The adjusted durations t_i' may then be used for the least squares estimation. We know of no studies investigating the efficiency of estimations obtained applying this approach.

Of special interest is the case in which the distribution of ω is not known. In the literature, one finds various suggestions for constructing least squares estimators, for example, Miller (1976), Buckley and James (1979), and Koul, Susarla, and van Ryzin (1981). Here we only deal with the estimator introduced by Buckley and James. Buckley and James define pseudo random variables

$$y_i^*(\boldsymbol{\beta}) = y_i\ \delta_i + E(y_i|y_i > c_i)\ (1 - \delta_i),$$

where c_i is the censoring time and δ_i is the censoring indicator.

For $\delta_i = 1$ it follows that $y_i^*(\beta) = y_i$. On the other hand, given $\delta_i = 0$ one must insert estimates. Buckley and James (1979) apply the Kaplan-Meier estimator to estimate the unknown distribution function F of ω. Ordering the residuals $\hat{e}_i(0,\hat{\boldsymbol{\beta}}) = y_i^* - \mathbf{x}_i'\hat{\boldsymbol{\beta}}$ according to size calculated at $\hat{\beta}_0 = 0$ as

$$\hat{e}_{(1)} < \ldots < \hat{e}_{(n)},$$

the Kaplan-Meier estimator is then (see also Section 3.2.4)

$$1 - \hat{F}(e;0,\hat{\boldsymbol{\beta}}) = \prod_{\hat{e}_{(i)} \leq e} \left(1 - \frac{d_{(i)}}{R_{(i)}}\right)^{\delta_{(i)}}.$$

$R_{(i)}$ represents the number of individuals in the risk set directly before $e_{(i)}$. $d_{(i)}$ is the number of persons who experience an event in $e_{(i)}$ and $\delta_{(i)} = 1$ for $d_{(i)} = 1$ and $\delta_{(i)} = 0$ otherwise. $y_i^*(\hat{\boldsymbol{\beta}})$ is then estimated by

$$\hat{y}_i^*(\hat{\boldsymbol{\beta}}) = \mathbf{x}_i'\hat{\boldsymbol{\beta}} + \frac{\sum_{j:\hat{e}_j > \hat{e}_i} w_j(\hat{\boldsymbol{\beta}})\ \hat{e}_j}{1 - \hat{F}(\hat{e}_i;0,\hat{\boldsymbol{\beta}})}.$$

The jumps of the Kaplan-Meier estimator are given by $w_j(\hat{\boldsymbol{\beta}})$. If the largest $\hat{e}_i$ value is censored, the remaining mass is usually given to this residual. The least squares estimator is then given by

$$\hat{\boldsymbol{\beta}} = [(\mathbf{X} - \overline{\mathbf{X}})'(\mathbf{X} - \overline{\mathbf{X}})]^{-1}\ (\mathbf{X} - \overline{\mathbf{X}})'\hat{\mathbf{y}}^*(\hat{\boldsymbol{\beta}}),$$

where the matrix $\overline{\mathbf{X}}$ possesses the elements $n^{-1}\sum_i x_{ij}$, and $\hat{\mathbf{y}}^*(\hat{\boldsymbol{\beta}}) = (\hat{y}_1^*(\hat{\boldsymbol{\beta}}), \ldots, \hat{y}_n^*(\hat{\boldsymbol{\beta}}))'$. The estimation is iterative since $\mathbf{y}^*$ is dependent upon $\hat{\boldsymbol{\beta}}$. A start value $\hat{\boldsymbol{\beta}}_0$ is obtained by considering all observations as being noncensored.

After deriving the estimator $\hat{\boldsymbol{\beta}}$, an estimate for β_0 is obtained as

$$\hat{\beta}_0 = \frac{1}{n} \sum_{i=1}^{n} (\hat{y}_i^*(\hat{\boldsymbol{\beta}}) - \mathbf{x}_i'\hat{\boldsymbol{\beta}}).$$

The described iterative least squares estimation procedure may be interpreted as a nonparametric version of the EM algorithm (Dempster et al., 1977; see Schneider and Weisfeld, 1986). The censored observations are replaced by their expected values and then the respective sum of squares is minimized.

Buckley and James use an estimator for the unknown variance σ^2 which ignores the censored observations. They apply

$$\hat{\sigma}_{BJ}^2 = \sum_{i=1}^{n} (\delta_i\ \hat{e}_i - \frac{1}{n_u} \sum_{k=1}^{n} \delta_k\ \hat{e}_k)^2 / (n_u - p).$$

Thereby, $n_u = \sum_{i=1}^{n} \delta_i$ is the number of uncensored observations. Schneider and

Weisfeld (1986) propose an alternative estimator which, in their simulation studies, exhibits a lower mean square error.

In summary, one can conclude that the least squares estimator may be implemented relatively simply given censored observations. However, if the distribution of ω_i is known, then application of the least squares method compared to the maximum likelihood method may lead to significant efficiency losses. The situation is different, if the distribution of ω_i is unknown. This is especially relevant in connection with the introduction of unobserved heterogeneity. In Section 3.9 we will deal with this in more detail.

3.7 Hypotheses Tests and Model Choice

Within the framework of a concrete data analysis, the determination of the general structure of the relationship between the covariates or prognostic factors and the durations or lifetimes is of special interest. After choosing appropriate covariates, the specification of the model which determines the hazard rate is especially important. With a step-by-step evaluation of the model, it is thus possible to test some of the assumptions of the model. In the simple situation with only one covariate x, it is usually possible to obtain information concerning the form of the relationships by means of graphic representation, graphing t or lnt on x. Furthermore, it is sometimes helpful to calculate the average duration or lifetime given varying x characteristics. These methods usually do not suffice if the model has many covariates.

Furthermore, many models rely on various additional assumptions. For example, the assumption of proportional hazards is asserted in the Cox or Weibull models given time invariant covariates. These presumptions must also be examined.

The following section presents a few methods of residual analysis as well as graphic model tests. Subsequently, special tests for the Cox model are dealt with, and the final section discusses the possibility of examining the significance of individual regression coefficients or model parts in addition to variable selection procedures.

3.7.1 Residual Analysis and Model Tests

Observing the relationship between the survivor function and the cumulative hazard rate (under a given covariate vector)

$$S(t|\mathbf{x}) = \exp(-\int_0^t \lambda(u|\mathbf{x})\, du) = \exp(-\Lambda(t|\mathbf{x})), \tag{3.7.1}$$

one sees that the random variable

$$r = \Lambda(T|\mathbf{x})$$

has an exponential distribution with parameter equal to 1.

Given a sample of noncensored data $(t_1, \mathbf{x}_1), \ldots, (t_n, \mathbf{x}_n)$, the $\Lambda(t_i|\mathbf{x}_i)$'s are independent realizations of standard exponentially distributed stochastic variables. It thus comes to mind (Cox and Snell, 1968; Kay, 1977), through

$$\hat{r}_i = \hat{\Lambda}(t_i|\mathbf{x}_i) = -\ln\hat{S}(t_i|\mathbf{x}_i) \tag{3.7.2}$$

to define "residuals," where $\hat{\Lambda}(t_i|\mathbf{x}_i)$ contains the estimates of the unknown parameters (see, e.g., (3.6.24) to (3.6.27) for the Cox model). In a first approximation $\hat{r}_1, \ldots, \hat{r}_n$ are also regarded as a random sample of some standard exponentially distributed variable. It must be observed, however, that the residuals defined in (3.7.2) generally are not independent nor do they possess identical distributions, so that the above described procedure is to be understood simply as an approximation.

Example

The survivor function of the Weibull regression model is

$$S(t|\mathbf{x}) = \exp(-(t\exp(-\mathbf{x}'\boldsymbol{\beta}))^{\delta}), \tag{3.7.3}$$

where the first component of $\mathbf{x}$ is always equal to 1. Consequently, the cumulative hazard rates are

$$r_i = \Lambda(t_i|\mathbf{x}_i) = (t_i\exp(-\mathbf{x}'_i\boldsymbol{\beta}))^{\delta} \qquad , i = 1, \ldots, n.$$

The r_i's are independent and distributed standard exponentially. Substituting the corresponding estimates in r_i, one obtains the residuals

$$\hat{r}_i = (t_i\exp(-\mathbf{x}'_i\hat{\boldsymbol{\beta}}))^{\hat{\delta}} \qquad , i = 1, \ldots, n. \tag{3.7.4}$$

Another way of defining the residuals is to apply the regression approach

$$y = \mathbf{x}'\boldsymbol{\beta} + \sigma\omega,$$

in which $y = \ln T$, $\sigma = 1/\delta$, and ω is characterized by a standard extreme value distribution. Analogous to traditional multiple regressions, the residuals are defined as

$$\hat{\omega}_i = \frac{y_i - \mathbf{x}'_i\hat{\boldsymbol{\beta}}}{\hat{\sigma}} \qquad i = 1, \ldots, n. \tag{3.7.5}$$

The residuals defined in (3.7.5) are related to (3.7.4) by $\hat{\omega}_i = \ln\hat{r}_i$, and depending on the goal of the analysis either (3.7.5) or (3.7.4) may be utilized.

The residuals may be calculated for various purposes. With the aid of residual plots the distributional assumptions may be examined graphically, or one can set the residuals of certain regression variables in relationship to another to check the fit of the model.

Given the survivor function of the Weibull regression model as presented in (3.7.3), by taking logarithms twice one obtains

$$\ln(-\ln S(t|\mathbf{x})) = \delta \ln t - \mathbf{x}'\boldsymbol{\beta}.$$

From this relationship, a model test for examining the validity of the Weibull model may be constructed. One categorizes the continuous covariates, such that for each covariate vector $\mathbf{x}_j$ multiple observations are given. Deriving the estimates $\hat{S}(t|\mathbf{x}_j)$ for each group $\mathbf{x}_j$ and plotting $\ln(-\ln\hat{S}(t|\mathbf{x}_j))$ against $\ln t$, they should be approximately linear and parallel. A further graphical model test will be presented in the next section in connection with the Cox model.

To this point, it has always been assumed that no censored observations were at hand. Given censored data, the procedure must be modified accordingly. One possible method is to apply the product-limit method (Kaplan-Meier estimator). An analogous method may be applied to the residuals. If t_i^* is a censored duration, the corresponding residual will also be censored and one may deal with the residuals $\hat{r}_1, \ldots, \hat{r}_n$ as a censored sample. As such, the product-limit estimator or the empirical hazard rate may be derived from the residuals and used to estimate the underlying survivor function. Graphical tests may then be applied to obtain information regarding the distribution of the residuals.

Another possible method (see Lawless, 1982, p. 281) given censored data is to calculate the average residual duration. The estimate is then added to the censored observations and the resulting value is included in the sample as a noncensored observation. This method is especially simple for an exponential distribution. If the duration T_i is distributed exponentially with a mean $1/\lambda_i = \exp(\mathbf{x}_i\boldsymbol{\beta})$ given $\mathbf{x}_i$, and t_i^* represents a censored observation, it is easy to calculate

$$E(T_i|T_i \geq t_i^*) = t_i^* + 1/\lambda_i.$$

Defining the residuals for noncensored observations as

$$\hat{r}_i = \hat{\lambda}_i t_i = t_i \exp(-\mathbf{x}'_i\hat{\boldsymbol{\beta}}), \qquad (3.7.6)$$

one obtains for a censored observation t_i^* the adjusted residual

$$\hat{r}_i = t_i^* \exp(-\mathbf{x}'_i\hat{\boldsymbol{\beta}}) + 1. \qquad (3.7.7)$$

Since $r_i = \Lambda(t_i|x_i)$ possesses a standard exponential distribution the residuals $\hat{r}_i = \hat{\Lambda}(t_i|x_i)$, $(i = 1, \ldots, n)$ defined in (3.7.2), can be approximately regarded as a sample taken from a standard exponentially distributed population. According to (3.7.7) for a censored observation t_i^*, the calculation of the residual must be adjusted as

$$\hat{r}_i = \hat{\Lambda}(t_i^*|x_i) + 1. \qquad (3.7.8)$$

It is then once again possible to construct a graphical model test. Plotting $-\ln\hat{S}(r)$ against r, a line with a slope of 1 should be approximately obtained, given the validity of the postulated model. $\hat{S}(r)$ thereby represents the product-limit estimate of the survivor function of r. For a further construction of the residuals and a model check for the Cox model, see Schoenfeld (1982).

3.7.2 Proportional Hazards Model Tests

Some of the methods presented in the last section may also be used to check the validity of the Cox model, especially the proportional hazards assumption. The Cox model possesses the hazard rate

$$\lambda(t|\mathbf{x}) = \lambda_0(t) \ \exp(\mathbf{x}'\boldsymbol{\beta})$$

with the unspecified baseline hazard rate $\lambda_0(t)$ and the survivor function (see (3.3.21))

$$S(t|\mathbf{x}) = S_0(t)^{\exp(\mathbf{x}'\boldsymbol{\beta})}. \tag{3.7.9}$$

The method presented in the previous section may also be utilized for the Cox model. Defining the residuals according to (3.7.2)

$$\hat{r}_i = \hat{\Lambda}(t_i|\mathbf{x}_i) = -\ln\hat{S}(t_i|\mathbf{x}_i),$$

for the Cox model one obtains from (3.7.9)

$$\hat{r}_i = -\ln\hat{S}_0(t_i) \exp(\mathbf{x}'_i\hat{\boldsymbol{\beta}}), \tag{3.7.10}$$

where $\hat{S}_0(t)$ represents an estimate of the "baseline" survivor function applying the method presented in Section 3.6.4. $\hat{\boldsymbol{\beta}}$ is the maximum partial likelihood estimator. If one calculates the product-limit estimator $\hat{S}(r)$ from the values in (3.7.10) and plots $\ln\hat{S}(r)$ against r, then approximately a line with slope −1 should be obtained.

However, the methods described here are only valid if a parametric model is formulated for the baseline hazard rate. If one proceeds to estimate $\boldsymbol{\beta}$ with the aid of the partial likelihood method and $\lambda_0(t)$ with a nonparametric estimator of the form (3.6.25) or (3.6.26), then the distribution of the obtained residuals may differ substantially from an exponential distribution. This has been demonstrated by Lagakos (1981). Therefore, according to Crowley and Storer (1983) the above described graphical residual test given unspecified baseline hazard rates is not recommended. Thus, Chapter 5 omits an application of the graphical residual test. Further goodness-of-fit tests for the Cox model are described in Schoenfeld (1980), Andersen (1982), and Kemény, Rothmeier, and Hamerle (1986).

Based upon the estimate $\hat{S}_0(t)$ of the baseline survivor function, a further possible graphical test of the Weibull model, which is a special proportional hazards model, may be constructed. Beginning with a Cox model, an estimate of $S_0(t)$ using the methods presented in Section 3.6.4 is obtained. The Weibull model is a special case of the Cox model with baseline hazard rate

$$\lambda_0(t) = \lambda\delta(\lambda t)^{\delta-1}$$

and baseline survivor function

$$S_0(t) = \exp(-(\lambda t)^{\delta}).$$

Taking the logarithm of $S_0(t)$ twice, we obtain

$$\ln(-\ln S_0(t)) = \delta \ln t + \delta \ln\lambda,$$

and a graphical representation of $\ln(-\ln S_0(t))$ against $\ln t$ should result approximately in a line graph. For further details, see Kay (1977) or Kemény, Rothmeier, and Hamerle (1986).

If one is especially interested in testing the proportionality assumption of the Cox model, this may be done by building classes and introducing appropriate time dependent covariates. For the sake of simplicity we assume that the population is divided only into two subsets (e.g., male and female) with regard to the proportionality assumption in question. Under proportional risks, the hazard rates of both subsets are

$$h_1(t|\mathbf{x}) = \lambda_0(t)\exp(\mathbf{x}'\boldsymbol{\beta}) \text{ and } h_2(t|\mathbf{x}) = \lambda_0(t)\exp(\alpha_1 + \mathbf{x}'\boldsymbol{\beta}), \quad (3.7.11)$$

where $\mathbf{x}$ represents the vector of the other included covariates. Introducing the dummy variable z_1, which may take on the value 1 and 0 for the individuals of both subsets, (3.7.11) may be reformulated as

$$h(t|\mathbf{x}, z_1) = \lambda_0(t)\exp(z_1\alpha_1 + \mathbf{x}'\boldsymbol{\beta}).$$

Now defining an additional covariate z_2

$$z_2 = z_1 \ln t$$

with the corresponding parameter α_2, one obtains

$$\begin{aligned} h(t|\mathbf{x}, z_1, z_2) &= \lambda_0(t)\exp(z_1\alpha_1 + z_2\alpha_2 + \mathbf{x}'\boldsymbol{\beta}) \\ &= \lambda_0(t)\, t^{z_1\alpha_2}\exp(z_1\alpha_1 + \mathbf{x}'\boldsymbol{\beta}). \end{aligned} \quad (3.7.12)$$

Examining $H_0: \alpha_2 = 0$ yields a test of the proportionality assumption. For that the methods based upon partial likelihood which are presented in the next section can be used.

Occasionally, it is proposed (Kalbfleisch and McIntosh, 1977) that the time dependent covariate should be fixed as

$$z_2 = z_1(\ln t - c), \quad (3.7.13)$$

where c designates the mean of $\ln t_i$. According to Kalbfleisch and McIntosh (1977), this avoids a high asymptotic correlation of the parameter estimates $\hat{\alpha}_1$ and $\hat{\alpha}_2$.

If the presumption of proportional risks with regard to one or more subgroups does not hold, a stratified Cox model may be applied. With regard to this aspect, see Section 3.3.4. Given that one has constructed s classes (subgroups), the group specific hazard rates are

$$\lambda_j(t|x) = \lambda_{0j}(t)\ \exp(x'\boldsymbol{\beta}).$$

The base hazard functions $\lambda_{01}(t), \ldots, \lambda_{0s}(t)$ must not be proportional to one another. The regression coefficients $\boldsymbol{\beta}$ remain the same for all strata.

3.7.3 Tests for Regression Coefficients or Model Parts

If one has already tested the general model structure, for example, with regard to the time dependency of the hazard rate and the relation between the covariates and the hazard rate, and a special model is chosen, one may also test the significance of individual regression coefficients or model parts. The tests are based upon the asymptotic properties, especially the asymptotic normality of the maximum likelihood estimator or of the maximum partial likelihood estimator of the Cox model. It must, however, be mentioned at this point that the asymptotic properties for all of the situations dealt with in this book have not been completely proven.
Designating $\boldsymbol{\theta}$ as the model parameter vector, consisting of the regression coefficient vector $\boldsymbol{\beta}$ as well as further parameters included in the model (e.g., the parameter δ in the Weibull model), hypotheses concerning individual regression coefficients or model parts may be concisely expressed in the general linear hypothesis

$$\mathbf{C}\,\boldsymbol{\theta} = \mathbf{0}, \qquad \text{rank}(\mathbf{C}) = m \tag{3.7.14}$$

with an appropriate matrix $\mathbf{C}$.

Examining $H_0: \beta_i = 0$, the matrix $\mathbf{C}$ consists of only one row, with 1 in the corresponding position and zeros otherwise. For $H_0: \beta_i = \beta_j = 0$, $\mathbf{C}$ consists of two rows which are constructed analogously. Usually the rank m of $\mathbf{C}$ is equal to the number of rows of $\mathbf{C}$.

In principle, the hypothesis

$$\mathbf{C}\,\boldsymbol{\theta} = \boldsymbol{\xi} \tag{3.7.15}$$

can also be tested where $\boldsymbol{\xi}$ is a fixed value. However, the hypothesis (3.7.14) with $\boldsymbol{\xi} = \mathbf{0}$ is usually of interest in practical applications. We limit our discussion to this case.

Examination of Individual Regression Coefficients

To test specifically the hypothesis $H_0: \beta_i = 0$, one may apply the estimator $\hat{\beta}_i$ and the estimated (asymptotic) variance $\widehat{\text{Var}}\,(\hat{\beta}_i)$. $\widehat{\text{Var}}\,(\hat{\beta}_i)$ is the respective diagonal element in the estimated covariance matrix of $\hat{\boldsymbol{\beta}}$ being the inverse of the observed information matrix (see Section 3.6.1, especially relationship

(3.6.8) and the following comments). The test statistic

$$\frac{\hat{\beta}_i}{\sqrt{\hat{\text{Var}}\,(\hat{\beta}_i)}} \tag{3.7.16}$$

has an asymptotic standard normal distribution when H_0 is true. If one wishes to test the general hypothesis $H_0 : \beta_i = \xi$, then the test statistic is modified to be

$$\frac{\hat{\beta}_i - \xi}{\sqrt{\hat{\text{Var}}\,(\hat{\beta}_i)}}. \tag{3.7.17}$$

Simultaneous Testing of Multiple Regression Coefficients or Parameters

In order to examine the hypothesis that multiple parameters are equal to zero, the hypothesis is formulated in the form of (3.7.14). For an explicit application of the test, various test statistics may be chosen.

a) The likelihood ratio statistic is

$$\text{Lq} = 2(\ln L(\hat{\theta}) - \ln L(\tilde{\theta})), \tag{3.7.18}$$

where $\tilde{\theta}$ represents the maximum likelihood estimates under the constraint $\mathbf{C}\theta = \mathbf{0}$ and $\hat{\theta}$ are maximum likelihood estimates without restrictions.

b) The Wald statistic is

$$W = (\mathbf{C}\hat{\theta})' \,[\mathbf{C}\ \mathbf{Cov}\,(\hat{\theta})\ \mathbf{C}']^{-1}\,(\mathbf{C}\hat{\theta}). \tag{3.7.19}$$

c) Instead of the likelihood ratio and the Wald statistic, the score statistic may also be used to test the general linear hypothesis. The score statistic is based upon the score function $\mathbf{s}(\theta)$, the vector of derivatives of the log likelihood function (see Section 3.6.1). One constructs

$$S = \mathbf{s}(\tilde{\theta})'\,\mathbf{I}(\tilde{\theta})^{-1}\,\mathbf{s}(\tilde{\theta}). \tag{3.7.20}$$

$\mathbf{I}(\tilde{\theta})^{-1}$ is the inverse of the observed information matrix calculated at $\tilde{\theta}$.

If H_0 is true, the asymptotic distribution for all three test statistics is the same. If H_0 is true, they have asymptotically a central χ^2 distribution with m degrees of freedom.

It must be noted that we are always concerned with asymptotic distributions of test statistics. Practical application of the tests thus require large sample sizes.

Selecting Variables

Applying stepwise regression to the model in order to select variables, one introduces or eliminates certain variables after each step according to the calculated significance probabilities. Under the likelihood or partial likelihood quotient method the introduction or elimination of covariates is determined through the likelihood or partial likelihood ratio test. The test statistic is analogous to (3.7.18):

$$2(\ln L(\hat{\beta}_\nu) - \ln L(\hat{\beta}_{\nu+1})) \qquad , \nu = 1, 2, \ldots,$$

where $\hat{\beta}_\nu$ designates the ML estimator calculated in the νth step. This test statistic is asymptotically $\chi^2(1)$ distributed. For a detailed discussion of the selection of variables in regression models, especially with regard to numerically efficient methods, see Fahrmeir and Hamerle (1984, Chapter 4).

3.8 Introducing Time-Dependent Covariates

To this point it has been assumed that the covariates were measured at the beginning of an episode and their values did not change over the course of an episode. In many applications, however, the covariates may depend upon time. Illustrative examples are age, income, family status, or a therapy which may be applied only during some specific time interval. Many possibilities exist with regard to the manner of covariate time dependency. A simple form of time dependence exists, for example, for the variables age and occupational experience. For these variables, the dependence on time may be expressed in a previously determined functional form. In this case Kalbfleisch and Prentice (1980, Chapter 5) speak of "defined" covariates. On the other hand, covariates exist which are realizations of some stochastic process. Kalbfleisch and Prentice (1980) differentiate here between *external* and *internal* time-dependent covariates. Given external time-dependent covariates the path of the covariate vector $\mathbf{x}(t)$ is not influenced by the duration, yet the covariates may very well affect the duration. Environmental factors, occupational or economic conditions are common examples of this type of external time-dependent covariate. On the other hand, an internal time-dependent covariate is the output of a stochastic process that is generated by the individual. This implies that the value of $\mathbf{x}(t)$ carries information about the duration time. For example, the decision to continue education is directly dependent upon the concrete occupational level one wishes to achieve (Andreß, 1985).

In the following, we initially assume external time-dependent covariates. The problems arising from internal time-dependent covariates is discussed in the final part of this section. We restrict our discussion here to the single-episode case, however, the methods are also easily applied to the multiepisode case or competing risks approach.

Designating $\mathbf{x}(t)$ as the covariate vector at time t, the hazard rate is defined by

$$\lambda(t|\mathbf{x}(t)) = \lim_{\Delta t \to 0} \frac{1}{\Delta t} P(t \leq T \leq t + \Delta t | T \geq t, \mathbf{x}(t)). \tag{3.8.1}$$

The survivor and density function given external time-dependent covariates may also be defined analogously as in the case of fixed covariates.

$$S(t|\mathbf{x}(t)) = P(T \geq t|\mathbf{x}(t)) = \exp(-\int_0^t \lambda(u|\mathbf{x}(u))\, du) \tag{3.8.2}$$

$$f(t|\mathbf{x}(t)) = \lambda(t|\mathbf{x}(t)) \cdot \exp(-\int_0^t \lambda(u|\mathbf{x}(u))\, du). \tag{3.8.3}$$

As may be seen from examining (3.8.2), the numerical calculation of the survivor function may be quite difficult due to the integration of the time paths of the covariates.

Given "defined" time-dependent covariates, or if the functional form of time dependency of the covariates is not too complicated, the derivation of (3.8.2) generally does not present great problems. The same is valid for covariates, whose paths follow a step function. They are piecewise constant and the cumulative hazard rate may be segmented into a sum of integrals. Designating $t_0 < t_1 \ldots < t_s$ as the transition time points of the covariate vectors in the interval [0, t) and given $t_{s+1} = t$, the survivor function is consequently

$$S(t|\mathbf{x}(t)) = \exp(-\sum_{r=1}^{s+1} \int_{t_{r-1}}^{t_r} \lambda(u|\mathbf{x}(t_{r-1}))\, du)$$

$$= \prod_{r=1}^{s+1} S(t_r|t_{r-1}, \mathbf{x}(t_{r-1})) \tag{3.8.4}$$

where

$$S(t_r|t_{r-1}, \mathbf{x}(t_{r-1})) = P(T \geq t_r|T \geq t_{r-1}, \mathbf{x}(t_{r-1})).$$

The likelihood function may be constructed with the aid of (3.8.1) and (3.8.4).

If the change of covariates over time is continuous, it is eventually possible to derive the cumulative hazard rate by numerical integration if the paths of the covariates are measured at many time points. If this is not the case, one may attempt to linearly approximate the covariates as step functions and to proceed as in (3.8.4). A further possibility is the episode splitting approach as recommended by Tuma et al. (1979) and Tuma and Hannan (1984) which was handled in Section 3.3.2. They substitute for each period the actual values of the time-dependent covariates. Finally, the discrete models discussed in Section 3.10 are applicable as an approximation.

A further possible way of including covariates which continuously change over time is to apply the Cox model with the hazard rate

$$\lambda(t|\mathbf{x}(t)) = \lambda_0(t) \exp(\mathbf{x}'(t)\boldsymbol{\beta}). \tag{3.8.5}$$

The derivation in Section 3.6.4 of the partial likelihood function may be applied here, thus

$$PL(\boldsymbol{\beta}) = \prod_{i=1}^{k} \frac{\exp(\mathbf{x}_i'(t_{(i)})\, \boldsymbol{\beta})}{\sum_{l \in R(t_{(i)})} \exp(\mathbf{x}_l'(t_{(i)})\, \boldsymbol{\beta})}. \tag{3.8.6}$$

$t_{(1)} < t_{(2)} < \ldots < t_{(k)}$ represent the event time points ($k \leq n$) and $R(t_{(i)})$ designates the risk set at time $t_{(i)}$. Given ties (3.8.6) must be modified according to (3.6.23). The application of the Cox model and the estimation on the basis of (3.8.6) require a relatively exact recording of the covariate path. The values of the covariates must be available in each event time point $t_{(i)}$, for all individuals of the risk set $R(t_{(i)})$ who up until this point in time had not experienced an event and were not censored. This will generally only be the case if the covariates are regularly measured. For example, medical studies often repeatedly measure and record quantities such as blood pressure. In other situations where less information concerning the covariates is available, a common procedure is to apply the value of the covariate (instead of $\mathbf{x}(t_{(i)})$) which is actually measured at some time point closest to $t_{(i)}$. If the covariate state space is discrete and one knows the state transition of the covariate in the observed time period, generally (3.8.6) may be applied without problem. A detailed discussion of the inclusion of time-dependent covariates is found in Petersen (1985).

If one is dealing with internal time-dependent covariates as is, for example, the case in medical survival studies, special problems may arise. Under these circumstances, the covariates always contain direct information concerning lifetime. If one has observed an interval covariate until time t, then the individual must have survived up to this time point and we have

$$P(T \geq t|\mathbf{x}(t)) = 1.$$

The above expression thus does not possess the interpretation of the survivor function as in (3.8.2) and the likelihood function may not be constructed in the manner presented in Section 3.6.2. The likelihood function may, however, be derived in another manner (see Kalbfleisch and Prentice, 1980, Section 5.3) such that it can be shown that especially for the PH model once again the partial likelihood (3.8.6) is obtained.

The problems that arise under the introduction of internal time-dependent covariates have not yet been fully investigated. Thus, interpretation should be careful, especially if one is dealing with a comparison of the effectiveness of various treatments, programs or campaigns where the covariate process achieves its value after the treatment assignment. For an illustrative example, see Kalbfleisch and Prentice (1980, p. 126).

Internal time-dependent covariates may also lead to the problem of mutual dependency of events, since the covariates and the durations of successive events are often processes that affect one another. That is, the occurrence of a specific event changes the risk of occurrence of another event and vice versa. Examples are family status and the duration of employment in a job. This problem of interdependent processes has not yet been sufficiently examined. See further Tuma and Hannan (1984, Chapters 9 and 16), Coleman (1984b), Clayton and Cuzick (1985), and Petersen (1985).

3.9 Introducing Unobserved Population Heterogeneity

3.9.1 Examples of Unobserved Heterogeneity

As a rule personal and environment specific characteristics, which are not recorded or unknown as well as the covariates included in the model will influence the transition or hazard rates. To this point it has always been assumed that the measured covariates fully determine the hazard rate. Important unobserved features were not considered. The implications of ignoring unobserved heterogeneity may be demonstrated by the following simple examples.

Example 1

Suppose the population is divided into two subsets x_1 and x_2. In each subpopulation the hazard rate is constant with

$$\lambda(t|x_1) = 0.1, \qquad \lambda(t|x_2) = 0.4,$$

and the probabilities that an individual belongs to one of the subpopulations are respectively p_1 and p_2. If the segmentation of the two subsets is disregarded, one obtains the following transition rate of the total population

$$\lambda(t) = \frac{\sum_i f(t|x_i)\, p_i}{\sum_i S(t|x_i)\, p_i} \qquad (3.9.1)$$

where $S(t|x_i)$ characterizes the survivor function of the subpopulation x_i. The graphical representation of the hazard rates, given the assumed values of $\lambda(t|x_1)$ and $\lambda(t|x_2)$ as well as $p_1 = p_2 = 0.5$, is presented below in Figure 3.12.

Not considering the population heterogeneity results in a decreasing hazard rate. This may be heuristically made plausible. First of all, on the average

Figure 3.12: Hazard Rate Path of the Subpopulations and the Total Population Given $\lambda(t|x_1) = 0.1$ and $\lambda(t|x_2) = 0.4$

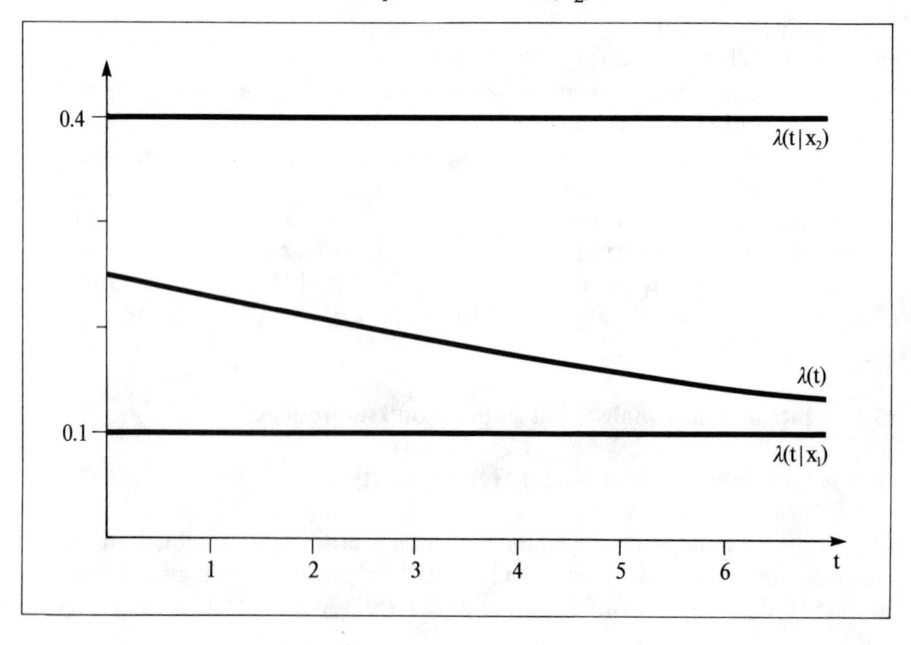

those individuals will experience an event whose transition rate is high and there is a tendency of those individuals characterized by low transition rates to remain in the risk set. Consequently, the aggregate time dependent hazard rate will decrease as time elapses.

Example 2

The original situation is the same as given in Example 1. The hazard rates of both subpopulations are characterized by

$$\lambda(t|x_1) = 3t^2 \qquad \text{and } \lambda(t|x_2) = 1.$$

$\lambda(t|x_1)$ is a Weibull hazard rate with $\alpha = 3$ and $\lambda = 1$. Furthermore, $p_1 = p_2 = 0.5$. The hazard rates are illustrated in Figure 3.13.

Whereas the hazard rate $\lambda(t|x_1)$ of the first subpopulation rapidly rises, the hazard rate $\lambda(t|x_2)$ of the second subpopulation is constant equal to one. The aggregate hazard rate $\lambda(t)$ possesses an entirely different profile. It increases at first, then decreases continuously and finally approaches the value 1.

Figure 3.13: Hazard Rate Path of the Subpopulations and the Total Population Given $\lambda(t|x_1) = 3t^2$ and $\lambda(t|x_2) = 1$

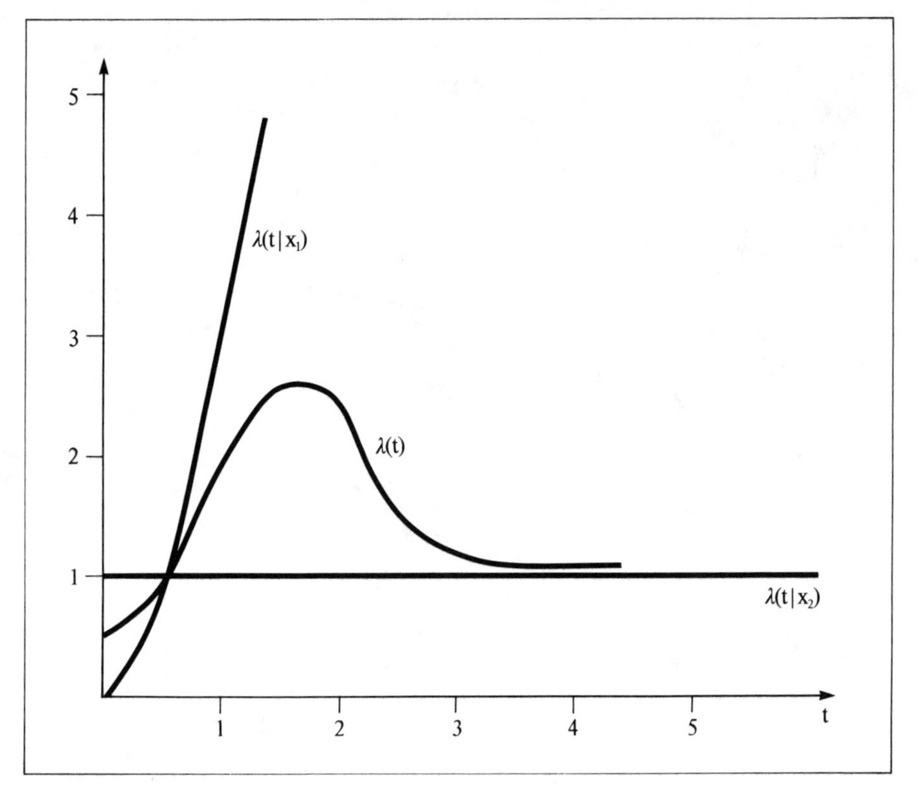

The tendencies demonstrated in these two examples are generally valid. If unobserved population heterogeneity is not considered, that is, if one aggregates the unobserved variables, then the hazard rate tends to exhibit negative time dependence, or in other words: a decreasing hazard rate arises. A proof of this statement may be found in Heckman and Singer (1984a). Observe, however, that the statement only concerns the tendency. It does not imply that aggregated hazard rates must always be monotonically decreasing. Consider the following example:

Example 3

The initial situation is as given in Examples 1 and 2. The subpopulation hazard rates are characterized by

$$\lambda(t|x_1) = 2t \qquad \text{and } \lambda(t|x_2) = 1.5t^{1/2}.$$

These are thus Weibull hazard rates with $\lambda_1 = 1$, $\alpha_1 = 2$ and $\lambda_2 = 1$, $\alpha_2 = 1.5$. The hazard rates are presented in Figure 3.14 (with $p_1 = p_2 = 0.5$).

The above mentioned tendency statement implies that over time t the aggregated hazard rate rises more slowly.

Figure 3.14: Hazard Rate Path of the Subpopulations and the Total Population Given $\lambda(t|x_1) = 2t$ and $\lambda(t|x_2) = 1.5t^{1/2}$

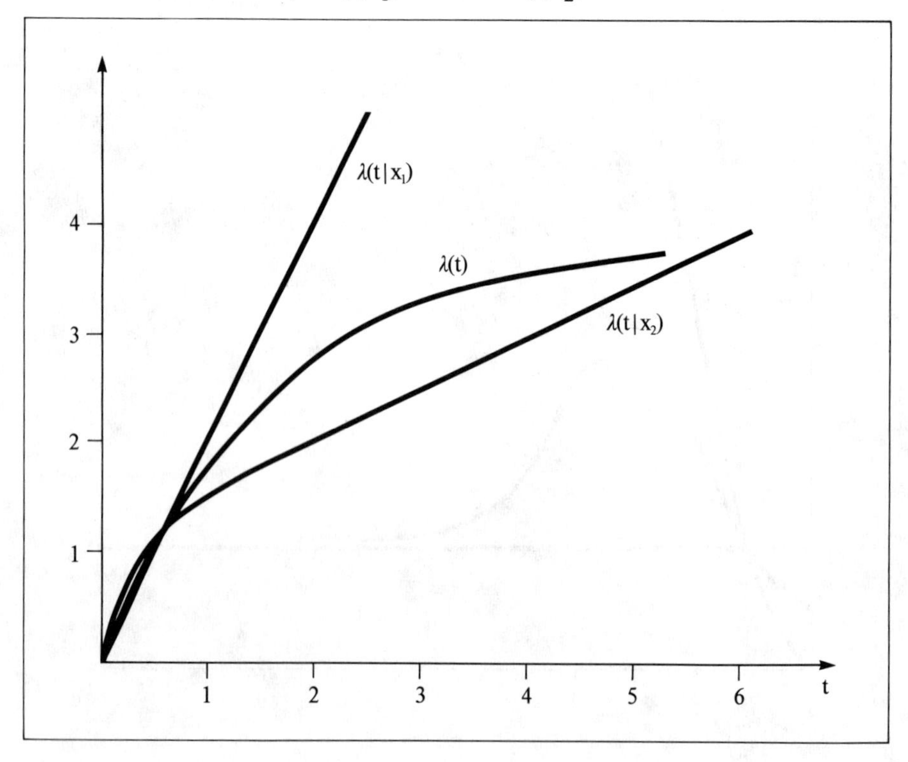

Models with unobserved heterogeneity have up to now almost exclusively dealt with one-episode case formulations. Exceptions are the works of Flinn and Heckman (1982) and Newman and McCulloch (1984). Especially from the theoretical perspective, the models have not been sufficiently investigated. In the following, we present possible methods for explicitly including unobserved heterogeneity into the model.

The unobserved heterogeneity will be represented by a (real valued) random variable ϵ, whose realization influences the hazard rate as a "disturbance." One apparent possible formulation is to introduce this deviation as a multiplicative term of the following form

$$\lambda(t|\mathbf{x}, \epsilon) = \lambda(t|\mathbf{x}) \cdot \epsilon. \tag{3.9.2}$$

Since the hazard rate is not negative, ϵ must be limited to positive values. One postulates

$$E(\epsilon) = 1, \tag{3.9.3}$$

which implies that on the average one obtains $\lambda(t|\mathbf{x})$. The "deviation" ϵ varies from individual to individual and is not directly observable. The procedure corresponds to the inclusion of a person parameter in the random effect model analysis of variance.

The distribution function of ϵ is designated as $G(\epsilon)$. The models (3.9.2) as common in demographic and medical studies are referred to as "frailty" models (see Vaupel et al., 1979; Spilerman, 1971). In event history analysis they have been introduced by Tuma (1978), applied to the case of an individual time invariant hazard rate $\lambda(t|\mathbf{x})$ in (3.9.2). With regard to the introduction of nonobserved population heterogeneity, see Heckman and Singer (1982, 1984a, 1984b), Flinn and Heckman (1982), Elbers and Ridder (1982), and Hougaard (1984, 1986).

3.9.2 Models and Parameter Estimation for Given Distributions of the Heterogeneity

Having a hazard rate with a given covariate vector and given heterogeneity $\lambda(t|\mathbf{x}, \epsilon)$, the density of T with a given covariate vector is given by

$$f(t|\mathbf{x}) = \int_0^\infty f(t|\mathbf{x}, \epsilon)\, dG(\epsilon) = \int_0^\infty \lambda(t|\mathbf{x}, \epsilon)\, S(t|\mathbf{x}, \epsilon)\, dG(\epsilon). \tag{3.9.4}$$

Distributions such as $f(t|\mathbf{x})$ are referred to as *mixture distributions.* $G(\epsilon)$ represents the mixing distribution.

(3.9.4) is the distribution of the observable quantities of the model and requires "integrating out" the disturbance ϵ. Thereby the distribution $G(\epsilon)$ must be fully specified. If a parametric distribution for $G(\epsilon)$ is asserted and $\lambda(t|\mathbf{x})$ is specified, then one can estimate the unknown parameters of $G(\epsilon)$

along with the parameters of $\lambda(t|x)$ with the maximum likelihood method on the basis of (3.9.4). Designating the censoring indicator by δ_i, one obtains a "marginal" log likelihood function (see Sections 3.6.2 and 3.6.3)

$$\ln L_M = \sum_{i=1}^{n} \ln \int_0^{\infty} \lambda(t_i|x_i, \epsilon)^{\delta_i} S(t_i|x_i, \epsilon)\, dG(\epsilon). \tag{3.9.5}$$

The maximization of (3.9.5) can cause numerical problems since in each step an integration with regard to the distribution of ϵ is necessary.

Under the assumption of a continuous heterogeneity with the density function $g(\epsilon)$, one obtains

$$\ln L_M = \sum_{i=1}^{n} \ln \int_0^{\infty} \lambda(t_i|x_i, \epsilon)^{\delta_i} S(t_i|x_i, \epsilon)\, g(\epsilon)\, d\epsilon. \tag{3.9.6}$$

In the following the procedure is demonstrated according to a simple specification of $\lambda(t|x)$ chosen by Tuma (1978) and Tuma and Hannan (1984, Section 6.3). The hazard rate is postulated to be time invariant

$$\lambda(t|x) = \phi(x'\beta)$$

with a (deterministic) nonnegative function ϕ. The distribution of the heterogeneity is assumed to be characterized by a gamma distribution with the density

$$g(\epsilon) = \frac{\theta}{\Gamma(\alpha)} (\theta\, \epsilon)^{\alpha-1} \exp(-\, \theta\, \epsilon), \quad \epsilon \geq 0,\ \alpha, \theta > 0, \tag{3.9.7}$$

where $\Gamma(\alpha)$ represents the gamma function

$$\Gamma(\alpha) = \int_0^{\infty} x^{\alpha-1}\, e^{-x}\, dx.$$

The expected value and variance of gamma distributed random variables are

$$E(\epsilon) = \frac{\alpha}{\theta} \text{ and } Var(\epsilon) = \frac{\alpha}{\theta^2}.$$

Thus, in order to meet the assumption (3.9.3) $E(\epsilon) = 1$, it is necessary that $\alpha = \theta$. As a result, the density of the unobserved heterogeneity is

$$g(\epsilon) = \frac{\alpha}{\Gamma(\alpha)} (\alpha\, \epsilon)^{\alpha-1} \exp(-\, \alpha\, \epsilon). \tag{3.9.8}$$

According to the assumption $\lambda(t|x) = \phi(x'\beta)$, the duration T given the heterogeneity and given the covariate vector is

$$f(t|x, \epsilon) = \phi(x'\beta)\, \epsilon \exp(-\, \phi(x'\beta)\, \epsilon\, t),$$

and from (3.9.4)

$$f(t|x) = \int_0^{\infty} \phi(x'\beta)\, \epsilon \exp(-\, \phi(x'\beta)\, \epsilon\, t)\, \frac{\alpha}{\Gamma(\alpha)} (\alpha\, \epsilon)^{\alpha-1} \exp(-\, \alpha\, \epsilon)\, d\epsilon. \tag{3.9.9}$$

Under appropriate substitution and application of the relationship $\Gamma(\alpha + 1) = \alpha\, \Gamma(\alpha)$, (3.9.9) may be integrated. One obtains

$$f(t|x) = \frac{\phi(x'\boldsymbol{\beta})\, \alpha^{\alpha+1}}{(\phi(x'\boldsymbol{\beta})\, t + \alpha)^{\alpha+1}}. \tag{3.9.10}$$

Going through the integration of the right-hand side of

$$S(t|x) = 1 - \int_0^t f(u|x)\, du,$$

the resulting survivor function is

$$S(t|x) = \left(\frac{\alpha}{\phi(x'\boldsymbol{\beta})\, t + \alpha}\right)^{\alpha}, \tag{3.9.11}$$

and due to $\lambda(t|x) = f(t|x) / S(t|x)$ one obtains the hazard rate

$$\lambda(t|x) = \frac{\phi(x'\boldsymbol{\beta})\, \alpha}{\phi(x'\boldsymbol{\beta})\, t + \alpha}. \tag{3.9.12}$$

Observe that the hazard rate in (3.9.12) does indeed depend upon the time t, although the individual hazard rates $\lambda(t|x, \epsilon) = \phi(x'\boldsymbol{\beta}) \cdot \epsilon$ are assumed to be time invariant. This is once again a consequence of the above mentioned result that unobserved heterogeneity shifts the hazard rate of the total population towards negative time dependence.

Observing (3.9.5) or (3.9.6), it is evident that the choice of $G(\epsilon)$ affects the form of the marginal likelihood function and may thus influence the estimation of the structural model parameter $\boldsymbol{\beta}$. Heckman and Singer (1982) have obtained in their analysis of unemployment spells for an assumed Weibull hazard rate given various $G(\epsilon)$ highly differing results for the $\boldsymbol{\beta}$ estimation. They therefore propose an alternative strategy which we present in the following section. Newman and McCulloch (1984) obtain still another result in their analysis of data regarding time segments between successive births. They chose as "mixing" distribution $G(\epsilon)$ various discrete approximations of the gamma distribution, as well as the lognormal distribution, and come to the result that the $\boldsymbol{\beta}$ estimates varied only minimally. One possible explanation of these different results compared to Heckman and Singer's analysis may be due to the possibility that the Weibull model does not adequately fit the data used by Heckman and Singer.

For multiple successive time periods, the state of affairs is more complicated. At present, we only sketch here the repeated event case. Assuming for each episode another heterogeneity ϵ_k one must necessarily also make an assumption about the joint distribution of ϵ_k since the individual components will not generally be invariant. A justification of some certain choice of the joint distribution is—as also is the case in choosing $G(\epsilon)$—usually quite difficult. Occasionally the simplifying assumption

$$\epsilon_k = \gamma_k\, \epsilon$$

is asserted, with γ_k representing an episode specific parameter.

If the hazard rate $\lambda^k(t|x_k, \gamma_k, \epsilon)$ as well as the distribution $G(\epsilon)$ are specified, a "marginal" likelihood function may be constructed applying the described likelihood functions of Section 3.6.6 through integration of ϵ. The marginal likelihood function may then be used to estimate the unknown model parameters.

At this point, a critical remark must be made. Usually the assumption is met that the heterogeneity is independent of the observed covariates. This is especially true in all of the cases cited in the following section. On the other hand, the motivation of introducing unobserved population heterogeneity as a rule stems from the assumption that in empirical applications it is never possible to grasp all relevant influences, and that if one does not attempt to account for unobserved features the results will be biased (omitted variable bias). However, the unobserved characteristics of an individual will not be independent of the gathered features. If one assumes independence of unobserved influences, the omitted variable bias remains. With regard to this issue, see Chamberlain (1980).

If one wishes to consider the dependence of the heterogeneity and the covariates, then not only the distribution $G(\epsilon)$ must be specified but also the distribution $G(\epsilon|\mathbf{x})$. This is difficult in practical application. One possibility is to formulate the regression approach

$$\epsilon = \mathbf{x}'\pi + u$$

and include it in the model. The parameter π must now also be estimated. Of course, this increases the number of unknown parameters greatly. Furthermore, identification problems may arise.

3.9.3 Simultaneous Estimation of the Structural Model Parameter and the Distribution of Heterogeneity

Due to possible sensitivity of estimates of the structural model parameters with regard to the choice of distribution of the unobserved heterogeneity Heckman and Singer (1982, 1984a, 1984b) propose a simultaneous estimation of the model parameters and the distribution of the heterogeneity, similar to the empirical Bayes estimation (see, e.g., Moritz, 1971).

In the following, the one-episode case is discussed and the covariate vector is presumed to be time invariant. Designating the conditional distribution of the duration given the covariates and heterogeneity as $f(t|\mathbf{x}, \epsilon; \boldsymbol{\theta})$, the dependence of the vector $\boldsymbol{\theta}$ of the structural parameter is explicitly formulated. $\boldsymbol{\theta}$ contains the regression coefficients as well as further parameters determining the hazard rate and thus $f(\cdot)$. The distribution of the heterogeneity ϵ, the mixing distribution is expressed as $G(\epsilon)$, and consequently one obtains for the density function of T given $\mathbf{x}$

$$h(t|\mathbf{x}; \boldsymbol{\theta}) = \int f(t|\mathbf{x}, \epsilon; \boldsymbol{\theta})\, dG(\epsilon). \tag{3.9.13}$$

The goal is to find not only a parameter estimate $\hat{\boldsymbol{\theta}}$, but also an estimate $\hat{G}(\epsilon)$ which possesses desirable properties such that with increasing sample size it converges to $G(\epsilon)$ in probability.

Before one may further examine this goal, various identification problems must be solved. If one entirely disregards the estimation problem, then the question arises if knowledge of the mixture distribution $h(t|\mathbf{x};\boldsymbol{\theta})$ is sufficient to insure that the integral equation (3.9.13) is fulfilled with uniquely determined functions $f(t|\mathbf{x},\epsilon;\boldsymbol{\theta})$ and $G(\epsilon)$. Without making additional assumptions this is definitely not the case. Also, in specifying the conditional distribution $f(t|\mathbf{x},\epsilon;\boldsymbol{\theta})$ it is not always identified. Two different distributions $G_1(\epsilon)$ and $G_2(\epsilon)$ may lead to the same mixed distribution. For example, see Heckman and Singer (1984b).

For continuous duration models (single duration models) Elbers and Ridder (1982) have demonstrated for the class of proportional hazards models with

$$\lambda(t|\mathbf{x}) = \lambda_0(t)\exp(\mathbf{x}'\boldsymbol{\beta})$$

the possibility of identification. An important condition thereby is that the covariate vector contains at least one continuous component. With regard to this, see Hougaard (1984) and Heckman and Singer (1984a, 1984b). A further central requirement that exists in the relevant literature on the subject concerns the independence of covariates and unobserved heterogeneity. As discussed above in the previous section, the problem of omitted variables is, however, generally not solved.

Kiefer and Wolfowitz (1956) have derived general conditions necessary for the existence of a consistent estimator of the mixing distribution and structural model parameters. Their analysis, however, does not contain a procedure of deriving numerically the estimates. Heckman and Singer (1984b) have verified the Kiefer and Wolfowitz conditions for proportional hazards models with time dependent covariates and possible censored data. They further propose a nonparametric maximum likelihood estimation algorithm founded upon the theoretical characterization of the ML estimation of mixture distributions from Lindsay (1983a, 1983b). They propose the application of the EM algorithm (see Dempster et al., 1977).

3.9.4 Comparison of Some Estimators in the Presence of Unobserved Heterogeneity

Heckman and Singer (1982, 1984) examined in their application to unemployment data the traditional approach and assumed a functional form for the distribution of unobserved heterogeneity. As outlined above, they found that the estimates of the regression coefficients are extremely sensitive to the parametric form chosen. This finding led to the development of their nonparametric approach. But a nonparametric representation of the distribution of the heterogeneity implies that one must impose a parametric form on the hazard rate conditionally on the covariates and the heterogeneity. This hazard rate determines the conditional duration distribution $f(t|\mathbf{x},\epsilon)$. However, Trussell and Richards (1985) demonstrate that the maximum likelihood estimator is severely affected if the conditional duration distribution is misspecified. The same is true if both $f(t|\mathbf{x},\epsilon)$ and $G(\epsilon)$ are misspecified. Often one cannot empirically distinguish misspecification of the duration distribution given the covariates and the heterogeneity from misspecification of the mixing distribution $G(\epsilon)$.

In summary we see that it would be desirable to have an estimator which does not require special assumptions about the functional form of the duration distribution, conditionally on covariates and heterogeneity, as well as the distribution of the heterogeneity. So the least squares estimator discussed in Section 3.6.7 comes to one's mind. The least squares estimator (LS estimator) can be calculated for a wide class of duration models when the hazard regression corresponds to a linear model after a logarithmic transformation of the duration. We consider log-duration $y = \log T$ and log-linear models of the form

$$y = \mathbf{x}'\boldsymbol{\beta} + \sigma\,\omega.$$

The first component of the vector $\mathbf{x}$ is 1, ω is a random disturbance term, and σ is a scale parameter in the distribution of y. A short description of such models can be found in Section 3.3.2. For a detailed discussion see Prentice (1975) or Lawless (1982). If the duration of T belongs to the class of generalized F (see Prentice, 1975), one obtains a linear model for logT. There are important special cases as exponential, gamma, Weibull, lognormal, and log-logistic. Only the exponential and the Weibull models are proportional hazards models.

Let us, for example, conditionally on the heterogeneity ϵ and the covariates, assume a Weibull model with hazard rate

$$\lambda(t|\mathbf{x},\ \epsilon) = \alpha\ t^{\alpha-1}\ \exp(\mathbf{x}'\boldsymbol{\beta})\ \epsilon,$$

where ϵ is a positive random variable with density $g(\epsilon)$ and expectation 1. The model can be written in a log-linear form

$$y = \mathbf{x}'\boldsymbol{\beta}^* + \sigma(\omega - \log\ \epsilon)$$

where $\boldsymbol{\beta}^* = -\boldsymbol{\beta}/\alpha$ and $\sigma = \alpha^{-1}$. ω has an extreme value distribution with expectation –0.5772 and variance $\pi^2/6$. Now suppose that ϵ is gamma distributed with expectation 1 and variance σ_ϵ^2. Integrating out the heterogeneity, it can be shown (see Prentice, 1975) that the marginal distribution of y is also log-linear. One obtains

$$y = \mathbf{x}'\boldsymbol{\beta}^* + \sigma\ \omega^*$$

where the distribution of ω^* is log-generalized $F(1, \sigma_\epsilon^{-2})$. The same is true for other models. In general, if $\lambda(t, \exp(x'\boldsymbol{\beta}))$ denotes the hazard rate for a member of the generalized F-distribution, the conditional specification $\lambda(t, \epsilon \exp(x'\boldsymbol{\beta}))$ yields a linear model for logT (Brännäs, 1986). Note that the often used specification $\epsilon\, \lambda(t, \exp(x'\boldsymbol{\beta}))$ yields a log-linear model only in the exponential and the Weibull cases.

For the wide class of linear regression models for logT a LS estimator can be calculated, perhaps after adjusting censored observations as outlined in Section 3.6.7. When the distribution of the disturbance ω or ω^* respectively is known and not normal, there may be a rather large loss of efficiency, compared to other estimators, for example, the maximum likelihood estimator. But when the distribution of the disturbance is not known or misspecified, the LS estimator may perform much better than the maximum likelihood estimator. See Buckley and James (1982) for a small Monte Carlo experiment.

Within the framework of duration models with unmeasured heterogeneity Brännäs (1986) compared the performance of the Heckman and Singer estimator, the LS estimator, and the maximum likelihood estimator for linear regression models after logarithmic transformation of the duration by Monte Carlo experimentation. The EM algorithm turned out to be extremely sensitive to the conditional density of the duration. This result coincides with the above mentioned findings of Trussell and Richards (1985). In the studied cases the EM algorithm was never better than the LS estimator, but often much worse. Furthermore, the bias and mean squared error of the EM algorithm were sensitive to the initial specifications for the heterogeneity. The performance of the maximum likelihood estimator strongly depends on a correct specification of the distribution of the heterogeneity. Although further research is needed here, it seems that the least squares estimator can be highly recommended for empirical application when unmeasured heterogeneity is important.

3.9.5 Testing for Neglected Heterogeneity

Most of the hazard rate models used in social sciences to analyze event history data do not account for unobserved heterogeneity, and nearly all available computer programs only estimate models without unobserved heterogeneity. But when estimating such a model we cannot be sure if the underlying model is correct or if we have misspecified the model neglecting important unobserved variables. Therefore, we need a test procedure which detects this kind of misspecification, and if possible, we should be able to calculate the test statistic with available computer programs. In addition, it should not use a special form for the distribution of the heterogeneity.

Our test for neglected heterogeneity combines the results of Lancaster (1985), Kiefer (1984), and Burdett et al. (1985). The derivation of the test statistic follows Lancaster. His approach of constructing diagnostics can be applied to any duration model, but he does not consider censored data. We shall generalize his approach to allow for censored durations, but we use the estimator of the test statistic proposed by Burdett et al. Burdett et al. (1985) derive a variance estimator which is conditional on the estimated parameter values, whereas Lancaster estimates the unconditional variance of the test statistic.

From (3.9.2) we have

$$\lambda(t|\mathbf{x},\epsilon) = \lambda(t|\mathbf{x})\ \epsilon.$$

We assume $E(\epsilon)=1$ and that ϵ and $\mathbf{x}$ are independent. Density and distribution function of ϵ are again denoted by $g(\epsilon)$ and $G(\epsilon)$ respectively.

In Section 3.7.1 we defined the cumulative hazard rate for a model without unmeasured heterogeneity

$$r(t|\mathbf{x}) = \int_0^t \lambda(u|\mathbf{x})du.$$

We saw that the random variable $r(t|\mathbf{x})$ has a standard exponential distribution which does not depend on the covariates. Therefore, we can consider the $r(t_i|\mathbf{x}_i)$, $i = 1, \ldots, n$, as "residuals," if we replace unknown parameters by maximum likelihood estimates. We discussed these residuals in Section 3.7.1. We shall see that the test statistic will only depend on these residuals which can be calculated from models that ignore unobserved heterogeneity. So common computer packages can be used. For example, the residuals of the exponential, the Weibull, and the log-logistic model are given by

$$r(t|\mathbf{x}) = t \exp(\mathbf{x}'\hat{\boldsymbol{\beta}}) \qquad \text{(exponential model)}$$

$$r(t|\mathbf{x}) = t^{\hat{\alpha}} \exp(\mathbf{x}'\hat{\boldsymbol{\beta}}) \qquad \text{(Weibull model)}$$

$$r(t|\mathbf{x}) = \log(1 + t^{\hat{\alpha}} \exp(\mathbf{x}'\hat{\boldsymbol{\beta}}) \qquad \text{(log-logistic model).}$$

Introducing the heterogeneity term ϵ, we obtain for the survivor function of T, given $\mathbf{x}$ and ϵ

$$S(t|\mathbf{x},\epsilon) = \exp(-\epsilon\ r(t|\mathbf{x}))$$

and on integrating this equation with regard to the distribution of ϵ, we find that

$$\begin{aligned} S(t|\mathbf{x}) &= \int \exp(-\epsilon\ r(t|\mathbf{x}))\ g(\epsilon)\ d\epsilon \\ &= E_\epsilon\ (\exp(-\epsilon\ r(t|\mathbf{x}))). \end{aligned}$$

The density function $f(t|\mathbf{x},\epsilon)$ is given by

$$\begin{aligned} f(t|\mathbf{x},\epsilon) &= \epsilon\ r'(t|\mathbf{x})\ \exp(-\epsilon\ r(t|\mathbf{x})) \\ &= \epsilon\ \lambda(t|\mathbf{x})\ \exp(-\epsilon\ r(t|\mathbf{x})). \end{aligned}$$

Now we develop the test statistic to examine neglected heterogeneity. The distribution of ϵ is not known, and hence we need a test statistic which does not require knowledge about the form of the distribution of ϵ. Let $\text{Var}(\epsilon)=\sigma^2$. The null hypothesis is that there is no neglected heterogeneity. This can be expressed as $\sigma^2 = 0$. With the alternative hypothesis of small σ^2 we approximate the density $f(t|\mathbf{x},\epsilon)$ and the survivor function $S(t|\mathbf{x},\epsilon)$ with a Taylor expansion around $\epsilon_o = 1$. Differentiating $S(t|\mathbf{x},\epsilon)$ with respect to ϵ we obtain

$$\frac{dS(t|\mathbf{x},\epsilon)}{d\epsilon} = -r(t|\mathbf{x}) \exp(-\epsilon \; r(t|\mathbf{x}))$$

$$\frac{d^2S(t|\mathbf{x},\epsilon)}{d\epsilon^2} = r^2(t|\mathbf{x}) \exp(-\epsilon \; r(t|\mathbf{x})),$$

and for the Taylor expansions we have

$$S(t|\mathbf{x},\epsilon) = \exp(-r(t|\mathbf{x})) - r(t|\mathbf{x}) \exp(-r(t|\mathbf{x})) (\epsilon - 1) + \\ + 1/2 \; r^2(t|\mathbf{x}) \exp(-r(t|\mathbf{x})) (\epsilon - 1)^2 + R.$$

Integrating $S(t|\mathbf{x},\epsilon)$ with respect to the distribution of ϵ and neglecting terms of order higher than two, we can write

$$S(t|\mathbf{x}) \approx \exp(-r(t|\mathbf{x})) + \sigma^2/2 \; r^2(t|\mathbf{x}) \exp(-r(t|\mathbf{x})) \\ = S(t|\mathbf{x},1) (1 + \sigma^2/2 \; r^2(t|\mathbf{x})). \qquad (3.9.14)$$

For the density $f(t|\mathbf{x})$ we obtain

$$f(t|\mathbf{x}) = -\frac{dS(t|\mathbf{x})}{dt} \approx r'(t|\mathbf{x}) \exp(-r(t|\mathbf{x})) - \sigma^2/2 \; [2r(t|\mathbf{x}) \; r'(t|\mathbf{x}) \\ \exp(-r(t|\mathbf{x}) + r^2(t|\mathbf{x}) \exp(-r(t|\mathbf{x})) (-r'(t|\mathbf{x}))] \\ = f(t|\mathbf{x},1) - \sigma^2 \; r(t|\mathbf{x}) \; f(t|\mathbf{x}) + \sigma^2/2 \; r^2(t|\mathbf{x}) \; f(t|\mathbf{x},1) \\ = f(t|\mathbf{x},1) \; [1 + \sigma^2/2 \; r^2(t|\mathbf{x}) - \sigma^2 \; r(t|\mathbf{x})] \qquad (3.9.15)$$

Now we assume that the functional form of $f(t|\mathbf{x})$ and $S(t|\mathbf{x})$ is known up to a finite dimensional parameter vector $\boldsymbol{\theta}$. $\boldsymbol{\theta}$ contains the regression coefficients $\boldsymbol{\beta}$ as well as further parameters determining time dependence. We write $f(t|\mathbf{x},\boldsymbol{\theta})$ and $S(t|\mathbf{x},\boldsymbol{\theta})$. We introduce a censoring indicator δ_i where $\delta_i = 0$ if the duration of individual i is right censored and $\delta_i = 1$ otherwise. The log likelihood contribution is given by

$$\log L = \delta \log f(t|\mathbf{x};\boldsymbol{\theta}) + (1 - \delta) \log S(t|\mathbf{x};\boldsymbol{\theta}),$$

where we dropped the subscript i to simplify notation. Using (3.9.14) and (3.9.15) as an approximation of the true survivor function and the true density function for small σ we have

$$\log L = \delta[\log f(t|x,1;\boldsymbol{\theta}) + \log(1 + \sigma^2(1/2 \; r^2(t|x;\boldsymbol{\theta}) - r(t|x;\boldsymbol{\theta}))] + \\ + (1 - \delta) [\log S(t|x,1;\boldsymbol{\theta}) (1 + \sigma^2/2 \; r^2(t|x;\boldsymbol{\theta})].$$

The proposed test for neglected heterogeneity is a score test for the hypothesis $\sigma^2 = 0$. Let $\hat{\boldsymbol{\theta}}$ denote the maximum likelihood estimates under the null hypothesis, that is, when σ^2 is known to be zero. To derive the test statistic we have to calculate the score function at the point $(\hat{\boldsymbol{\theta}}, 0)$. We obtain

$$\left.\frac{\partial \log L}{\partial \theta_j}\right|_{(\boldsymbol{\theta},\sigma^2) = (\hat{\boldsymbol{\theta}},0)} = 0 \qquad \text{for all j,}$$

$$\frac{\partial \log L}{\partial \sigma^2} = \delta \frac{1/2\, r^2(t|\mathbf{x};\boldsymbol{\theta}) - r(t|\mathbf{x};\boldsymbol{\theta})}{1 + \sigma^2(1/2\, r^2(t|\mathbf{x};\boldsymbol{\theta}) - r(t|\mathbf{x};\boldsymbol{\theta}))} + (1-\delta) \frac{r^2(t|\mathbf{x};\boldsymbol{\theta})/2}{1 + \sigma^2/2\, r^2(t|\mathbf{x};\boldsymbol{\theta})}.$$

Evaluating $\frac{\partial \log L}{\partial \sigma^2}$ at $(\hat{\boldsymbol{\theta}},0)$ and collecting terms we see that

$$\frac{\partial \log L}{\partial \sigma^2}\Big|_{(\boldsymbol{\theta},\sigma^2) = (\hat{\boldsymbol{\theta}},0)} = 1/2\, r^2(t|\mathbf{x};\hat{\boldsymbol{\theta}}) - \delta r(t|\mathbf{x};\hat{\boldsymbol{\theta}}).$$

Using the total likelihood function the score test for the null hypothesis $\sigma^2 = 0$ is based on the quantity

$$S = \frac{1}{2n} \sum_{i=1}^{n} (r^2(t_i|\mathbf{x}_i;\hat{\boldsymbol{\theta}}) - 2\delta_i\, r(t_i|\mathbf{x}_i;\hat{\boldsymbol{\theta}})). \tag{3.9.16}$$

The $r(t_i|\mathbf{x}_i;\hat{\boldsymbol{\theta}})$ in (3.9.16) are the maximum likelihood estimates of the integrated hazards for the model without heterogeneity. Therefore, common computer programs for duration models (e.g., SAS, GLIM, BMDP, RATE) can be used for the computation of S.

The variance of S can be estimated by

$$\hat{Var}(S) = \frac{1}{4n^2} \sum_{i=1}^{n} (s_i - \bar{s})^2$$

where $s_i = r^2(t_i|\mathbf{x}_i;\hat{\boldsymbol{\theta}}) - 2\delta_i r(t_i|\mathbf{x}_i;\hat{\boldsymbol{\theta}})$ and $\bar{s}$ is the sample mean of s_i. This conditional estimate of the variance of the test statistic was proposed by Burdett et al. (1985), whereas Lancaster (1985) uses an unconditional variance estimator which is more difficult to obtain in the case of censored durations.

The test statistic

$$\frac{S}{\sqrt{\hat{Var}(S)}} \tag{3.9.17}$$

is asymptotically standard normal, conditionally on the estimated value of $\hat{\boldsymbol{\theta}}$. We use a one-tailed test because the alternative hypothesis is $\sigma^2 > 0$.

The proposed test statistic applies quite general provided that the general residuals can be calculated. It is a score test for neglected heterogeneity when the variance of the heterogeneity is small.

For an illustration of the test for unobserved heterogeneity see Blossfeld and Hamerle (1988b).

3.10 Discrete Hazard Rate Regression Models

In the previous sections, it has been assumed that the points of time in which state transitions or events occur could be stated explicitly. In many cases, however, the determination of the exact point of time of transition or occurrence of some event is not possible. It may only be possible to state intervals of time in which state transitions or event occurrences may take place. This applies in particular to longitudinal panels where event histories, for example, about employment status and other important qualitative changes between the successive panel waves are registered retrospectively in fixed-length periods. One of the new panel studies of this kind is the socio-economic panel of the "Sonderforschungsbereich 3" (see Hanefeld, 1984). Other applications are in medical work when patients are followed up and detailed information on each patient is collected at fixed intervals, or in sociological research when attention is given to qualitative changes that occur in specific time intervals.

If there are only a few time intervals or if the time units are large then many failures are reported at the same time and the number of ties becomes high. Then, strictly speaking, continuous time techniques are inappropriate (Cox and Oakes, 1984, p. 99 f.). Some continuous time methods, especially the partial likelihood estimation procedure for Cox's proportional hazards model (see, e. g., Kalbfleisch and Prentice, 1980, Chapter 4) make use of the temporal order in which the failures or events occur and they cannot be applied directly when the data include tied observations. In the presence of ties an approximate partial likelihood function is widely used (Breslow, 1974). But when the number of ties becomes high, this approximation yields severely biased estimates (Cox and Oakes, 1984, p. 103; Kalbfleisch and Prentice, 1980, p. 74 f.). Kalbfleisch and Prentice (1980, p. 75) emphasize that there is some asymptotic bias in both the estimation of the regression coefficients and in the estimation of its covariance matrix. This applies not only to the Cox model but also to fully parametrized specifications of the hazard rate. Moreover, the papers which deal with the derivation of the asymptotic properties of the estimators in hazard rate models (see, e.g., Andersen and Gill, 1982; Borgan, 1984) assume that ties only occur with zero probability.

In such situations discrete time models are more suited for the analysis of failure time data. Several authors, including Thompson (1977), Prentice and Gloeckler (1978), Mantel and Hankey (1978), Allison (1982), Aranda-Ordaz (1983), Laird and Olivier (1981), Hamerle (1985), and others have studied discrete time regression models for failure time data.

In this section, we present only a brief introduction of models of discrete hazard rates. Further, Chapters 4 to 6 present no examples, since in life course studies the point of time of state transition may be determined with sufficient accuracy. We limit ourselves to the most important statistical concepts regarding a single episode and without competing risks. A detailed analysis of discrete hazard rate models for the single or multiepisode case may be found in Hamerle and Tutz (1988) upon which the following discussion is based.

The time axis is divided into q + 1 intervals

$$[a_0, a_1), [a_1, a_2), \ldots, [a_{q-1}, a_q), [a_q, \infty),$$

where, as a rule $a_0 = 0$ and the assumed terminal period of observation is a_q. The time interval $[a_{t-1}, a_t)$ is expressed in short form simply as t.

The duration time or lifetime is represented by a nonnegative stochastic variable T. T takes on only integer values, and $T = t$ means that in the interval $[a_{t-1}, a_t)$ a transition or change of state has occurred.

In addition to the duration time or lifetime for each individual or subject, a p-dimensional vector **x** of covariates or prognostic factors is collected. The covariates are postulated to be time invariant. Introducing external and internal time dependent covariates is possible as illustrated in Hamerle and Tutz (1988).

As in Sections 3.2 and 3.3, hazard rates and survivor functions may be defined. The hazard rate is

$$\lambda(t|\mathbf{x}) = P(T = t|T \geq t, \mathbf{x}). \tag{3.10.1}$$

(3.10.1) represents the conditional probability that an individual in time interval t experiences an event given the covariates and the fact that the individual has reached the beginning of the time interval.

The conditional probability of "surviving" the time interval t is then

$$P(T > t|T \geq t, \mathbf{x}) = 1 - \lambda(t|\mathbf{x}). \tag{3.10.2}$$

The survivor function is

$$S(t|\mathbf{x}) = P(T \geq t|\mathbf{x}), \tag{3.10.3}$$

the (unconditional) probability of "surviving" the time interval t, that is, up till the beginning of this interval an event has not occurred. The relationship between the survivor function and the hazard rate is obtained by successive application of

$$P(T \geq s|\mathbf{x}) = P(T \geq s|T \geq s-1, \mathbf{x}) \cdot P(T \geq s-1|\mathbf{x})$$

and with (3.10.2). Consequently,

$$S(t|\mathbf{x}) = \prod_{s=1}^{t-1} (1 - \lambda(s|\mathbf{x})). \tag{3.10.4}$$

Finally, one obtains for the unconditional event probability or, in other words, the probability of experiencing an event in time interval t given the covariates

$$P(T = t|\mathbf{x}) = P(T = t|T \geq t, \mathbf{x}) \cdot P(T \geq t|\mathbf{x}) = \lambda(t|\mathbf{x}) \prod_{s=1}^{t-1} (1 - \lambda(t|\mathbf{x})). \tag{3.10.5}$$

With the aid of (3.10.5), $P(T = t|\mathbf{x})$ may also be expressed by the hazard rate, and as in models with continuously measured time we parameterize the dependence of the hazard rate upon the covariates. Various possibilities exist in which one may accomplish this. The most important models are discussed in detail in Hamerle and Tutz (1988). Here we sketch only two specifications, the logistic model and the grouped Cox model.

The hazard rate of the logistic model is

$$\lambda(t|\mathbf{x}) = P(T = t|T \geq t, \mathbf{x}) = \frac{\exp(\beta_{0t} + \mathbf{x}'\boldsymbol{\beta})}{1 + \exp(\beta_{0t} + \mathbf{x}'\boldsymbol{\beta})} \quad t = 1, \ldots, q. \tag{3.10.6}$$

An equivalent formulation of the model is obtained as

$$\ln \frac{P(T = t|x)}{P(T > t|x)} = \beta_{0t} + \mathbf{x}'\boldsymbol{\beta}. \tag{3.10.7}$$

The parameters $\beta_{01}, \ldots, \beta_{0q}$ represent, as in the Cox model, given continuously measured time, a "baseline hazard rate" which is common to all individuals.

The hazard rate of the grouped Cox model is

$$\lambda(t|\mathbf{x}) = 1 - \exp(-\exp(\beta_{0t} + \mathbf{x}'\boldsymbol{\beta})). \tag{3.10.8}$$

Model (3.10.8) may be regarded as a Cox model with discrete observations. If the Cox model is valid for the continuous durations, but we can only observe grouped data, then (3.10.8) is valid for the grouped observations. The parameter vector $\boldsymbol{\beta}$, which regulates the influence of the covariates, remains unchanged. It is identical with the corresponding weighting vector of the continuous time Cox model.

Maximum Likelihood Estimation

Also for discrete models, in analogy to the discussion of Section 3.6.2, the contribution of an individual i to the likelihood function may be derived.

One obtains

$$L_i = P(T_i = t_i|\mathbf{x}_i)^{\delta_i} P(T_i \geq t_i|\mathbf{x}_i)^{1-\delta_i} = \lambda(t_i|\mathbf{x}_i)^{\delta_i} P(T_i \geq t_i|\mathbf{x}_i), \tag{3.10.9}$$

where δ_i once again designates the censoring indicator. Thus given (3.10.4),

$$L_i = \lambda(t_i|\mathbf{x}_i)^{\delta_i} \prod_{s=1}^{t_i-1} (1-\lambda(s|\mathbf{x}_i)). \tag{3.10.10}$$

Most discrete duration models may be estimated on the basis of (3.10.9) within the framework of generalized linear models. Generalized linear models are discussed in detail in Fahrmeir and Hamerle (1984, Chapter 7) or McCullagh and Nelder (1983). One observes t_i independent dichotomous random variables $Y_{i1}, \ldots, Y_{it_i}$. Then it follows for the corresponding likelihood function

$$L_i = \prod_{r=1}^{t_i} P(Y_{ir} = 1)^{y_{ir}} (1 - P(Y_{ir} = 1))^{1-y_{ir}}.$$

In the case in which the conditional random variables are dependent upon the regressor variables $\mathbf{x}_i$, one obtains

$$L_i = \prod_{r=1}^{t_i} P(Y_{ir} = 1|\mathbf{x}_i)^{y_{ir}} (1 - P(Y_{ir} = 1|\mathbf{x}_i))^{1-y_{ir}}. \tag{3.10.11}$$

This is equivalent to the likelihood of t_i observations of a generalized linear model, in which $P(Y_{ir} = 1|\mathbf{x}_i)$ with the aid of a response function, namely, $\lambda(r|\mathbf{x}_i)$ may be specified dependent upon the parameter $\boldsymbol{\beta}$. Setting for the observed vector $\mathbf{Y}_i = (Y_{i1}, \ldots, Y_{it_i}) = (0, \ldots, 0, 1)$, one obtains from (3.10.11) the contribution of the ith person to the likelihood function in the duration model for the case $\delta_i = 1$. Observing only $t_i - 1$ random variables, one obtains

$$L_i = \prod_{r=1}^{t_i-1} P(Y_{ir} = 1|\mathbf{x}_i)^{y_{ir}} (1 - P(Y_{ir} = 1|\mathbf{x}_i))^{1-y_{ir}},$$

the contribution (3.10.10) to the likelihood function for $\delta_i = 0$, where once again $P(Y_{ir} = 1|\mathbf{x}_i) = \lambda(r|\mathbf{x}_i)$ and the observation vector $\mathbf{Y}_i = (y_{i1}, \ldots, y_{i,t_i-1}) = (0, \ldots, 0)$ is given.

The possibility of estimating duration models within the framework of generalized linear models offers the advantage of working with existing program packages such as BMDP (for the logistic model), GLIM or SAS.

The discrete models may be further modified to include external as well as internal time dependent covariates; however, the likelihood function must be derived alternatively (see Hamerle and Tutz, 1988). Also, one may generalize the model to include unobserved population heterogeneity (see Hamerle, 1986).

Chapter 4: Data Organization and Descriptive Methods

In general, the main goal of data analysis is to find an economical statistical model that approximates reality as closely as possible and permits a substantive interpretation. However, one does not always know which variables are the most important to examine at the beginning of an analysis. Therefore, normally one attempts in a first step to gain an impression of the distributions of all possible variables with the aid of descriptive and graphical methods. Additional comparisons and tests among subgroups may then be made in a second step in order to obtain additional information regarding relationships among the variables. Finally, in a third step, theoretically-derived causal factors may be included in order to develop an appropriate explanatory model.

This chapter discusses ways to handle files of event oriented data (Section 4.1) as well as various ways of presenting such data graphically (Section 4.2). This is then followed by a description of the application of the life table method and Kaplan-Meier estimator, using data from the German Life History Study (GLHS) as an illustration (Section 4.3). Special attention is given to considering the specific problems and advantages of event history analysis and to documenting this discussion with illustrative program examples, tests, and suggestions about how to interpret results.

4.1 Managing Event Oriented Data Structures

The first substantial problem that arises in analyzing event history data is the question of appropriate data storage in preparation for statistical evaluation. *Event oriented data structures are much more complex* than cross-sectional data because, given the *state information,* one needs to simultaneously consider both the exact *beginning and end time points.* Furthermore, in economics and the social sciences it is common to find *repeated events the recurrence of which may vary greatly over the period of observation.* This leads to the question of an appropriate method of data storage. For example, if, as in the GLHS, the employment history of three birth cohorts is recorded continuously, large differences may occur in the number of job episodes among persons ranging from 0 (never having been employed) to as many as 19 different job episodes. If one stores this data for each individual according to rows in square form, then a derived matrix will have many empty cells and the available storage capacity will not be used economically. A *small change in the statement of a problem* which thus results in a *new definition of the state space,* usually requires a fundamental reorganization of the data file because program packages such as SPSS, BMDP, RATE, SAS, or GLIM use only certain episodes as the units of analysis.

In practical application, these problems of data management of event history data are solved with the aid of a data bank system which stores the data in an hierarchical and economical manner permitting flexible retrieval. The program package SIR (Robinson et al., 1980) exemplifies such a data bank system. We used this system to store the GLHS data. Given such a data bank system it is then possible to produce event oriented data sets according to a specific question at hand, which may be analyzed through the program packages SPSS, BMDP, RATE, SAS, or GLIM.

One speaks of an event oriented data set when *each record of a file corresponds exactly to one event or one episode.* If only one episode is observed for each subject or sample member, then the number of records in the file corresponds exactly to the number of sample members. On the other hand, given repeated events (e.g., jobs) whose number may vary among persons, then the summation of these person-specific episodes represents the number of records in the file.

Each episode is fully characterized by a beginning time (TS) and an end time (TF) as well as by a beginning (SS) and an end state (SF). Commonly, instead of analyzing the beginning and end times, the duration (TD) is used as the actual dependent variable in event history analysis. If the beginning and end state have the same value, then one must deal with *right censored observations.* The episode is not terminated by a state transition but rather is cut off at the time point of some retrospective interview. Sometimes a specific *censoring (or indicator) variable* (CEN), which takes the value "1" if an episode is terminated by a regular state transition or otherwise the value "0," is introduced to indicate right censored observations. For each such episode, *covariates* may be introduced, which can either be independent of time (**x**) or can vary over time (**x**(t)), taking into consideration the *history of the process* or *other parallel processes.*

Table 4.1 gives an example of an event oriented data set, prepared for analyzing employment trajectories. Each record in the file represents an employment episode and the successive records of a person's employment are stored successively. For each episode the exact beginning (TS) and end (TF) of employment as well as the resulting duration of employment (TD) are given. In order to be able to simply compare and calculate dates for the three birth

Table 4.1: Example for an Event Oriented Data Set for the Study of Occupational Careers

Case No.	Beginning Time (TS)	End Time (TF)	Duration (TD)	State at the Beginning of the Spell (SS)	State at the End of the Spell (SF)	Censoring Variable (CEN)	Number of the Episode	Time Constant Covariates							Time Varying Covariates		
								Sex	Social Origin	...	Cohort	Labor Force Experience at Entry into the Job	...	Number of Previously Held Jobs	Time of First Marriage	...	Time of Leaving Home
								x_1	x_2	...	.	....	...	x_m	$x_{m+1}(t)$	...	$x_p(t)$
1	590	626	36	1	2	1	1	1	2	...	1	0	...	0	660	...	665
1	626	698	72	2	3	1	2	1	2	...	1	36	...	1	660	...	665
1	698	981	313	3	3	0	3	1	2	...	1	108	...	2	660	...	665
2	610	680	70	10	11	1	1	2	1	...	1	0	...	0	705	...	700
.	.	.	.	.	.	.	.	.	.	...	.	.	...	.	.	...	...
.	.	.	.	.	.	.	.	.	.	...	.	.	...	.	.	...	...
.	.	.	.	.	.	.	.	.	.	...	.	.	...	.	.	...	...

cohorts of the GLHS, we coded episode begin and end times according to the number of months from the beginning of this century. For example, the number 590 represents the second month (February) in the year 1949 ($49 \cdot 12 + 2 = 590$). The state space in this example consists of twelve occupations (see Table 4.4). Thus, person 1 was first employed in occupation 1 ($SS_1 = 1$) until he changed to occupation 2 ($SF_1 = 2$ and $SS_2 = 2$) and finally worked in occupation 3 ($SF_2 = 3$ and $SS_3 = 3$). In occupation 3 this person was employed ($SF_3 = 3$) at the point of time at which the interview occurred. Whereas the first two episodes of the first individual ended with an event (CEN = 1), the third episode has been terminated by the interview (right censored (CEN = 0)). Finally, in the remaining columns for each episode, next to the number of the employment episode, various covariates are recorded. Illustrative of variables that remain constant over the employment history are sex, social origin, and cohort membership. Illustrative of time varying variables that express the history of the process being analyzed are the number of months of occupational experience at the beginning of each episode or the number of previously practiced occupations. An example of a parallel process that changes over time (and therefore within an episode) would be marital status. For example, one could examine whether or not the course of employment after marriage becomes more stable.

Occupational changes may be analyzed in this event oriented form as a function of covariates by applying one of the program packages SPSS, BMDP, RATE, GLIM, or SAS. The episodes (or observation units) in this case are the individual's occupations, regardless of whether a state transition implies promotion, demotion, or simply horizontal mobility. If one wishes, however, to analyze how long it takes for an individual to get a better job (Sørensen, 1984; Sørensen and Blossfeld, 1988) it is necessary to construct new duration times or spells. All recorded job episodes up to the time of promotion must then be combined to form one spell. Whether or not one considers possible unemployment periods depends upon the definition of the process and theoretical interest. If no better jobs were found, the spell is censored at the time of the interview, since up to this point in time no event occurred. The individual, of course, may obtain a better job after the interview. The event oriented file in Table 4.1 is, however, no longer applicable for this sort of modified analysis since each occupational transition was regarded as an episode. A newly structured data file must be constructed.

In practical application of event oriented data structures the method of data storage in data bank systems (e.g., SIR) or program packages with data bank characteristics (e.g., SAS) has proved to be successful. With these programs, it is possible to work in a very flexible way with extremely complicated event oriented data sets and to restructure data sets to answer specific questions.

4.2 Graphical Presentation of Event History Data

Due to the high density of information in event history data, it is not only necessary (compared with cross-sectional analyses) to use new methods of data storage and data management, but also to find new ways to present such time dependent data lucidly and understandably. In particular methods are required that are appropriate for illustrating the multistate and multiepisode processes commonly found in the economic and social sciences. In this section, three possible ways of graphically presenting event history data are discussed: the illustration of individual histories, event sequences, and cumulative distributions.

The most comprehensive presentation of the high information density of event history data is exemplified in the *graphical illustration of individual histories.* For each individual or unit of examination, the entire history is graphed (see Figure 4.1). The x axis represents historical time or age, and the y axis records discrete states. The goal of this illustrative procedure is to build a typology via the simultaneous presentation of many individual histories.

Figure 4.1, taken from Müller (1978), illustrates this way of presenting event history data. The horizontal lines of the graph symbolize duration in various states of the educational and employment system. The vertical lines represent state transitions. With the aid of this rather highly differentiated state space one may follow in detail the employment history of 30 randomly chosen males born in 1946 who where employed in agriculture in 1971. It is easy to see that this occupational group consists mainly of individuals who after elementary school, completed agricultural or vocational training, and subsequently worked on the family farm.

Although the development of plot programs (see Carr-Hill and Macdonald, 1973; Müller, 1978) has significantly facilitated the use of this method, the principle disadvantage of constructing multiple plot graphs still remains: With an increasing number of states along an increasing number of histories, the complexity of the plots increases very rapidly, and the clarity of the graph decreases. Often one cannot identify the path and history of an individual. Although it is possible to select and plot specific subgroups (e.g., selecting individuals with common beginning and end states) and thereby increase the clarity of the diagram, this method is in principle still quite limited due to its complexity.

One attempt to deal with this problem is the method of presenting event history data by *event sequences.* A greater clarity of individual sample paths is achieved by plotting only subpopulations with a specific sequence of events. Time or age is again plotted on the abscissa and the ordinate characterizes a specific sequence pattern. The diagram thus contains only those event sequences which may be assigned to some specific sequence pattern. Figure 4.2 exemplifies this method. Here Schulz and Strohmeier (1985) plot in conjunction the events of various biographical careers (partnerships, family, employ-

Figure 4.1: Occupational Life Histories of 30 Randomly Selected Men of the Birth Cohorts 1946, Who Were Farmers at Age 25

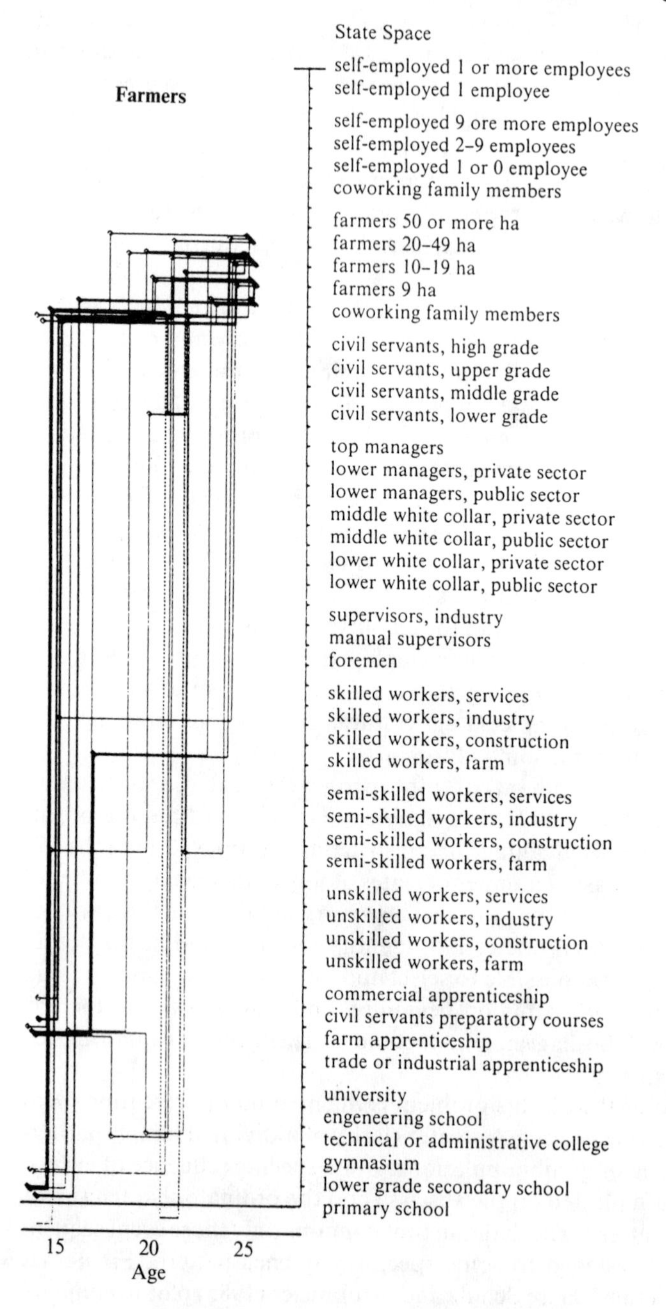

Source: Müller, W.: Klassenlage und Lebenslauf. Habilitationsschrift. Mannheim 1978.

Figure 4.2: An Event Sequence Derived From 61 Cases With the Same Course Pattern

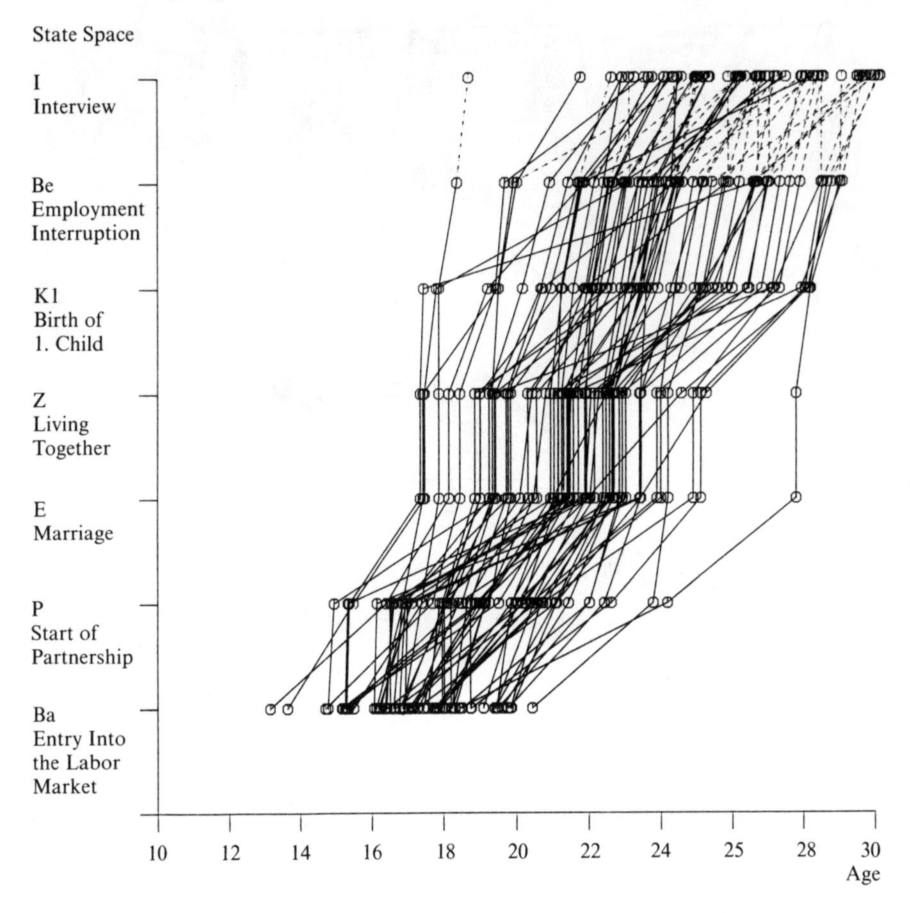

Source: Schulz, M., and Strohmeier, K.: "Familienkarriere und Berufskarriere" In: Franz, H.-W. (Ed.): 22. Deutscher Soziologentag 1984. Sektions- und Ad-hoc-Gruppen. Opladen 1985.

ment). The steepness of the plotted curve paths illustrates the speed with which an individual has undergone a specific sequence pattern. The spread or width of the plotted curves offers information about the degree of age homogeneity. Duration heterogeneity is represented in the diagram by a large number of criss-crossing lines.

The advantage of this method is that much detailed information about individual histories remains relatively accessible. One can observe how certain consecutive events are distributed over the life course. On the other hand, this method requires an a priori choice of sequences. It is therefore not possible to

Figure 4.3: Educational and Occupational History of Men and Women of the Birth Cohort 1929-31

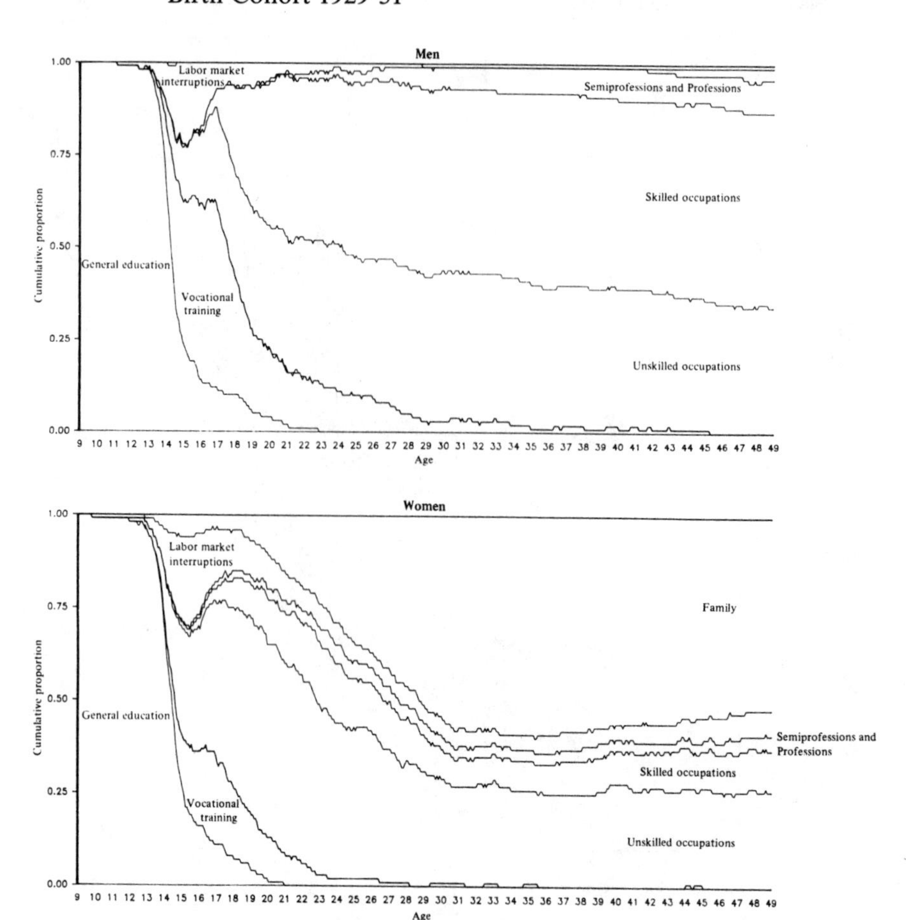

see how typical or atypical a specific pattern generally is. The lucidity of this graph again decreases with the number of individual histories plotted and an increase in overlapping curves.

However, the case is entirely different when event history data are used not to graph individual histories but to *describe aggregates* with cumulated distributions. In this case, the diagram becomes clearer as the number of individuals increases, because this enforces the stability of the distributions. For each point in time the distribution of the individuals is calculated for a given number of states and then cumulated. When these points are connected, a picture is produced of the structural changes that have occurred over time.

Figure 4.3 demonstrates the use of cumulated distributions to illustrate the educational and occupational history of men and women of the birth cohort

1929–31 (see Blossfeld, 1987a, 1987b). The x axis designates the age of the cohort (the smallest unit being one month), and the y axis reflects the cumulative distribution of these persons (in percent) on selected states of the educational and employment system as well as on labor market interruptions (e. g., unemployment, family work, etc.). Along the x axis one can follow structural changes over the life course. The various opportunities, which potentially occur for men and women of the 1929-31 cohort over the 40-year (maximal) observation period, can be compared in a simple manner.

Although this method may be used to obtain important information about the course of a process given multistate and multiepisode models, its main disadvantage is the fact that individual histories are not retrievable. Therefore, the explanatory power of the diagram strongly depends on individual movements, which underlie the structure, because at each time point only the balance of the individual movements is plotted. Hence, stability evident in the aggregate may not correspond to stability at the individual level. Despite this disadvantage, researchers have preferred this approach, and presentations using the aggregate plotting method are becoming more prominent.

In general, one can say that the three methods of graphical presentation offer a valuable first impression of the process being analyzed, even when complexly structured multistate and multiepisode processes are studied. These plot diagrams are capable of giving information about important relationships between variables. Another important function involves the use of graphic presentations to check the plausibility of results indicated by statistical models in subsequent phases of the analysis process. The graphs help to avoid false interpretations and methodological artefacts.

4.3 Life Table Method and Kaplan-Meier Estimator

Whereas the previous discussion of methods of presenting event history data was exclusively based on the entire course of the process, this section discusses *nonparametric estimation methods* that concentrate on each of the episodes in such histories. These methods of analysis may, however, also be applied to the multistate and multiepisode cases as will be shown in a later discussion.

Let us first examine the *one-episode case* and limit the discussion at the same time to *one specific type of event.* A typical application of such analysis is, for example, commonly found in medical duration studies, which examine the length of time before individuals suffering from some specific illness die. The method is also used in technical studies that analyze the industrial reliability of newly installed machines. In the economic and social sciences there also exist applications for this special case. Consider, for example, the time period between the introduction of a new product and the time when the customer purchase decision is made (marketing), or the period between the birth of a child and the point at which the child decides to leave the parents' home (family sociology). Repeated events such as employment episodes can also be considered as one-episode cases, for example, if one is interested only in the transition to the first job, or in the period between entry to employment and the first job change.

For estimating the survivor functions of such models, the life table method (such as may be implemented in the program packages SPSS and BMDP) and the Kaplan-Meier estimator (e.g., implemented in the BMDP and SAS program packages) are available. Both methods are nonparametric estimation techniques that make no assumption about the distribution of the process to be analyzed and therefore are especially suited for first exploratory analyses of the data. Especially helpful are the graphical presentations of the estimated survivor function and their logarithm, the estimated hazard function, and the estimated duration density function that may permit a detailed insight into the course of the process.

Life Table Method

The life table method, as has already been explained in Section 3.2.3, divides duration until the occurrence of an event or until censoring into fixed intervals of any length. For each of these intervals the number of individuals who are exposed to the risk of the event at the beginning of the interval, who experience an event during the interval, and who are censored are counted. Based on this, the duration density function, the survivor function, and the hazard function are estimated. The estimation is made under the assumption that the censored cases in each interval are equally distributed (see Section 3.2.3).

As a concrete example of the application of the method of life table analysis with the program package SPSS, we examine the length of time between entry to employment and the first job change occurs. Here we were interested in determining whether or not differences arise between men and women, as well as between the three birth cohorts 1929–31, 1939–41, and 1949–51 of the GLHS.

The data are organized according to the event oriented data structure presented in Table 4.1. In addition to various interesting variables, it contains the beginning and end points for each employment episode, coded in relation to the number of months from the beginning of this century. With the following SPSS setup, the life tables for men and women as well as for the three birth cohorts are separately calculated and the paths of the various estimated functions plotted:

Program Example 4.1:

```
GET FILE        DATA
COMPUTE         DUR = M51 - M50 + 1
COMPUTE         CEN = 1
IF              (M51 EQ M47) CEN = 0
```

```
IF              (M48 GE 348 AND LE 384) COHO = 1
IF              (M48 GE 468 AND LE 504) COHO = 2
IF              (M48 GE 588 AND LE 624) COHO = 3
*SELECT IF      (M5 EQ 1)
SURVIVAL        TABLES = DUR BY M3(1,2),COHO(1,3)/
                INTERVALS=THRU 24 BY 3, THRU 300 BY 12/
                STATUS = CEN(1)/
                PLOTS (ALL)/
                COMPARE
OPTIONS         5,8
FINISH
```

After reading in the SPSS system file, the second line of the program calculates the employment duration (DUR), by subtracting the beginning time point (M50) from the terminal time point of each employment episode (M51) (see Appendix 1). The employment duration for each episode is consequently expressed in monthly units. With the next two commands a censoring variable is generated that receives the value "1," given a regular employment termination, or the value "0," if the end of the spell is identical with the time point of the interview (M47). The next three commands are used to construct a new variable (COHO) to differentiate the three birth cohorts. This variable is based on the birth dates (M48), which are coded according to the number of months from the beginning of the century. Because we are only interested in the first episode of the employment history, we filter out each of the first jobs with the aid of the sequence number of the episodes (M5). The life table analysis is initiated with the command SURVIVAL. The TABLES statement contains the duration variable DUR, the group variables sex (M3) and cohort (COHO). The values (in parentheses) in this statement group the total sample into subsamples.

Commands concerning the period of time to be considered in the life tables and the division into intervals are given with the code word INTERVALS. The duration period in this example during the first 24 months is divided into 3-month intervals and thereafter up until the 300th month (or 25th employment year) into 12-month intervals. This is done because the probability that a job change will occur within the first year of employment may be especially high (probationary period, worker, or employer dissatisfaction) and in later phases should decrease. In the STATUS command the program assigns the censoring variable (CEN). The value in brackets means that regular events are characterized by 1, whereas all others are considered as censored. The PLOT command, along with the code word ALL, is used to graph the survivor function and its logarithm, the hazard function and the duration density function. Implementing the command COMPARE results in a comparison of the survivor functions of the respective subgroups and a statistical test of this comparison. Finally, the OPTIONS statement requests that, if there is not enough core to do exact comparisons, only approximate comparisons should be done (option 5) and that all survival tables should be printed out in a raw data file, which may then be used for a plot program (option 8).

With this SPSS program five life tables (for men and women as well as for the three birth cohorts) are calculated. Here our discussion is limited to men (Table 4.2). The first column of this life table contains the lower limits of the chosen intervals (a_{k-1}). In the second column, for each individual the number of men still exposed to the risk of a change of employment (risk set 1: R_k) is registered at the beginning of an interval. For example, in the first interval this is $R_1 = 1077$ men and in the second interval this reduces to $R_2 = 1063$ men.

Table 4.2: Illustration of a Life Table

LIFE TABLE
SURVIVAL VARIABLE DUR
FOR M3

INTVL START TIME	NUMBER ENTRNG THIS INTVL	NUMBER WDRAWN DURING INTVL	NUMBER EXPOSD TO RISK	NUMBER OF TERMNL EVENTS	PROPN TERMI- NATING	PROPN SURVI- VING	CUMUL PROPN SURV AT END	PROBA- BILITY DENSTY	HAZARD RATE	SE OF CUMUL SURV- IVING	SE OF PROB- ABILTY DENS	SE OF HAZRD RATE
0.0	1077.0	0.0	1077.0	14.0	0.0130	0.9870	0.9870	0.0043	0.0044	0.003	0.001	0.001
3.0	1063.0	1.0	1062.5	44.0	0.0414	0.9586	0.9461	0.0136	0.0141	0.007	0.002	0.002
6.0	1018.0	1.0	1017.5	64.0	0.0629	0.9371	0.8866	0.0198	0.0216	0.010	0.002	0.003
9.0	953.0	0.0	953.0	46.0	0.0483	0.9517	0.8438	0.0143	0.0165	0.011	0.002	0.002
12.0	907.0	1.0	906.5	85.0	0.0938	0.9062	0.7647	0.0264	0.0328	0.013	0.003	0.004
15.0	821.0	0.0	821.0	36.0	0.0438	0.9562	0.7312	0.0112	0.0149	0.014	0.002	0.002
18.0	785.0	2.0	784.0	45.0	0.0574	0.9426	0.6892	0.0140	0.0197	0.014	0.002	0.003
21.0	738.0	0.0	738.0	42.0	0.0569	0.9431	0.6500	0.0131	0.0195	0.015	0.002	0.003
24.0	696.0	5.0	693.5	183.0	0.2639	0.7361	0.4785	0.0143	0.0253	0.015	0.001	0.002
36.0	508.0	0.0	508.0	92.0	0.1811	0.8189	0.3918	0.0072	0.0166	0.015	0.001	0.002
48.0	416.0	3.0	414.5	75.0	0.1809	0.8191	0.3209	0.0059	0.0166	0.014	0.001	0.002
60.0	338.0	3.0	336.5	47.0	0.1397	0.8603	0.2761	0.0037	0.0125	0.014	0.001	0.002
72.0	288.0	1.0	287.5	40.0	0.1391	0.8609	0.2377	0.0032	0.0125	0.013	0.000	0.002
84.0	247.0	2.0	246.0	27.0	0.1098	0.8902	0.2116	0.0022	0.0097	0.013	0.000	0.002
96.0	218.0	3.0	216.5	21.0	0.0970	0.9030	0.1911	0.0017	0.0085	0.012	0.000	0.002
108.0	194.0	3.0	192.5	20.0	0.1039	0.8961	0.1712	0.0017	0.0091	0.012	0.000	0.002
120.0	171.0	2.0	170.0	21.0	0.1235	0.8765	0.1501	0.0018	0.0110	0.011	0.000	0.002
132.0	148.0	4.0	146.0	13.0	0.0890	0.9110	0.1367	0.0011	0.0078	0.011	0.000	0.002
144.0	131.0	7.0	127.5	9.0	0.0706	0.9294	0.1271	0.0008	0.0061	0.010	0.000	0.002
156.0	115.0	5.0	112.5	8.0	0.0711	0.9289	0.1180	0.0008	0.0061	0.010	0.000	0.002
168.0	102.0	7.0	98.5	5.0	0.0508	0.9492	0.1120	0.0005	0.0043	0.010	0.000	0.002
180.0	90.0	2.0	89.0	11.0	0.1236	0.8764	0.0982	0.0012	0.0110	0.010	0.000	0.003
192.0	77.0	2.0	76.0	4.0	0.0526	0.9474	0.0930	0.0004	0.0045	0.009	0.000	0.002
204.0	71.0	3.0	69.5	3.0	0.0432	0.9568	0.0890	0.0003	0.0037	0.009	0.000	0.002
216.0	65.0	2.0	64.0	1.0	0.0156	0.9844	0.0876	0.0001	0.0013	0.009	0.000	0.001
228.0	62.0	1.0	61.5	6.0	0.0976	0.9024	0.0791	0.0007	0.0085	0.009	0.000	0.003
240.0	55.0	2.0	54.0	2.0	0.0370	0.9630	0.0761	0.0002	0.0031	0.009	0.000	0.002
252.0	51.0	2.0	50.0	3.0	0.0600	0.9400	0.0716	0.0004	0.0052	0.009	0.000	0.003
264.0	46.0	5.0	43.5	3.0	0.0690	0.9310	0.0666	0.0004	0.0060	0.009	0.000	0.003
276.0	38.0	3.0	36.5	1.0	0.0274	0.9726	0.0648	0.0002	0.0023	0.009	0.000	0.002
288.0	34.0	4.0	32.0	0.0	0.0000	1.0000	0.0648	0.0000	0.0000	0.009	0.000	0.000
300.0+	30.0	24.0	18.0	6.0	0.3333	0.6667	0.0432	**	**	0.009	**	**

** THESE CALCULATIONS FOR THE LAST INTERVAL ARE MEANINGLESS.

THE MEDIAN SURVIVAL TIME FOR THESE DATA IS 34.49

Column 3 records the number of men whose employment episode in the respective interval has been censored (w_k). No censoring occurred in the first interval ($w_1 = 0$), and only one episode is censored in the second episode ($w_2 = 1$). Column 4 contains the number of men remaining in the population at risk, taking censoring into account (risk set 2: $R_k - \frac{w_k}{2}$). For the second interval one obtains a value of the risk set 2 of $1063 - \frac{1}{2} = 1062.5$.

Finally, the events for each interval are expressed in column 5 (d_k). For example, $d_2 = 44$ is obtained for the second interval.

Based on this data, the conditional probability of changes in employment in each interval (column 6)

$$\hat{\lambda}_k = \frac{d_k}{R_k - \frac{w_k}{2}},$$

the conditional probability to experience no event for each interval (column 7)

$$\hat{p}_k = 1 - \hat{\lambda}_k,$$

the survivor function (column 8)

$$\hat{P}_k = \hat{p}_k \cdot \ldots \cdot \hat{p}_1,$$

the duration density function (column 9)

$$\hat{f}_k = \frac{\hat{P}_{k-1} - \hat{P}_k}{h_k},$$

the hazard function (column 10)

$$\hat{\lambda}(m_k) = \frac{2\hat{\lambda}_k}{h_k\,(1 + \hat{p}_k)},$$

as well as the standard errors (columns 11–13) are estimated (see Section 3.2.3).

The second interval is characterized, for example, by an estimated conditional probability of a change in employment equal to

$$\hat{\lambda}_2 = \frac{44}{1063 - \frac{1}{2}} = 0.0414.$$

The conditional probability that no event occurs in the second interval is

$$\hat{p}_2 = 1 - 0.0414 = 0.9586.$$

The survivor function is estimated to be

$$\hat{P}_2 = 0.9870 \cdot 0.9586 = 0.9461,$$

the estimated density function is calculated as

$$\hat{f}_2 = \frac{0.9870 - 0.9461}{3} = 0.0136,$$

and the estimated hazard function is

$$\hat{\lambda}(m_2) = \frac{2 \cdot 0.0414}{3(1+0.9586)} = 0.0141 \;.$$

The corresponding standard error of the survivor function is 0.007, for the duration density function 0.002, and for the hazard function 0.002. Finally, at the bottom of the life table an estimation of the median of the first employment duration is given (a value equal to 34.49 here means approximately two years and ten months).

The life table, which describes in great detail the process of leaving the first job, is naturally quite complex and is not easy to be used for a comparison of subgroups. On the other hand, the SPSS plots available are of great value when comparing subpopulations.

Let us first examine the *duration density functions* (see equation 3.2.2) of the first employment of men (1) and women (2) (see Figure 4.4). It is clear to see, that in the case of both sexes, density of job changes at the beginning of the process rises steeply and levels off asymptotically with increasing duration to the x axis. The density function exhibits for both sexes the same right-skewed path, and the differences between both processes are at first glance purely minimal.

Another presentation of the process is demonstrated by plotting the survivor function (Figure 4.5). In this case the survivor function expresses how many men (1) or women (2) have not changed their occupation up to a particular point of time. Examination of the plot reveals that after 48 months (or 4 years) approximately 70 percent of all men and women have already changed their first occupation. For this period of time, the process is quite similar for both sexes. However, the differences between the sexes become larger and larger thereafter. Women after this point, tend to leave their first job faster than males do. The survivor function of women thus falls beneath that of males.

Based on the estimation of the *hazard function* (see equation 3.2.5) one obtains indications as to whether the process of occupational mobility out of the first employment position is time dependent or not. In Figure 4.6, it is easy to see that the probability of changing jobs in the first nine months increases for both men and women and then decreases with increasing duration. For females, one observes once again a significant rise in the risk of changing jobs between the 60th and 110th month of employment duration. After approximately the 180th month (or 15 years), plots of the hazard function become difficult to interpret due to the small number of persons still exposed to risk. As such, the estimated path of the hazard function becomes quite unstable and large variances are possible.

Finally, on the basis of a *log scale graph of the survivor function* in Figure 4.7, another way of demonstrating the process of changing the first occupation is illustrated. If the process is constant over time, then this function

Figure 4.4: Example of a Duration Density Function Plot Estimated According to the Life Table Method

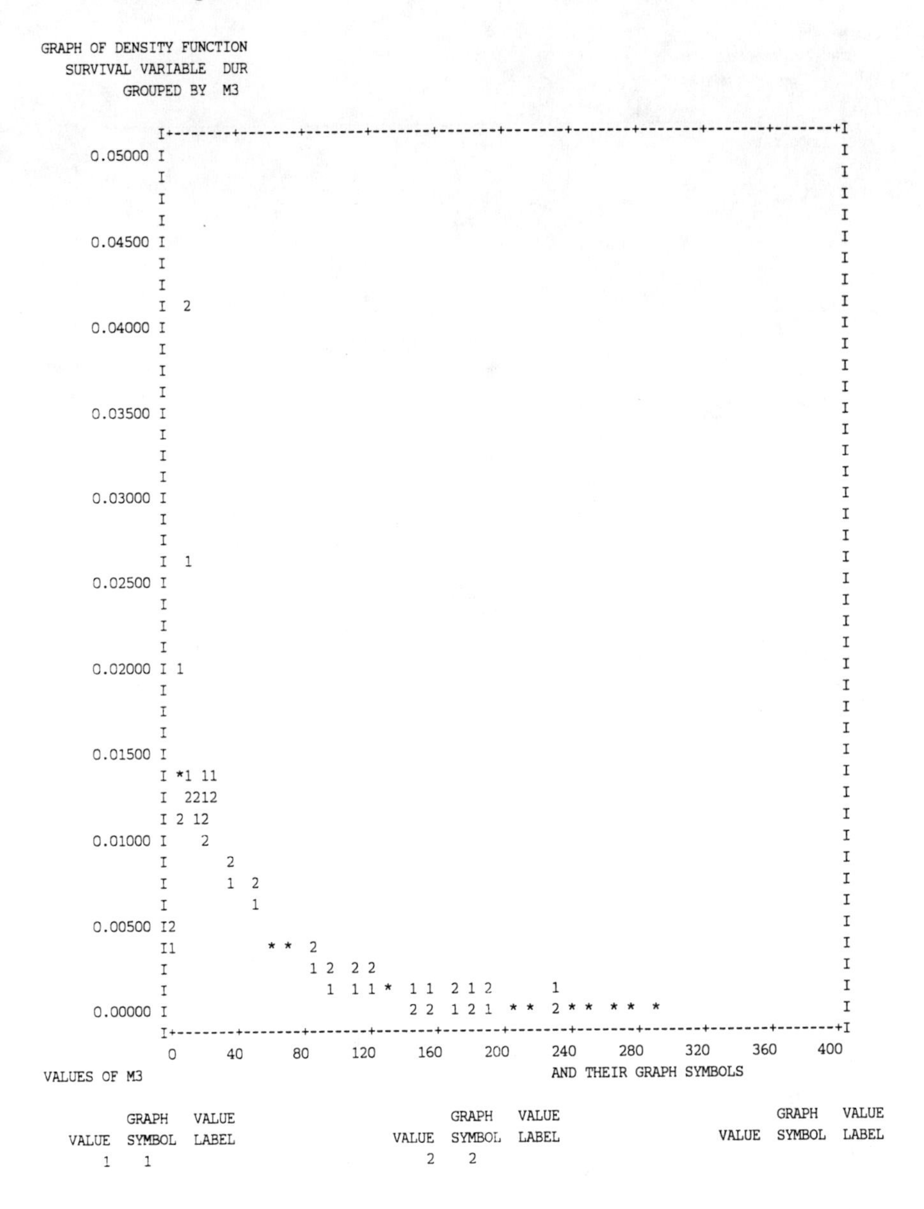

should fall linearly with a constant slope. If the risk decreases, as is the case for men (1) after the 120th month, then the curve bends upward. On the other hand, women (2) after the 80th month are characterized by a downward bending curve.

Figure 4.5: Example of a Survivor Function Plot Estimated According to the Life Table Method

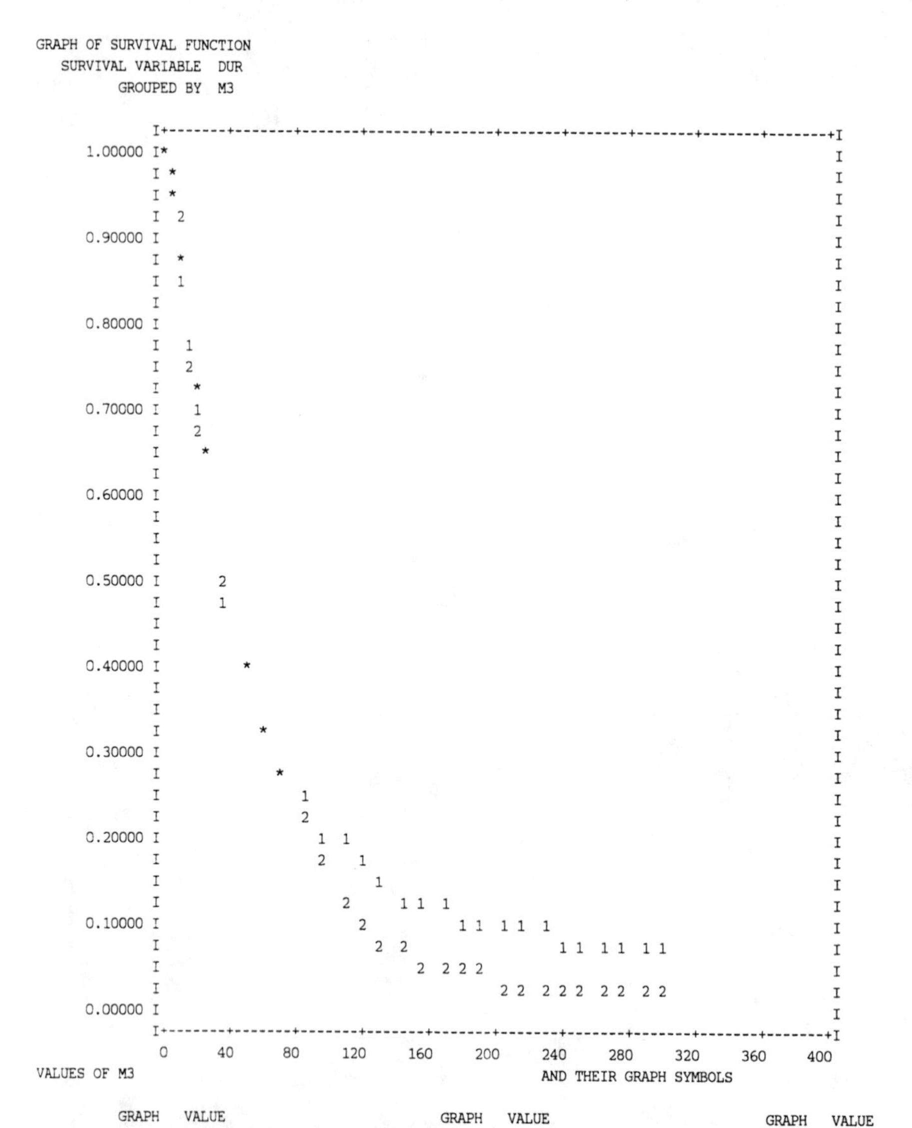

In addition to plots, one can use SPSS to examine whether or not the survivor functions of men and women differ significantly on the basis of the *Lee-Desu test statistic* (Lee and Desu, 1972), which is a modified Wilcoxon test (see Section 3.2.5). Prerequisite for this rank test is that the survivor functions

Figure 4.6: Example of a Hazard Function Plot Estimated According to the Life Table Method

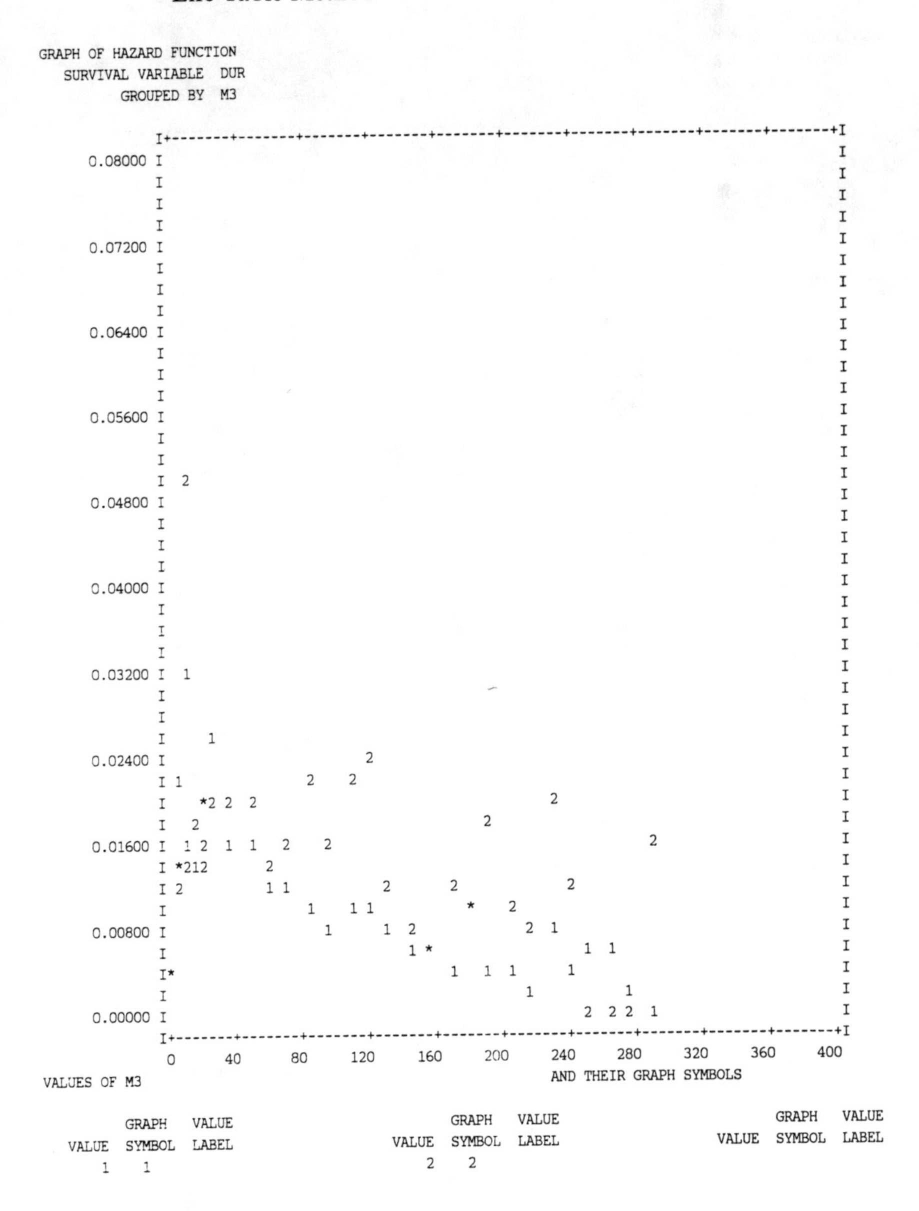

of the subpopulations do not intersect (i.e., the ranking may not be turned around) and that within the subpopulations identical censoring patterns are given. Under the null hypothesis that the survivor functions of the subgroups do not differ, the test statistic is asymptotically χ^2 distributed with k–1 degrees

Figure 4.7: Example of a Log Scale Graph of the Survivor Function Estimated According to the Life Table Method

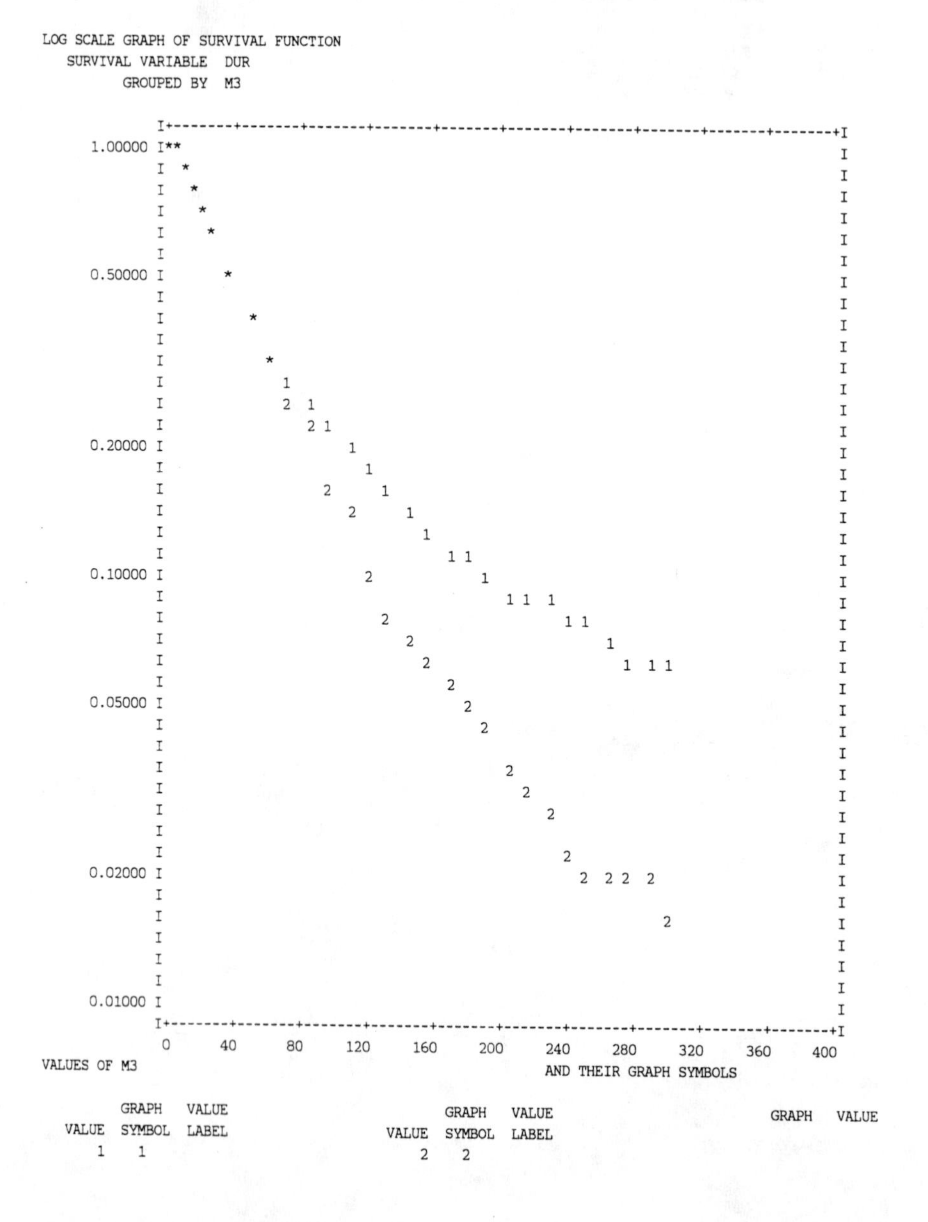

of freedom (k = number of subgroups). A comparison of the survivor functions between men and women in the previous example resulted in a value of the Lee-Desu test statistic of 0.048 with 1 d.f. The survivor functions of men and women do not significantly vary from one another at the 0.05 significance level. One must consider, however, that this test reacts sensitively to differences in the survivor functions at the beginning of the process. Because the differences in first occupational experiences of men and women only crystallize after the 48th month, it is advisable to use the Cox-Mantel test statistic offered in the BMDP program package. This test statistic is especially sensitive to differences toward the end of the process. It is presented in more detail below in connection with the Kaplan-Meier estimator.

Kaplan-Meier Estimator

Although the life table method is a very useful method to analyze event history data, especially for larger samples, the estimated results largely depend on the chosen intervals. As a rule, the larger the interval, the poorer and more inaccurate are the estimations obtained for the functions. A further problem is that when the choice of interval length varies, different estimations are normally to be expected. Thus, for samples that are not very large, one should apply the Kaplan-Meier estimator (or the product-limit estimator), which is available in the BMDP program package. This method of estimation, which has already been discussed in detail in Section 3.2.4, does not use data grouped according to intervals. Instead it applies the actually measured event and censoring times. The basic idea of this estimator is that by segmenting the durations into successively smaller units a point will finally be reached in which each event or censoring time falls within only one specific interval. Indeed the life table and the Kaplan-Meier estimator are more or less identical, if one chooses a corresponding small interval in the life table method. The recorded event and censoring times are then ordered according to size. As a rule, censored observations that have occurred at the same time point as events are regarded as somewhat lagged. Based on such a ranking of event and censoring times, estimations of survivor functions are made, whereby the censored times only decrease the risk set of the later occurring events. Thus, under the Kaplan-Meier method of estimating the survivor function, a step function with discrete jumping points at the event time points is obtained. However, a problem arises with this method if there exist censored times in this ordered ranking that are larger than the largest event time. Under these circumstances, the estimated survivor function can no longer approach zero and the average duration will be underestimated. In the practical application of such cases, an interpretation only considers the length of time until the largest event occurs.

In illustrating the application of the Kaplan-Meier estimator we once again apply the example of time until a change of the first job occurs. The event history data set is the same; however, it is now read in the PL1 program of BMDP.

Program Example 4.2:

```
/INPUT UNIT IS 30.
       CODE IS DATA.
/VARIABLE NAMES ARE (63) DUR,(64)CEN,(65)COHO.
          ADD = 3.
/TRANSFORM DUR = M51 - M50 + 1.
           CEN = 1.
           IF (M51 EQ M47) THEN CEN = 0.
           IF (M48 GE 348 AND M48 LE 384) THEN COHO = 1.
           IF (M48 GE 468 AND M48 LE 504) THEN COHO = 2.
           IF (M48 GE 588 AND M48 LE 624) THEN COHO = 3.
           USE = M5 EQ 1.
/GROUP CODES (65) ARE 1,2,3.
       NAMES (65) ARE COHO1,COHO2,COHO3.
       CODES (3) ARE 1,2.
       NAMES (3) ARE MEN,WOMEN.
/FORM TIME IS DUR.
      STATUS IS CEN.
      RESPONSE IS 1.
/ESTIMATE METHOD IS PROD.
          GROUP IS COHO.
          PLOTS ARE SURV,LOG.
          STATISTICS ARE BRESLOW,MANTEL.
/ESTIMATE METHOD IS PROD.
          GROUP IS M3.
          PLOTS ARE SURV,LOG.
          STATISTICS ARE BRESLOW,MANTEL.
/END
```

In the BMDP program following the input of the BMDP system file (see Appendix 1) that contains 62 variables, the paragraph VARIABLE defines three additional variables required for the run. In the TRANSFORM paragraph, as was the case in the previous SPSS example, first the duration (DUR) is calculated, then a censoring variable (CEN) is constructed, a cohort variable (COHO) is created and finally with the command USE only the first occupations are selected.

In the FORM paragraph the BMDP program assigns the duration variable (TIME IS DUR) and the censoring variable (STATUS IS CEN). Given an event, the latter has the value "1" (RESPONSE IS 1). The estimation is then made separately according to the cohort variable (COHO) and the sex variable (M3), whereby the product-limit estimator (Kaplan-Meier estimator) (METHOD IS PROD) is applied. The respective survivor functions (SURV) and the logarithmic survivor functions (LOG) are then plotted. In addition, the Breslow (BRESLOW) and the Cox-Mantel statistic (MANTEL) are also applied.

Table 4.3 demonstrates how the Kaplan-Meier method is applied to estimate the survivor function regarding the first change of employment for men. In the column CASE NUMBER, the case numbers of males are ordered according to their duration in the TIME column. The STATUS column identifies whether one is dealing with a regular event (DEAD) or a censored observation (CENSORED). The case with the number 6432, which is also ranked after the event times with three months, exemplifies the rule that given the same durations of event time and censoring time, the censoring time is to be regarded as lagged. Because in the GLHS the employment history is recorded monthly, it is quite possible that various cases have the same duration. In Table 4.3, for example, six cases have a one-month duration. According to equation 3.2.41 with a correction for ties,

$$\hat{S}(t) = \begin{cases} 1 & \text{for } t \leq t_{(1)} \\ \prod_{k|t_{(k)}<t} \left(1 - \frac{1}{R_k}\right) & \text{for } t > t_{(i)} \end{cases}$$

the survivor function in the column CUMULATIVE SURVIVAL jumps after the first month by a value $\frac{6}{1077} = 0.0056$ from 1 to 0.9944 and in the column STANDARD ERROR the standard error evaluated at this point is printed (see equation 3.2.42). If the portion of such ties due to measurement of durations is large, then the life table method may be the more efficient and robust approach. Finally, in the column CUM DEATHS, the number of regular events is counted, and in the column REMAIN AT RISK, the number of men who are still exposed to the risk of occupational change is recorded.

Case number 6675 possesses the longest event time with 410 months. Nine additional censored cases follow, which have not been included in the estimation. Hence, the average duration (MEAN SURVIVAL TIME) of the first job of 73.83 months is somewhat underestimated (see equation 3.2.43).

Generally, the printout of the detailed calculation scheme of the survivor function in accordance with the Kaplan-Meier method (Table 4.3) is of little help. The printout is usually very large (in the previous example for males only it is already 1,077 lines) and may be suppressed by the command NO PRINT. It does, however, give an impression as to how large a sorting and storage capacity is needed to calculate the Kaplan-Meier estimator. For large samples, the life table method is therefore more economical to compute than the Kaplan-Meier estimator.

The plots available with the BMDP program are extremely useful for comparing subpopulations. Figure 4.8 illustrates the *path of the survivor function* for the cohorts 1929–31 (A), 1939–41 (B), and 1949–51 (C). It is

Table 4.3: Example of the Kaplan-Meier Estimator

TIME VARIABLE IS DUR

CASE LABEL	CASE NUMBER	TIME	STATUS	CUMULATIVE SURVIVAL	STANDARD ERROR	CUM DEATHS	CUM LOST	REMAIN AT RISK
	1788	1.00	DEAD			1	0	1077
	1944	1.00	DEAD			2	0	1076
	2396	1.00	DEAD			3	0	1075
	3632	1.00	DEAD			4	0	1074
	4644	1.00	DEAD			5	0	1073
	4770	1.00	DEAD	0.9944	0.0023	6	0	1072
	1843	2.00	DEAD			7	0	1071
	2645	2.00	DEAD			8	0	1070
	2660	2.00	DEAD			9	0	1069
	2936	2.00	DEAD			10	0	1068
	2951	2.00	DEAD			11	0	1067
	3339	2.00	DEAD			12	0	1066
	4875	2.00	DEAD			13	0	1065
	6023	2.00	DEAD			14	0	1064
	6653	2.00	DEAD	0.9861	0.0036	15	0	1063
	69	3.00	DEAD			16	0	1062
	758	3.00	DEAD			17	0	1061
	871	3.00	DEAD			18	0	1060
	1413	3.00	DEAD			19	0	1059
	1468	3.00	DEAD			20	0	1058
	2484	3.00	DEAD			21	0	1057
	3155	3.00	DEAD			22	0	1056
	3465	3.00	DEAD			23	0	1055
	4448	3.00	DEAD			24	0	1054
	4527	3.00	DEAD			25	0	1053
	4662	3.00	DEAD			26	0	1052
	5113	3.00	DEAD			27	0	1051
	5298	3.00	DEAD			28	0	1050
	5509	3.00	DEAD			29	0	1049
	6370	3.00	DEAD			30	0	1048
	6391	3.00	DEAD	0.9712	0.0051	31	0	1047
	6432	3.00	CENSORED			31	0	1046
	136	4.00	DEAD			32	0	1045
	313	4.00	DEAD			33	0	1044
	855	4.00	DEAD			34	0	1043
	897	4.00	DEAD			35	0	1042
	1270	4.00	DEAD			36	0	1041
	1389	4.00	DEAD			37	0	1040
	.	.	.	.	.	.	.	.
	.	.	.	.	.	.	.	.
	.	.	.	.	.	.	.	.
	4198	331.00	DEAD	0.0551	0.0086	976	0	21
	2866	333.00	CENSORED			976	0	20
	4319	334.00	CENSORED			976	0	19
	1612	363.00	DEAD	0.0522	0.0086	977	0	18
	180	367.00	CENSORED			977	0	17
	4689	368.00	CENSORED			977	0	16
	3119	377.00	CENSORED			977	0	15
	244	388.00	CENSORED			977	0	14
	4346	392.00	CENSORED			977	0	13
	372	397.00	CENSORED			977	0	12
	5194	398.00	CENSORED			977	0	11
	553	404.00	CENSORED			977	0	10
	6675	410.00	DEAD	0.0470	0.0092	978	0	9
	3565	413.00	CENSORED			978	0	8
	6590	413.00	CENSORED			978	0	7
	375	414.00	CENSORED			978	0	6
	1986	416.00	CENSORED			978	0	5
	6161	416.00	CENSORED			978	0	4
	5217	422.00	CENSORED			978	0	3
	6186	426.00	CENSORED			978	0	2
	1	428.00	CENSORED			978	0	1
	3458	429.00	CENSORED			978	0	0

MEAN SURVIVAL TIME = 73.83 LIMITED TO 429.00 S.E. = 3.291

Figure 4.8: Example of a Survivor Function Plot Based on the Kaplan-Meier Estimator

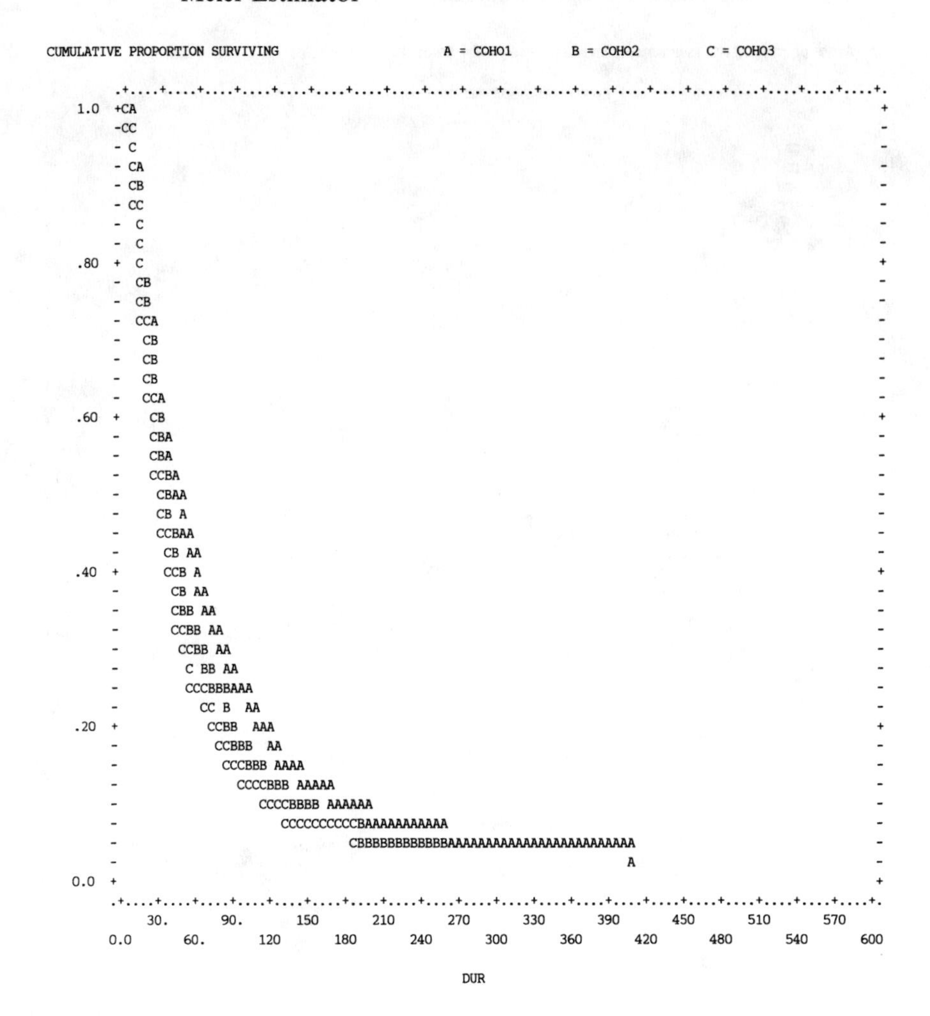

evident that the survivor functions have relatively similar paths up until a duration period of some 30 months. Thereafter, differences between the cohorts occur. The younger the cohort is, the faster the first job is changed. This is also observable from the *log scale graph of the survivor function* (Figure 4.9; see equation 3.2.9), which bends increasingly downward from the oldest to the youngest cohorts. This implies that the risk of changing the first occupation increases from the oldest to the youngest cohorts.

In addition to the purely visual examination of differences between the subgroups based on graphical plots, the BMDP program also offers the possibility of supplementing the graphical analysis with statistical tests (see

Figure 4.9: Example of a Log Scale Graph of the Survivor Function Based on the Kaplan-Meier Estimator

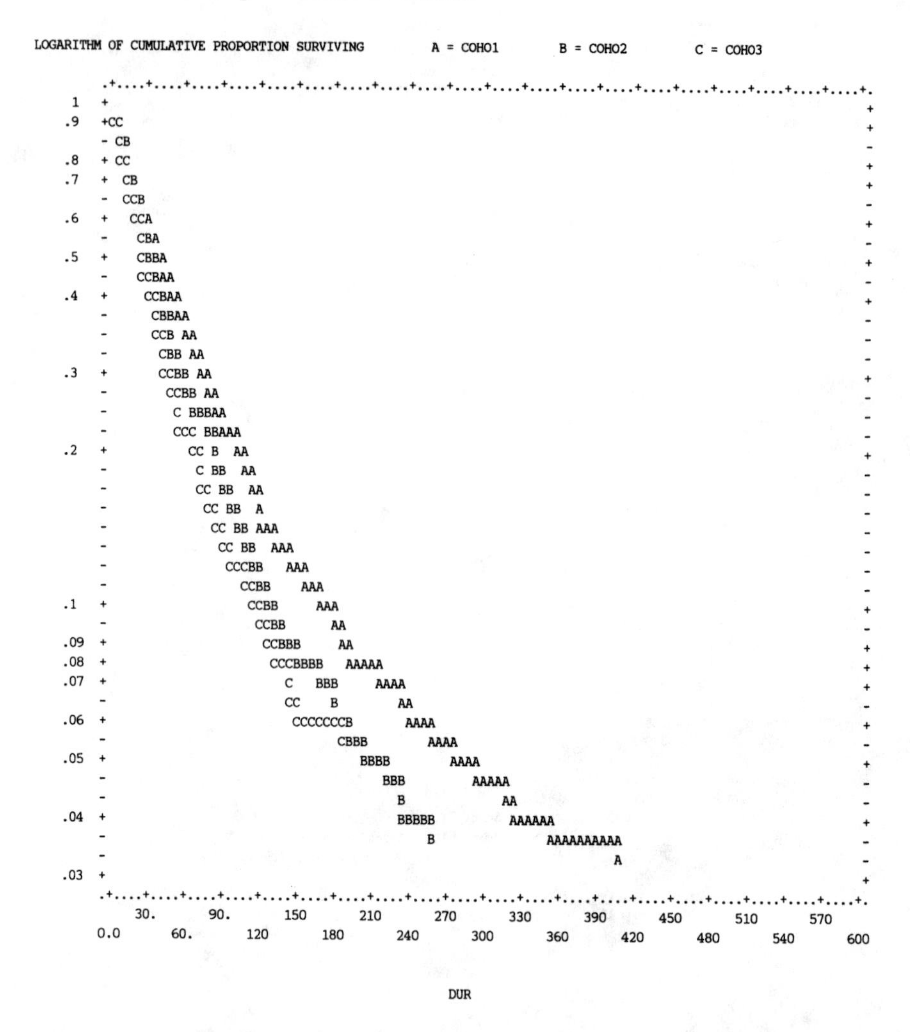

Section 3.2.5). Applying BMDP, one can calculate the Breslow version (1970) of the Wilcoxon statistic and a log-rank statistic according to the method described by Cox and Mantel (1966). Both test statistics presuppose that the survivor functions do not intersect and that the subgroups are characterized by similar censoring patterns. The second assumption may be judged quite simply on the basis of the *censoring and event pattern plot* of BMDP (Figure 4.10). For our birth cohort example, various censoring and event patterns are obtained for each different birth cohort. An explanation can be found in the fact that up until the time of the interview, the older cohorts had a longer

Figure 4.10: Illustration of a Printout for Event and Censoring Patterns

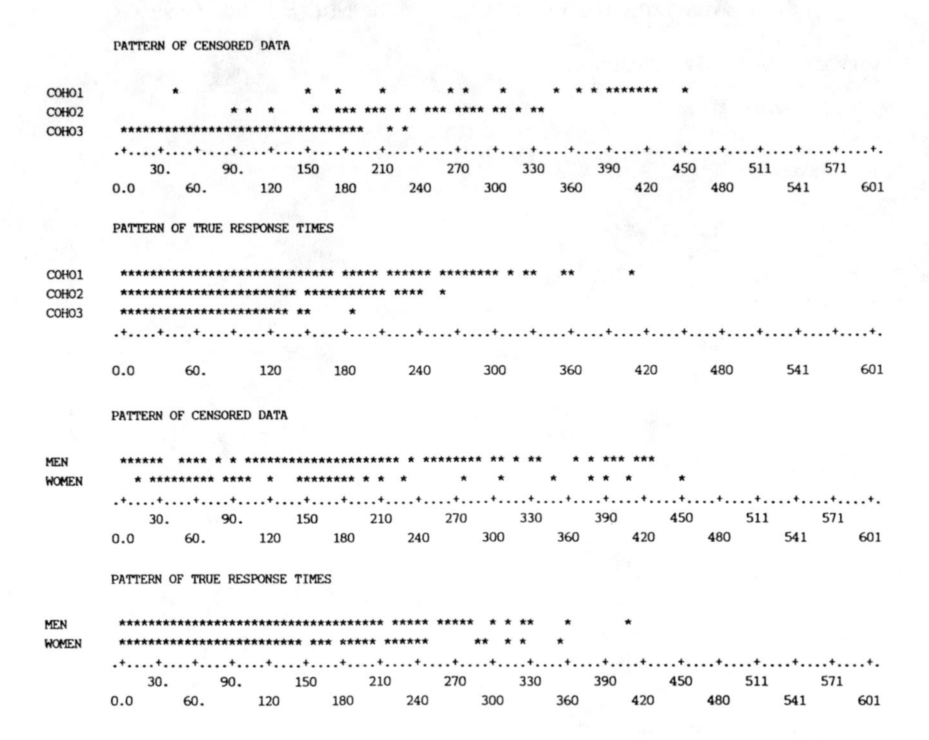

chance to be employed and consequently also have longer censoring and event times. However, the incidence of censoring is thus correlated with cohort membership, and the presumptions made in the tests for analyzing differences among cohorts are not fulfilled. Let us therefore only interpret the results of a test for men and women, in which the censoring times over the entire observation period are distributed quite similarly and where one may postulate that the assumption of a constant censoring pattern is fulfilled (Figure 4.10). Both tests are based on the assumption that the survivor functions of the subgroups to be analyzed do not differ and are asymptotically χ^2 distributed with k–1 degrees of freedom (k = number of subgroups). The Breslow statistic, which along with the Lee-Desu test statistic in SPSS stresses the differences of the survivor functions at the beginning of the process, has a value of 0.079 (with 1 d.f.) and is not significant for a significance level of 0.05. The Cox-Mantel statistic on the other hand, which stresses increasing differences at the end of the process, is significant with the value of 8.745 and one degree of freedom (significance level 0.05). Various test statistics may thus lead to different conclusions for the same survivor functions (see Figure 4.5). Thus, when assessing the test results, one should check that the censoring patterns are somewhat similar and that the survivor functions do not intersect. Based on

Figure 4.11: Example of a Log Scale Graph of the Survivor Function in Analyzing the Distribution of the Multiepisode Case

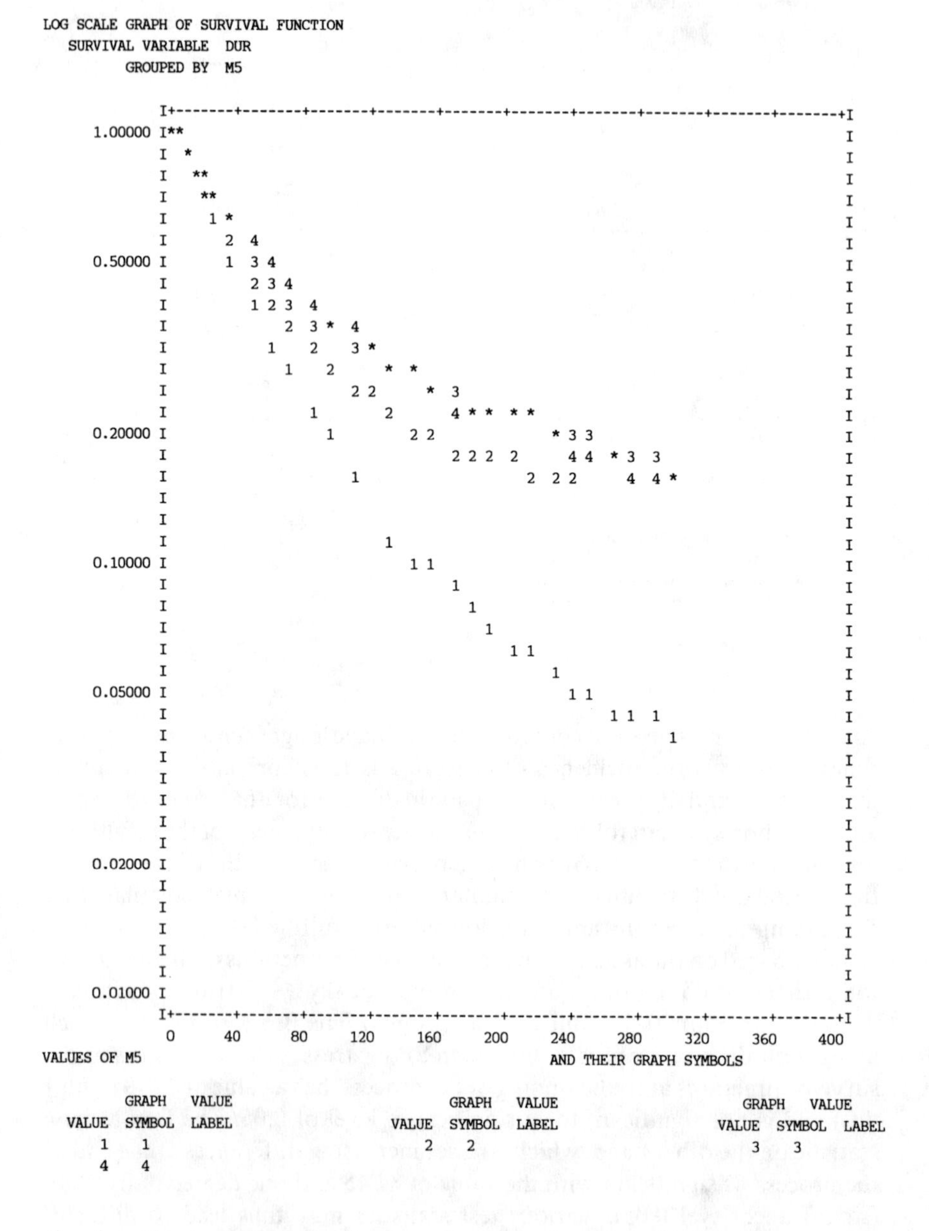

the visual inspection of the survivor plots, one should decide in advance which of the test statistics is more sensitive and thus more appropriate for the situation.

Multiepisode Case

Up until now the presentation of the distribution free methods has been limited to cases where only one episode (e. g., the duration in the first occupation) and a specific type of event (e. g., leaving the first occupation) was given. This type of situation is, however, only a special case. Usually one encounters more than one episode per unit of analysis, as is the case in job shifts during an employment career or residency episodes in migration studies. Consequently, the question arises as to whether or not these methods may be applied to analyze the multiepisode case, and the most practiced manner in which to do this.

Some researchers regard the various episodes of an object being analyzed as independent units, store them in an event oriented data set (Table 4.1), and apply the methods of analysis previously discussed to all episodes. The crucial question in so doing is, however, whether one may work with the different episodes of an individual in this manner or not. For example, individuals with a greater number of occupations would be represented in an event oriented data set more often. This would not pose a problem if one were dealing with a homogeneous population, that is, if the individuals possessed the same characteristics with regard to the process being analyzed. If, however, one is confronted with heterogeneous subpopulations then this would result in a mixing that may lead to time dependency and incorrect conclusions (see Section 3.9.1). In principle, it is the same assumption of homogeneity that has been implicitly made in the one-episode case. Generally, in dealing with questions in economics and the social sciences, which are as a rule characterized by high levels of interdependency between variables, the homogeneity assumption is quite improbable. Sometimes it is possible to avoid this problem by disaggregating the samples according to theoretically important variables. Unfortunately, this approach is only applicable to a limited extent, due to the limitations arising from the *nonavailability of variables in the data set* (problem of unobserved heterogeneity), or *the sample size* that would be necessary for very small subpopulations in order to estimate survivor functions. Thus, in the process of data analysis, the methods just presented offer only a first heuristic step for obtaining information about the structure of the data.

In the multiepisode case of occupational careers the survivor function may, for example, provide information as to whether or not the duration in one occupation is dependent on past history. In order to see this, one has to check whether the survivor functions, for example, in the first four employment activities (which contain approximately 85% of the employment episodes of the GLHS data) differ. In doing so, one can apply the following SPSS program, which compared to program example 4.1 uses the sequence number of the employment episode (M5) as the grouping variable (see Appendix 1) instead of the variables sex (M3) and cohort (COHO).

Program Example 4.3:

```
GET FILE        DATA
COMPUTE         DUR = M51 - M50 + 1
COMPUTE         CEN = 1
IF              (M51 EQ M47) CEN = 0
SURVIVAL        TABLES = DUR BY M5 (1,4)/
                INTERVALS = THRU 24 BY 3,THRU 300 BY 12/
                STATUS = CEN (1)/
                PLOTS (ALL)/
                COMPARE
OPTIONS         5,8
FINISH
```

The estimated log scale graph of the survivor functions based on this program demonstrate that, especially between the first occupation and later occupations, large differences exist (Figure 4.11). The path of the curve representing the first occupation bends sharply downward and indicates that the risk of leaving the first job given increasing duration is significantly higher than for the following employment episodes. However, when an occupational activity occurs more frequently, the differences in any previous occupational level decrease. Generally, based on the course patterns illustrated in Figure 4.11, one must assume that the process of various occupational episodes is different and this should be considered when regression models (Cox or parametric rate models) are used.

Multistate Case

A final extension of the application of the methods presented here consists of the treatment of the *multistate case.* Under such circumstances not only one type of event (e.g., a change of occupation as such) exists, but rather different types of events (e.g., a change from an unskilled worker status to a white-collar employee, or a transition from employee to self-employed). Moreover, these events may be observed as competing with one another ("competing risks"). In research in economics and the social sciences, the case, in which more than one possible state may occur, is of special importance. A concrete example of the multistate model is obtained by observing a specific type of event (or a specific state transition) and regarding competing events as being censored.

The following example demonstrates the program procedure for the multistate case. Once again, the event oriented occupation data set is used. The occupational activities of the employment histories have been classified according to 12 occupational groups (see Table 4.4). For each employment episode, the respective occupation group code (initial state) has been recorded, as well as the occupation group code of the respective successive employment episode (end state). If the last employment episode is considered, then the end state is coded by 0 (censored). The following SPSS run (see Appendix 1) now demonstrates for the specific transition from an unskilled manual occupation (code 2) to a skilled manual occupation (code 3) how one may estimate the life tables for men and women separately:

Program Example 4.4:

```
GET FILE        DATA
COMPUTE         DUR = M51 - M50 + 1
COMPUTE         CEN = 1
IF              (M51 EQ M47) CEN = 0
IF              (M62 NE 3) CEN = 0
*SELECT IF      (M61 EQ 2)
SURVIVAL        TABLES = DUR BY M3(1,2)/
                INTERVALS = THRU 36 BY 3,THRU 180 BY 12/
                STATUS = CEN(1) FOR DUR/
                PLOTS(LOGSURV,SURVIVAL)/
                COMPARE
FINISH
```

After calculation of the duration DUR, the censoring variable CEN is set equal to "1" for all episodes. If the end of the employment episode (M51) corresponds to the time of the interview (M47), then one is dealing with a censored observation, and CEN is set equal to "0." In addition, the program treats all episodes, which by means of the transition into the end state of interest (M62) "skilled manual occupation" (with code 3) have not ended, as being censored. The idea behind this is that these individuals are exposed to the risk of a change to the skilled manual occupation up to the point in time where one of the competing events occurs (transition to one of the other occupational groups). It is therefore necessary to include these durations in the estimation of the life table. With the SELECT IF statement only those episodes are chosen in which an individual actually is in the beginning state "unskilled manual occupation" (with code 2). At the beginning of this process the actual risk set is, therefore, defined as employees who are working in an unskilled manual occupation. For men and women, the life table for the specific transition from an unskilled manual occupation to a skilled manual occupation is separately estimated.

In examining the survivor functions of men (1) and women (2) presented in Figure 4.12 one observes that the transition from an unskilled manual occupation to a skilled manual occupation occurs sluggishly. After a duration of approximately 120 months (or 10 years of employment) in an unskilled manual occupation, only 25 percent of the males and about 8 percent of the

Table 4.4: Classification of the Occupations

Name of the occupational group	Description of the occupational group	Composition of the occupational groups according to the German occupational classification (1970)	Composition of the occupational groups according to the international classification of occupations (ISCO)	Examples
Production				
Agricultural occupations (AGR)	Occupations with a dominant agricultural orientation	011-022,041-051,053-062	6-11 to 6-49	Farmers, agricultural workers, gardeners, workers in the forest economy, fishermen, etc.
Unskilled manual occupations (EMB)	All manual occupations that showed at least 60 percent unskilled workers in 1970	071-133,135-141,143,151-162,164, 176-193,203-213,222-244,252,263, 301,313,321-323,332-346,352-371, 373,375-377,402-403,412,423-433, 442,452-463,465-472,482,486,504, 512-531,543-549	3-92,7-11 to 7-23,7-26 to 7-34,7-51 to 7-61,7-71,7-72,7-74,7-75,7-77,7-79 to 7-89,7-93 to 7-95,7-99,8-02,8-12,8-20,8-35,8-39,8-72,8-91 to 9-01, 9-10,9-25,9-39,9-42 to 9-49,9-52,9-53,9-56,9-59,9-69,9-72 to 9-74,9-82, 9-99	Miners, rockbreakers, papermakers, wood industry occupations, printing industry occupations, welders, riverters, unskilled workers, road and railroad construction workers, etc.
Skilled manual occupations (QMB)	All manual occupations that showed at most 40 percent unskilled workers in 1970	134,142,144,163,171-175,201-202, 221,251,261-262,270-291,302,305-312,314-315,331,351,372,374,378-401,411,421-422,441,451,464,481, 483-485,491-503,511,541-542	7-41 to 7-49,9-02,9-26 to 9-31,7-24, 7-25,8-31 to 8-34,8-71,8-73 to 8-80, 8-41 to 8-59,9-41,7-91,8-01,8-03,7-62,7-92,7-76,7-73,5-31,7-78,9-51,9-54,9-55,9-57,7-96,8-11,8-19,9-61	Glassblowers, bookbinders, typesetters, locksmiths, precision instrument makers, electrical mechanics, coopers, brewers, carpenters, etc.
Technicians (TEC)	All technically trained specialists	303,304,621-635,721-722,733,857	0-75,0-32 to 0-39,0-14,7-00,0-54,0-84,9-27,0-42,0-43,0-62,0-66,0-69	Machinery technicians, electrical technicians, construction technicians, mining technicians, etc.
Engineers (ING)	Highly trained specialists who solve technical and natural science problems	032,052,601-612,726,883	0-21 to 0-31,0-11 to 0-13,0-82,0-83, 0-41,0-51 to 0-53	Construction engineering, electrical engineers, production designers, chemical engineers, physicists, mathematicians, etc.
Service				
Unskilled services (EDB)	All unskilled personal services	685-686,688,706,713-716,723-725, 741-744,791-794,805,838,911-913, 923-937	4-52,4-90,3-59,9-81,9-85 to 9-89,3-91,9-71,9-79,5-89,5-51,5-92, 1-75, 1-80,3-94,5-00,5-10,5-32,5-40,5-52, 5-60	Cleaners, waiters, servers, etc.
Skilled services (QDB)	Essentially order and security occupations as well as skilled service occupations	684,704-705,711-712,801-804,812, 814,831,837,851-852,854-857,892-902,921-922	4-43,9-83,9-84,3-51,3-60,5-82,1-71, 1-63,0-79,0-76,0-72,5-99,1-49,5-70, 5-20,4-31,5-81	Policemen, firemen, locomotive engineers, photographs, hairdressers, etc.
Semiprofessions (SEMI)	Service positions which are characterized by professional specialization	821-823,853,861-864,873-877	1-51,1-59,1-79,1-91,1-93,1-95,0-71, 0-73,0-74,0-77,1-94,1-33 to 1-39	Nurses, educators, elementary school teachers, Kindergarten teachers, etc.
Professions (PROF)	All liberal professions and service positions which require a university degree	811,813,841-844,871-872,881-882, 891	1-21 to 1-29,0-61,0-63 to 0-65,0-67, 0-68,1-31,1-32,0-81,0-90,1-92,1-99, 1-41	Dentists, doctors, pharmacists, judges, secondary education teachers, university professors, etc.
Administration				
Unskilled commercial and administrational occupations (EVB)	Relatively unskilled office and commerce occupations	682,687,731-732,734,782-784,773	4-51,4-32,3-52,3-70,3-80,3-21,3-22, 3-99	Postal occupations, shop assistants, typists, etc.
Skilled commercial and administrational occupations (QVB)	Occupations with medium and higher administrative and distributive functions	031,681,683,691-703,771-772,774-781	6-00,4-00 to 4-22,3-39,4-41,4-42,5-91,3-31,3-41,3-42,3-00,3-10,3-93, 3-95	Credit and financial assistants, foreign trade assistants, data processing operators, bookkeepers, goods traffic assistants, etc.
Managers (MAN)	Occupations which control factors of production as well as funtionaires of organizations	751-763	2-01 to 2-19,1-10	Managers, business administrators, deputies, ministers, social organization leaders, etc.

Figure 4.12: Example of Survivor Function Plots for the Transition From an Unskilled Manual Occupation to a Skilled Manual Occupation

GRAPH OF SURVIVAL FUNCTION
SURVIVAL VARIABLE DUR
GROUPED BY M3

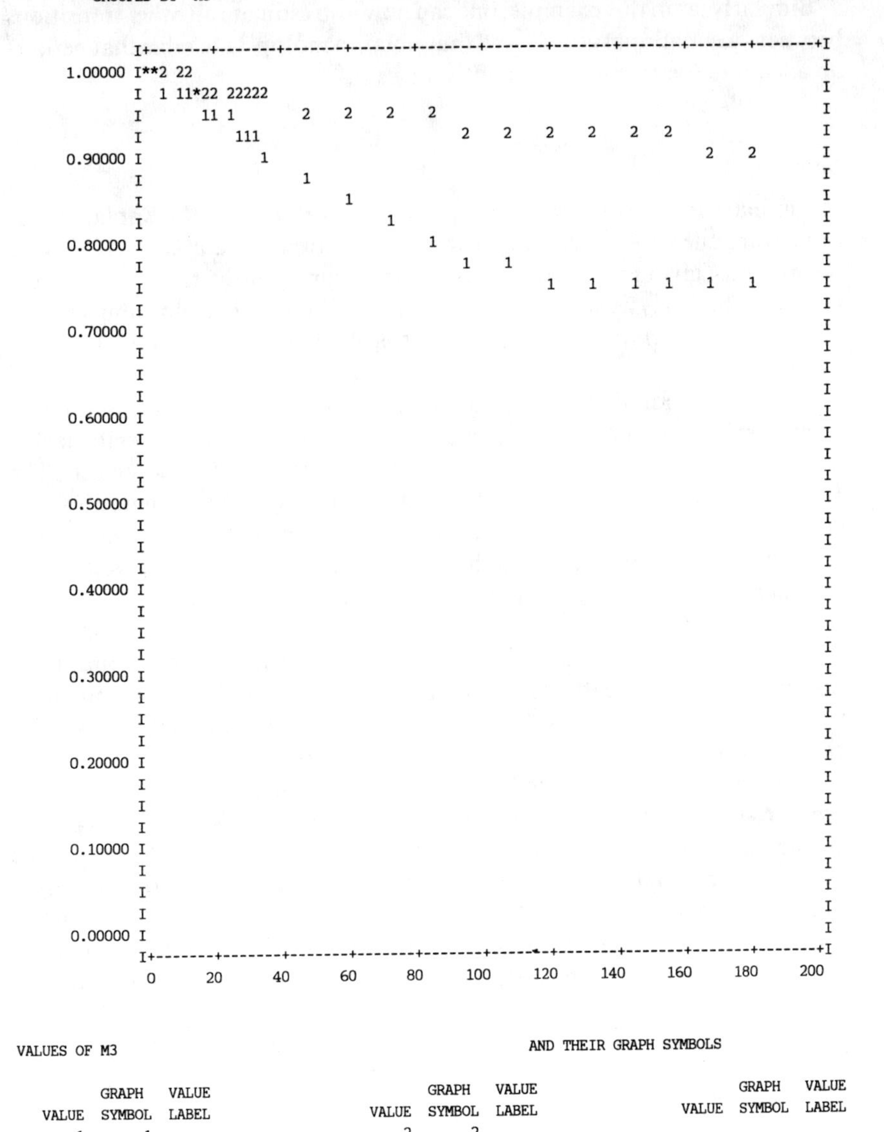

females have actually changed and made the transition to a skilled manual occupation. Thereafter, especially for men, the survivor functions remain largely unchanged. That is, no further occupational change to a skilled manual occupation occurs. Since the survivor function for males is somewhat steeper up until the 120th month than it is for women, the opportunity of this specific change of occupation is higher for men.

Similarly, as in this example, one can now also estimate all other transitions between occupational groups and thus obtain a multiple life table that permits an adequate treatment of the multistate case.

Summary

In summarizing the discussion of the life table method and the Kaplan-Meier estimator, one can assert that their main function in data analysis may be regarded as giving a rather heuristic first impression of the course of the process. The analysis is focused for the most part on subgroup comparisons, which can offer initial explanations regarding the importance of certain variables.

The assumption that in such comparisons only homogeneous subpopulations exist is, however, not very plausible for economics and the social sciences because a high degree of intercorrelation between variables usually exists. The alternative—disaggregating until only very specific subgroups are compared—is quite often not possible due to the limits of sample size in many data sets. Therefore comparisons between the subgroups and the possibility of time dependency as judged by the survivor functions or cumulated survivor functions usually possess a more heuristic character. One must be especially careful when examining the multiepisode case, in which the distributions of the episodes differ greatly between individuals. This implies that individuals with certain characteristics will be overrepresented in any analysis. In most cases the problem of homogeneous subpopulations may be solved by applying regression models as discussed in the following chapters, whereby covariates are introduced not only to differentiate among subgroups but also to permit consideration of the process history. It has been demonstrated that the life table method and the Kaplan-Meier estimator may also be applied to analyze multistate models. Although it must be stressed that one must be careful when interpreting the results of the presented methods, these techniques offer an important means of establishing the relevancy of specific variables affecting the process being analyzed.

Chapter 5: Semi-Parametric Regression Models: The Cox Proportional Hazards Model

When applying the life table method or the Kaplan-Meier estimator to a data set, with an increasing number of subgroups a point is rapidly reached at which it is no longer sensible to estimate survivor functions due to a small sample bias. Often one is also confronted with the problem that *subgroup comparisons become more and more complex as the number of subgroups increase,* and interpretation thus becomes highly difficult. In the past years, therefore, regression methods have been increasingly applied to the analysis of event history data. Here the hazard function, which describes the process, is modeled as dependent on the duration in a state and different covariates. These regression methods permit not only an easy way of considering *time-constant differences in subgroups* (such as differences in sex, birth cohorts, etc.) and the *influence of an individual's previous history* (e.g., occupational experience), but also permit a way to analyze the *impact of one or more parallel processes* by the introduction of time-dependent covariates.

Compared to the life table method or the Kaplan-Meier estimator, in which *continuous characteristics must be classified under a loss of information,* regression models can handle qualitative as well as quantitative covariates. Where measurement permits it, there is a possibility of controlling various effects by introducing their metric form as proxies in the analysis. Well-known examples are the inclusion of social classes in the form of status scores (Treiman, 1977; Handl, Mayer, and Müller, 1977; Wegener, 1985) or the inclusion of qualification levels in terms of the average number of school years necessary to obtain a specific educational degree (Helberger, 1980; Blossfeld, 1985c). Previous history may also be taken into account in the analysis; for example, the number of months one has worked in the labor force. With the use of these variables, the effects in the regression models may be studied, without significantly increasing the number of parameters to be estimated.

In this chapter the main point of interest is the use of the method of partial likelihood proposed by Cox. After an examination of the proportionality assumption (Section 5.1), a detailed interpretation of a Cox model is presented (Section 5.2). Model selection with the aid of stepwise regression is then demonstrated in Section 5.3. Especially relevant for the application of event history analysis in the economic and social sciences are examples of how time-dependent covariates are included in the Cox model (Section 5.4) and of how multistate models are handled (Section 5.5).

In general, the use of a Cox model is appropriate if the researcher is interested in the influence of the covariates without additional assumptions about the time dependency (see Section 3.3.3): It may be that one has no previous information on the time path of the hazard function, or a path that is already known cannot be adequately grasped by a parametric model, or one may only be interested in the magnitude and direction of the covariate effects, controlling for time dependency. The Cox model is thus extremely flexible and may be applied to many situations.

The duration dependency is introduced in the Cox model as the so-called baseline hazard rate $\lambda_0(t)$, which is not further specified. The covariates are included in a log-linear form in the model $\exp(\mathbf{x}'\boldsymbol{\beta})$. The baseline hazard rate and the log-linear covariate vector are then multiplicatively related:

$$\lambda(t|\mathbf{x}) = \lambda_0(t) \exp(\mathbf{x}'\boldsymbol{\beta}).$$

Due to the fact that the baseline hazard rate is unspecified, Cox has proposed a partial likelihood method to estimate the β coefficients of the model (see Section 3.6.4). This method considers the probabilities of events that are experienced under a given risk set. The censored observations diminish the risk set for future events. Analogous to the Kaplan-Meier estimator, durations must be sorted in ascending order. The estimation of the β's are finally obtained with iterative methods included in program packages such as BMDP, SAS, RATE, or GLIM. For further explanation of the iterative methods of estimation, see the discussion presented in Section 3.6.1.

In a comparison of the available program packages that estimate the Cox model in terms of their user friendliness or their capabilities of model selection and evaluation, the BMDP program package is highly recommended. The following examples of the partial likelihood estimation will thus only refer to the subprogram P2L from BMDP.

5.1 Testing the Proportionality Assumption

Despite the wide range of possible applications of the Cox model, the assumption of proportional risks is somewhat limiting in the sense that the ratio of the hazard functions for two members of the population should be constant throughout the entire observation period. This means, if two individuals are characterized by time invariant covariate vectors $\mathbf{x}_i$ and $\mathbf{x}_j$ the ratio of the hazard functions

$$\frac{\lambda(t|\mathbf{x}_i)}{\lambda(t|\mathbf{x}_j)} = \frac{\lambda_0(t) \exp(\mathbf{x}_i'\boldsymbol{\beta})}{\lambda_0(t) \exp(\mathbf{x}_j'\boldsymbol{\beta})} = \exp(\boldsymbol{\beta}'(\mathbf{x}_i - \mathbf{x}_j))$$

is independent of time. An illustration of two proportional hazard functions is given in Figure 3.10.

Checking the Proportionality Assumption Graphically

A first hint whether or not the proportionality assumption is fulfilled may be obtained by stratifying the sample according to the categories of a variable. The subgroup specific survivor functions are then estimated under the assumption that the influences of the other covariates in the model are identical for all categories (see the discussion in Section 3.7.2). After a log-minus-log transformation of the survivor function, the plotted curves should, over the entire observation period, only differ by a constant factor. The logic of this procedure is demonstrated below using the variable sex. If the unspecified baseline survivor function of the Cox model is designated as $S_0(t)$, then the survivor functions of men and women may be written as:

men $\quad S_M(t|\mathbf{x}) = S_0(t)^{\exp(\mathbf{x}'\boldsymbol{\beta})\exp(\beta_g 1)}$

women $\quad S_F(t|\mathbf{x}) = S_0(t)^{\exp(\mathbf{x}'\boldsymbol{\beta})}$

Through a log-minus-log transformation of these equations one obtains:

men $\quad \ln(-\ln S_M(t|\mathbf{x})) = \ln(-\ln S_0(t)) + \mathbf{x}'\boldsymbol{\beta} + \beta_g$

women $\quad \ln(-\ln S_F(t|\mathbf{x})) = \ln(-\ln S_0(t)) + \mathbf{x}'\boldsymbol{\beta}$

and finally:

$$\ln(-\ln S_M(t|\mathbf{x})) = \ln(-\ln S_F(t|\mathbf{x})) + \beta_g.$$

Given proportionality, the plotted survivor functions of men and women should only differ by the term β_g.

In the following, we always assume a hazard function in the form

$$\lambda^k(t|\mathbf{x}_k) = \lambda_0(t - t_{k-1})\exp(\mathbf{x}_k'\boldsymbol{\beta}) \qquad k = 1, 2, \ldots .$$

This implies that the hazard function is only dependent upon the duration $v = t - t_{k-1}$ and that the baseline hazard rate as well as the coefficients $\boldsymbol{\beta}$ are the same for all episodes (see Blossfeld and Hamerle, 1988a). At the end of Section 3.6.6 we demonstrated how this model may be reformulated for the one-episode case, in which the successive episodes of an individual are independent. The fact that one is dealing with the *k*th episode of an individual is expressed by an index for the covariates. In the following, we always write $\lambda^k(v|\mathbf{x}_k)$ instead of $\lambda^k(t|\mathbf{x}_k)$.

The following P2L BMDP program run demonstrates how to estimate a Cox model. We use the previously introduced event oriented employment data (see Appendix 1). After the variables education (EDU), labor force experience (LFX), number of previously held occupations (NOJ), and the cohorts (COHO2, COHO3) have been introduced into the Cox model, all employment episodes (multiepisode case) are used to determine whether or not the proportionality assumption holds for men and women.

Program Example 5.1:

```
/INPUT UNIT IS 30.
       CODE IS DATA.
/VARIABLE NAMES ARE (63)DUR,(64)CEN,(65)EDU,(66)LFX,
                    (67)NOJ,(68)COHO2,(69)COHO3.
          ADD IS 7.
/TRANSFORM DUR = M51 - M50 + 1.
           CEN = 1.
           IF (M51 EQ M47) THEN CEN = 0.
           IF (M41 EQ 1 AND M42 EQ 1) THEN EDU = 9.
           IF (M41 EQ 1 AND (M42 EQ 2 OR M42 EQ 3)) THEN EDU = 11.
           IF (M41 EQ 2 AND M42 EQ 1) THEN EDU = 10.
           IF (M41 EQ 2 AND (M42 EQ 2 OR M42 EQ 3)) THEN EDU = 12.
           IF (M41 EQ 3 AND (M42 EQ 1 OR M42 EQ 2 OR M42 EQ 3))
           THEN EDU = 13.
           IF (M42 EQ 4) THEN EDU = 17.
           IF (M42 EQ 5) THEN EDU = 19.
           COHO2 = 0.
           COHO3 = 0.
           IF(M48 GE 468 AND M48 LE 504) THEN COHO2 = 1.
           IF(M48 GE 588 AND M48 LE 624) THEN COHO3 = 1.
           LFX = M50 - M43.
           NOJ = M5 - 1.
/FORM TIME IS DUR.
      STATUS IS CEN.
      RESPONSE IS 1.
/REGRESSION COVARIATES ARE EDU,M59,NOJ,LFX,
                           COHO2,COHO3.
            STRATA IS M3.
/GROUP CODES (3) ARE 1,2.
       NAMES (3) ARE 'MEN','WOMEN'.
/TEST ELIMINATE = EDU,M59,NOJ,LFX,COHO2,COHO3.
      STATISTICS = WALD,LRATIO,SCORE.
/PLOT TYPE = LOG.
/PRINT CASES ARE 0.
/END
```

In the TRANSFORM paragraph in Program Example 5.1, first the duration (DUR) of each occupational episode is calculated and then the censoring information (CEN) is generated. To include the individual's education into the Cox model as a metric proxy (EDU), we use the average number of school years necessary to obtain a school degree. In assigning school years to school degrees the following values have been assumed: Lower secondary school qualification *(Hauptschule)* without vocational training is equivalent to 9 years, middle school qualification *(Mittlere Reife)* is equivalent to 10 years, lower secondary school qualification *(Hauptschule)* with vocational training is equivalent to 11 years, middle school qualification *(Mittlere Reife)* with vocational training is equivalent to 12 years, Abitur is equivalent to 13 years, professional college qualification is equivalent to 17 years, and a university degree is equivalent to 19 years. The distinction between the three birth cohorts in the Cox model is included through dummy variables. The oldest cohort (those born between 1929–31) is chosen as the reference category, whereas the cohorts 1939–41 and 1949–51 are designated by the variables COHO2 and COHO3:

	COHO2	COHO3
cohort 1929–31	0	0
cohort 1939–41	1	0
cohort 1949–51	0	1

At the start of each employment episode, the length of time spent in the labor force (LFX) is measured in months. The number of occupations prior to the current employment episode is then finally measured by the variable NOJ.

In the FORM paragraph of the BMDP program the duration variable (TIME IS DUR) and the censoring variable (STATUS IS CEN) are assigned. If an event occurs, the censoring variable has the value "1" (RESPONSE IS 1).

The REGRESSION paragraph specifies a Cox model in which the variables EDU, M59 (prestige) (see Wegener, 1985), NOJ, LFX, COHO2, and COHO3 are included. Because the variable sex (M3) should be checked with regard to proportionality, it is introduced as a stratification variable (STRATA IS M3). In the GROUP paragraph respective labels are assigned. The command to plot the log-minus-log survivor functions for men and women is given in the PLOT paragraph by the command TYPE=LOG. Finally, in the TEST paragraph a partial likelihood ratio test (LRATIO), a Wald test (WALD) and a score test (SCORE) are requested in order to check the hypothesis that none of the introduced covariates have a significant influence upon the rate.

Let us first interpret Figure 5.1, in which the paths of the log-minus-log survivor function of men (A) and women (B) are plotted against the duration (DUR). If the proportionality assumption is valid, then throughout the observation period the vertical (i.e., parallel to the y axis) distance between both

Figure 5.1: An Illustrative Graph of the Log-Minus-Log Survivor Function to Check the Proportionality Assumption

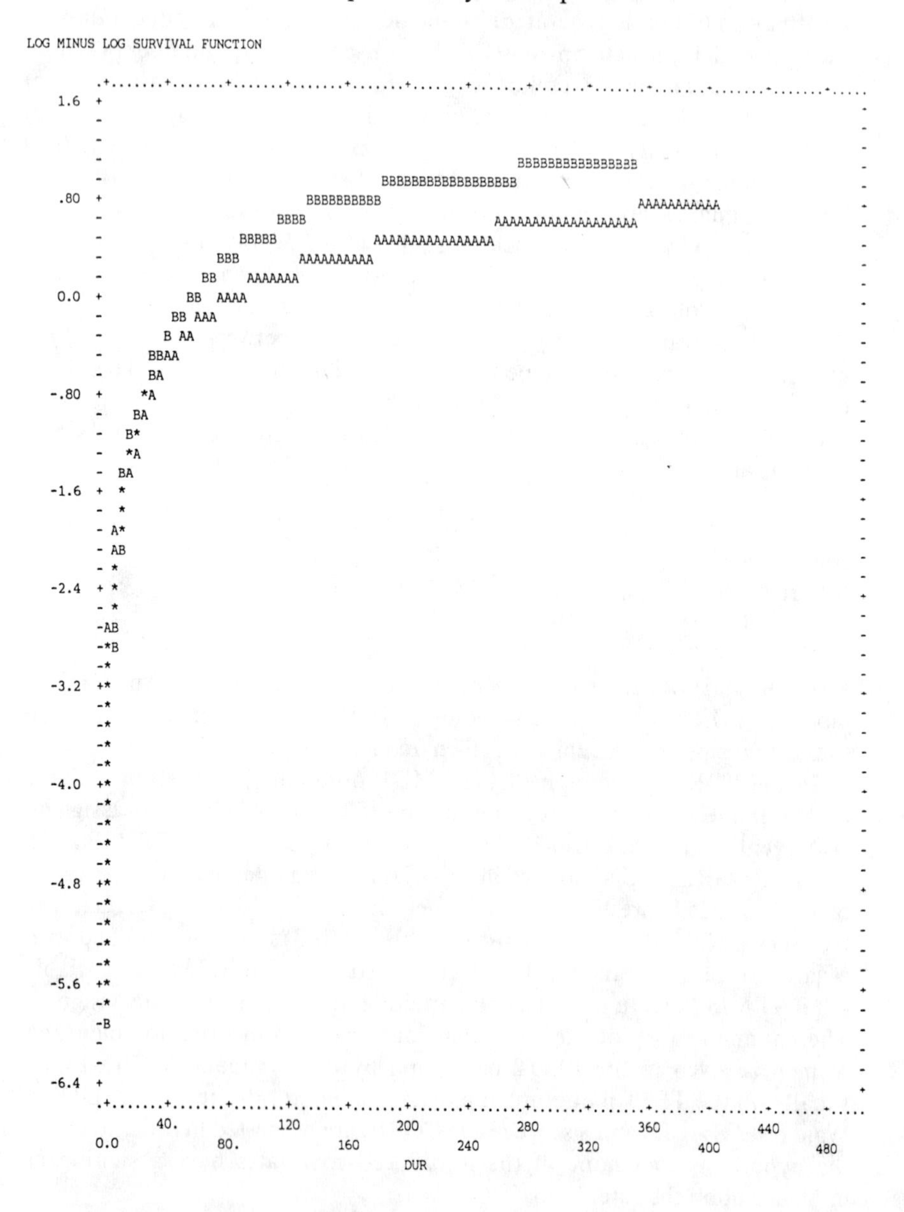

curves should remain constant. In Figure 5.1, the first 40 months are characterized by steep slopes of the survivor functions, making the plotted distance between the two curves relatively unrecognizable. Thereafter, however, with increasing duration the distance between both curves increases. But because the increase in distance is not very great, the question arises as to whether or not the proportionality assumption is already violated. Because visual inspection of the plotted curves can not answer this question, we apply a statistical test.

Statistical Test of the Proportionality Assumption

In the BMDP program package, the proportionality assumption may be tested quite elegantly by introducing time-dependent variables. As already demonstrated in Section 3.7.2, there should be no significant interaction effect between the variable to be tested and the duration when the ratio of the hazard functions of any two individuals characterized by the covariate vectors $\mathbf{x}_i$ and $\mathbf{x}_j$ are independent of the duration v. In the case of sex, which is introduced into the regression as a dummy variable, we define the following interaction variable:[1]

$$z = x_g\,(\ln v - \ln c) \qquad \text{with } x_g = \begin{cases} 1 & \text{for men} \\ 0 & \text{for women} \end{cases}$$

The Cox model is then:

$$\lambda^k(v|\mathbf{x}) = \lambda_0(v)\exp(\mathbf{x}_k'\boldsymbol{\beta} + \beta_g x_g + \beta_z z), \qquad , k = 1, 2, \ldots,$$

or:

$$\lambda^k(v|\mathbf{x}) = \lambda_0(v)\exp(\mathbf{x}_k'\boldsymbol{\beta} + \beta_g x_g)\left(\frac{v}{c}\right)^{\beta_z x_g} \qquad , k = 1, 2, \ldots .$$

If the proportionality assumption is valid, then the coefficient β_z should not be significantly different from zero and the hazard functions of men and women should only differ by the constant factor $\exp(\beta_g)$.

A realization of this test can be obtained in BMDP through a simple modification of the Program Example 5.1:

Program Example 5.2:

```
/INPUT UNIT IS 30.
       CODE IS DATA.
/VARIABLE NAMES ARE (63)DUR,(64)CEN,(65)EDU,
                    (66)LFX,(67)NOJ,(68)COHO2,
                    (69)COHO3,(70)SEX.
         ADD IS 8.
```

[1]In order to obtain a more accurate estimation of the parameters β_g and β_z, the logarithmic duration as well as the logarithmic mean of the duration is considered. An estimation of the mean of the duration (c) is obtained through the BMDP subprogram P1L (see Table 4.3).

```
/TRANSFORM DUR = M51 - M50 + 1.
           CEN = 1.
           IF (M51 EQ M47) THEN CEN = 0.
           IF (M41 EQ 1 AND M42 EQ 1) THEN EDU = 9.
           IF (M41 EQ 1 AND (M42 EQ 2 OR M42 EQ 3)) THEN EDU = 11.
           IF (M41 EQ 2 AND M42 EQ 1) THEN EDU = 10.
           IF (M41 EQ 2 AND (M42 EQ 2 OR M42 EQ 3)) THEN EDU = 12.
           IF (M41 EQ 3 AND (M42 EQ 1 OR M42 EQ 2 OR M42 EQ 3))
           THEN EDU = 13.
           IF (M42 EQ 4) THEN EDU = 17.
           IF (M42 EQ 5) THEN EDU = 19.
           COHO2 = 0.
           COHO3 = 0.
           IF (M48 GE 468 AND M48 LE 504) THEN COHO2 = 1.
           IF (M48 GE 588 AND M48 LE 624) THEN COHO3 = 1.
           LFX = M50 - M43.
           NOJ = M5 - 1.
           SEX = 0.
           IF (M3 EQ 1) THEN SEX = 1.
/FORM TIME IS DUR.
      STATUS IS CEN.
      RESPONSE IS 1.
/REGRESSION COVARIATES ARE EDU,M59,NOJ,LFX,COHO2,
                           COHO3,SEX.
           ADD IS Z2.
/FUNCTION Z2 = SEX * (LN(TIME)-LN(73.83)).
/PRINT CASES ARE 0.
/END.
```

After defining the new variable SEX in the VARIABLES paragraph, in the TRANSFORM paragraph of the Program Example 5.2, the dummy variable SEX is constructed from the variable M3. The dummy variable has the value "1" for men and "2" for women. The variable SEX takes on the value "1" when dealing with men and otherwise the value "0." Thus, women are the reference group in this example.

In the REGRESSION paragraph, a Cox model is specified which, compared to the Program Example 5.1, is characterized by an additional covariate, the dummy variable SEX. The ADD statement permits the inclusion of the interaction variable Z2 as a time-dependent covariate. Z2 is constructed in the FUNCTION paragraph and represents the interaction between sex (SEX) and duration (TIME), whereby the duration is additionally weighted by the logarithm of the average duration, 73.83 months.

After running the program[2] one obtains a significant effect ($\hat{\beta}_z/SE(\hat{\beta}_z) = -0.2071/0.0282 = -7.3548$) for the interaction variable Z2. The ratio of the hazard functions of men and women, which initially had the value $\exp(\hat{\beta}_g) = 0.5803$, is not constant over the duration period, but rather changes over time. Therefore, the proportionality assumption is not fulfilled, and the variable sex may not be included in this case as a covariate. The variable sex must be introduced in the Cox model as a stratification variable.

5.2 Interpretation of the Estimated Results

Let us now interpret the results of the Program Example 5.1 (Table 5.1), in which the Cox model has been estimated under the assumption of group specific baseline functions $\lambda_{0j}(v)$ for men (j = 1) and women (j = 2):

$$\lambda_j^k(v|\mathbf{x}_k) = \lambda_{0j}(v)\exp(\mathbf{x}_k'\boldsymbol{\beta}) \qquad j = 1,2 \qquad k = 1, 2, \ldots .$$

First of all, the GLOBAL CHI-SQUARE statistic offers a criterion of how well the estimated model, compared with the model without covariates, explains reality. Under the null hypothesis that none of the covariates influences the rate of job change, the test values are asymptotically χ^2 distributed with p (= number of estimated parameters) degrees of freedom:

$$\text{GLOBAL CHI-SQUARE} = s(\hat{\boldsymbol{\beta}}_0)'\, \mathbf{I}(\hat{\boldsymbol{\beta}}_0)^{-1}\, s(\hat{\boldsymbol{\beta}}_0),$$

where $\boldsymbol{\beta}_0$ is set equal to zero. This test statistic is equivalent to the test statistic (3.7.20) of the score test presented in Section 3.7.3. A χ^2 value of 667.39 with six degrees of freedom (see Table 5.1), indicates that the null hypothesis (none of the introduced variables are significant) must be rejected (P-VALUE = 0.0000).

The other test statistics, which were explicitly implemented in the TEST paragraph and with which the entire model was tested in this example, lead to the same conclusion. The partial likelihood ratio test (see Section 3.7.3)

---

[2]Results of Cox model as presented in Program Example 5.2:

```
  LOG LIKELIHOOD = -37109.8619
GLOBAL CHI-SQUARE =     845.12  D.F.=   8  P-VALUE =0.0000
```

VARIABLE	COEFFICIENT	STANDARD ERROR	COEFF./S.E.	EXP(COEFF.)
65 EDU	0.0261	0.0100	2.6030	1.0265
59 M59	-0.0078	0.0010	-7.5614	0.9922
67 NOJ	0.1470	0.0100	14.7182	1.1584
66 LFX	-0.0070	0.0003	-22.2626	0.9930
68 COHO2	0.1031	0.0350	2.9477	1.1086
69 COHO3	0.1780	0.0391	4.5561	1.1948
70 SEX	-0.5442	0.0442	-12.3025	0.5803
71 Z2	-0.2071	0.0282	-7.3548	0.8129

$$\text{LRATIO} = -2\ln\left(\frac{\text{L (model without covariates)}}{\text{L (present model)}}\right)$$

has a χ^2 value of LRATIO = 821.35.

The Wald statistic

$$\text{WALD} = \hat{\boldsymbol{\beta}}' \, \text{Cov}(\hat{\beta})^{-1} \, \hat{\boldsymbol{\beta}},$$

which is derived from the general Wald test statistic (3.7.19) has a value of WALD = 653.40 with six degrees of freedom.

Those variables which actually have a significant influence may be determined by examining the column of standardized coefficients (COEFF./S.E.). According to the discussion presented in Section 3.7.3, the estimated values have an asymptotical normal distribution, which, given a significance level of $\alpha = 0.05$, are significantly different from 0 when their absolute value is larger than 1.96. Applying this criterion, all of the variables introduced in the model have a significant effect (see Table 5.1).

The effect of a covariate x_i can be easily interpreted, when one examines the percent of change in the rate, given an increase in the covariate x_i:

$$\zeta_{\Delta x_i} = \frac{\lambda_0(v)\exp(\mathbf{x}'\boldsymbol{\beta} + \beta_i\,(x_i + \Delta x_i)) - \lambda_0(v)\exp(\mathbf{x}'\boldsymbol{\beta} + \beta_i x_i)}{\lambda_0(v)\exp(\mathbf{x}'\boldsymbol{\beta} + \beta_i x_i)} \cdot 100\%$$

$$\zeta_{\Delta x_i} = \frac{\exp(\beta_i\,(x_i + \Delta x_i)) - \exp(\beta_i x_i)}{\exp(\beta_i x_i)} \cdot 100\%$$

$$\zeta_{\Delta x_i} = (\exp(\beta_i \Delta x_i) - 1) \cdot 100\%$$

or:

$$\zeta_{\Delta x_i} = (\exp(\beta_i)^{\Delta x_i} - 1) \cdot 100\%.$$

The antilogarithms $\exp(\beta_i)$ of the β_i coefficients are given in the EXP (COEFF.) column of Table 5.1 and are referred to in the literature as α_i effects. They take the value 1 when the variable has no effect upon the rate ($\beta_i = 0$) and are otherwise smaller than or greater than 1 when the variable exhibits a diminishing ($\beta_i < 0$) or an increasing ($\beta_i > 0$) influence. If the value of the variable x_i is increased by just one unit, then the rate changes by

$$\zeta_1 = (\exp(\beta_i) - 1) \cdot 100\%.$$

The $\hat{\beta}$ coefficient of the variable education (EDU) has a positive sign (see Table 5.1). Each additional school year increases the tendency toward job change by 2.61 percent [$(\exp(\hat{\beta}_{EDU}) - 1) \cdot 100\% = (1.0261 - 1) \cdot 100\% = 2.61\%$]. The number of previously held occupations (NOJ) also has a positive effect on the rate. Each of these occupational episodes increases the individual inclination to leave the present employment position by 15.84 percent. Finally, the $\hat{\beta}$ coefficients of the birth cohort dummy variables have a positive effect. Compared to the oldest cohort (those born between 1929 and 1931) the tendency to change jobs for the cohort born between 1939 and 1941

(COHO2) is 10.94 percent and for the cohort born between 1949 and 1951 (COHO3) it is 19.32 percent higher.

On the other hand, the $\hat{\beta}$ coefficient of the variable prestige (M59) is negative. The higher a job is located in the hierarchy of the occupational pyramid, the less likely one is to leave such a position. The variable labor force experience (LFX) has a similar effect of decreasing the rate by 0.70 percent per month.

If the effect of labor force experience were to be expressed in years instead of months, special caution would be necessary due to the log-linear formulation of the Cox model. The calculation of the effect upon the rate by a yearly occupational experience is not the same as that of the linear regression model in which one would obtain $12 \cdot (-0.7\%) = -8.4\%$. Rather, in applying the aforementioned formula for the relation, one obtains $((\exp(\hat{\beta}_{LFX}))^{12} - 1) \cdot 100\% = (0.9923^{12} - 1) \cdot 100\% = -8.86\%$.

Finally, it must also be noted that, given log-linear covariate influences, a simultaneous change in two or more covariates has a multiplicative and not an additive effect:

$$\gamma_{\Delta x} = \frac{\lambda_0(v)\exp(\beta_1(x_1 + \Delta x_1) + \ldots + \beta_p(x_p + \Delta x_p)) - \lambda_0(v)\exp(\beta_1 x_1 + \ldots + \beta_p x_p)}{\lambda_0(v)\exp(\beta_1 x_1 + \ldots + \beta_p x_p)} \cdot 100\%$$

$$\gamma_{\Delta x} = \frac{\exp(\beta_1(x_1 + \Delta x_1) + \ldots + \beta_p(x_p + \Delta x_p)) - \exp(\beta_1 x_1 + \ldots + \beta_p x_p)}{\exp(\beta_1 x_1 + \ldots + \beta_p x_p)} \cdot 100\%$$

$$\gamma_{\Delta x} = (\exp(\beta_1)^{\Delta x_1} \cdot \ldots \cdot \exp(\beta_p)^{\Delta x_p} - 1) \cdot 100\%.$$

Table 5.1: Results of the Cox Model From Program Example 5.1

```
  LOG LIKELIHOOD = -33877.3618
GLOBAL CHI-SQUARE =     667.39  D.F.=   6  P-VALUE =0.0000
```

VARIABLE	COEFFICIENT	STANDARD ERROR	COEFF./S.E.	EXP(COEFF.)
65 EDU	0.0258	0.0100	2.5628	1.0261
59 M59	-0.0077	0.0010	-7.5078	0.9923
67 NOJ	0.1470	0.0100	14.7181	1.1584
66 LFX	-0.0070	0.0003	-22.2621	0.9930
68 COHO2	0.1038	0.0350	2.9690	1.1094
69 COHO3	0.1766	0.0391	4.5218	1.1932

STATISTIC	CHI-SQUARE	D.F.	P-VALUE
LRATIO	821.35	6	0.0000
SCORE	667.39	6	0.0000
WALD	653.40	6	0.0000

Accordingly, if one increases the number of previously held occupations (NOJ) by one unit and the prestige variable (M59) by 20 units, which would imply an occupational promotion, then the risk of leaving a job changes by $(1.1584^1 \cdot 0.9923^{20} - 1) \cdot 100\% = -0.75\%$.

5.3 Stepwise Regression Applied to the Cox Model

Although the selected variables and the interpretation of results are usually based on theoretical considerations and substantive hypotheses, situations may arise in which one possesses only a vague idea about the causality of variables and the relationships present in the data. Under such circumstances an important aid in model building using empirical data is the stepwise regression method. This method has already been described in Section 3.7.3 and is also implemented in the BMDP subprogram P2L. Applying this method to the previously used pool of variables, the individual variables are systematically tested to determine whether or not they should be included or excluded from the Cox model. The result of this procedure is an optimal model for the given pool of variables and the respectively chosen statistical criteria.

In the application of BMDP one may choose between two methods that control the step-by-step process of variable selection: the **m**aximum **p**artial **l**ikelihood **r**atio **t**est (MPLR) and the **P**eduzzy, **H**ardy, and **H**olford (PHH) test statistic. The MPLR method is statistically better and more exact; however, it requires more calculation time than the PHH method. The application of each method to the same pool of variables may lead to different results. Given a large number of variables, it is thus advisable to first apply the PHH method and eliminate those variables with little or no effect on the hazard rate and then, in a second step, to formulate an appropriate model based on the MPLR method. One must, however, remember that a multiple test problem exists which may lead to an increase in the α error.

In the application of BMDP each variable may be specified to determine whether it should be included in the first step of the model and how often each of the variables should be included or excluded during the stepwise regression. In addition, the selection process permits one to state initial values of the parameters to be estimated, as well as to determine various stop criteria for the applied Newton-Raphson algorithm (see Section 3.6).

A simple example for the application of stepwise regression in BMDP is demonstrated below. Here again the Program Example 5.1 has been slightly modified.

Program Example 5.3:

```
/INPUT UNIT IS 30.
      CODE IS DATA.
/VARIABLE NAMES ARE (63)DUR,(64)CEN,(65)EDU,(66)LFX,
                    (67)NOJ,(68)COHO2,(69)COHO3.
         ADD IS 7.
/TRANSFORM DUR = M51 - M50 + 1.
          CEN = 1.
          IF (M51 EQ M47) THEN CEN = 0.
          IF (M41 EQ 1 AND M42 EQ 1) THEN EDU = 9.
          IF (M41 EQ 1 AND (M42 EQ 2 OR M42 EQ 3)) THEN EDU = 11.
          IF (M41 EQ 2 AND M42 EQ 1) THEN EDU = 10.
          IF (M41 EQ 2 AND (M42 EQ 2 OR M42 EQ 3)) THEN EDU = 12.
          IF (M41 EQ 3 AND (M42 EQ 1 OR M42 EQ 2 OR M42 EQ 3))
          THEN EDU = 13.
          IF (M42 EQ 4) THEN EDU = 17.
          IF (M42 EQ 5) THEN EDU = 19.
          COHO2 = 0.
          COHO3 = 0.
          IF(M48 GE 468 AND M48 LE 504) THEN COHO2 = 1.
          IF(M48 GE 588 AND M48 LE 624) THEN COHO3 = 1.
          LFX = M50 - M43.
          NOJ = M5 - 1.
/FORM TIME IS DUR.
      STATUS IS CEN.
      RESPONSE IS 1.
/REGRESSION COVARIATES ARE EDU,M59,NOJ,LFX,
                           COHO2,COHO3.
           STRATA IS M3.
           STEPWISE = MPLR.
/GROUP CODES (3) ARE 1,2.
      NAMES (3) ARE 'MEN','WOMEN'.
/PRINT CASES ARE 0.
/END
```

Compared to the Program Example 5.1, in the REGRESSION paragraph of the previous run no specific Cox model is specified. Instead, only a pool of variables is stated from which a Cox model will be built up step-by-step according to the used statistical criteria. The inclusion or exclusion of variables is based on the maximum partial likelihood ratio test (STEPWISE = MPLR) stated as

$$\chi^2_{\mathrm{MPLR}} = 2(\ln L(\hat{\boldsymbol{\beta}}_\nu) - \ln L(\hat{\boldsymbol{\beta}}_{\nu+1})) \qquad , \nu = 1, 2, \ldots,$$

where $\hat{\beta}_\nu$ represents the MPL estimator of the corresponding νth step. In this program run, a number of additional decisions regarding the selection procedure have been implicitly determined by default values (see Dixon et al., 1983, p. 589). For example, none of the variables are presupposed to be included in the first regression step and each variable may only be included or excluded twice.

The result of this program is presented in detail in Table 5.2. In STEP NUMBER 0 for all variables from the given variable pool, based on a partial likelihood ratio test, the χ^2 values and the corresponding significance probabilities (P-VALUE) have been calculated assuming that each of these variables are individually introduced as the first in the Cox model. Based on these calculations, the variable labor force experience (LFX), with a χ^2 value of 541.11, is the variable that can explain the most about the job shift rate. As such, this variable is introduced in the Cox model in STEP NUMBER 1 as a covariate. The $\hat{\beta}$ coefficient is –0.0046, which is highly significant (COEFFICIENT/STANDARD ERROR –0.0046/0.0002 = –21.2017).

Table 5.2: An Example of Variable Selection With the Aid of Stepwise Cox Regression

```
STEP NUMBER   0              THERE ARE NO TERMS IN THE MODEL AT THIS STEP.
---------------

STATISTICS TO ENTER OR REMOVE VARIABLES
----------------------------------------

                           APPROX.          APPROX.
    VARIABLE               CHI-SQ.          CHI-SQ.                         LOG
NO. N A M E                ENTER            REMOVE          P-VALUE      LIKELIHOOD

 65 EDU                       5.86                          0.0155      -34285.1049
 59 M59                      68.79                          0.0000      -34253.6425
 67 NOJ                      34.78                          0.0000      -34270.6435
 66 LFX                     541.11                          0.0000      -34017.4808
 68 COHO2                     1.99                          0.1579      -34287.0382
 69 COHO3                    69.79                          0.0000      -34253.1410

STEP NUMBER   1                 LFX        IS ENTERED
---------------

                           LOG LIKELIHOOD = -34017.4808
IMPROVEMENT CHI-SQ   ( 2*(LN(MPLR) ) =      541.11  D.F.=    1  P = 0.0000
                        GLOBAL CHI-SQUARE =      462.24  D.F.=    1  P = 0.0000

                                              STANDARD
    VARIABLE            COEFFICIENT              ERROR    COEFF./S.E.    EXP(COEFF.)
    --------            -----------           --------   -----------    -----------
 66 LFX                     -0.0046             0.0002       -21.2017         0.9954
```

Table 5.2 continued

STATISTICS TO ENTER OR REMOVE VARIABLES

VARIABLE NO. NAME	APPROX. CHI-SQ. ENTER	APPROX. CHI-SQ. REMOVE	P-VALUE	LOG LIKELIHOOD
65 EDU	10.88		0.0010	-34012.0421
59 M59	56.60		0.0000	-33989.1796
67 NOJ	194.76		0.0000	-33920.0995
66 LFX		541.11	0.0000	-34288.0354
68 COHO2	2.01		0.1566	-34016.4775
69 COHO3	9.49		0.0021	-34012.7363

STEP NUMBER 2 NOJ IS ENTERED

LOG LIKELIHOOD = -33920.0995
IMPROVEMENT CHI-SQ (2*(LN(MPLR)) = 194.76 D.F.= 1 P = 0.0000
GLOBAL CHI-SQUARE = 583.41 D.F.= 2 P = 0.0000

VARIABLE	COEFFICIENT	STANDARD ERROR	COEFF./S.E.	EXP(COEFF.)
67 NOJ	0.1520	0.0100	15.2560	1.1642
66 LFX	-0.0074	0.0003	-23.8515	0.9926

STATISTICS TO ENTER OR REMOVE VARIABLES

VARIABLE NO. NAME	APPROX. CHI-SQ. ENTER	APPROX. CHI-SQ. REMOVE	P-VALUE	LOG LIKELIHOOD
65 EDU	10.81		0.0010	-33914.6965
59 M59	55.89		0.0000	-33892.1523
67 NOJ		194.76	0.0000	-34017.4808
66 LFX		701.09	0.0000	-34270.6435
68 COHO2	0.51		0.4737	-33919.8428
69 COHO3	6.54		0.0105	-33916.8291

STEP NUMBER 3 M59 IS ENTERED

LOG LIKELIHOOD = -33892.1523
IMPROVEMENT CHI-SQ (2*(LN(MPLR)) = 55.89 D.F.= 1 P = 0.0000
GLOBAL CHI-SQUARE = 643.02 D.F.= 3 P = 0.0000

Table 5.2 continued

VARIABLE	COEFFICIENT	STANDARD ERROR	COEFF./S.E.	EXP(COEFF.)
59 M59	-0.0053	0.0007	-7.2320	0.9947
67 NOJ	0.1505	0.0099	15.2358	1.1624
66 LFX	-0.0073	0.0003	-23.6724	0.9927

STATISTICS TO ENTER OR REMOVE VARIABLES

VARIABLE NO. N A M E	APPROX. CHI-SQ. ENTER	APPROX. CHI-SQ. REMOVE	P-VALUE	LOG LIKELIHOOD
65 EDU	8.32		0.0039	-33887.9899
59 M59		55.89	0.0000	-33920.0995
67 NOJ		194.05	0.0000	-33989.1796
66 LFX		691.27	0.0000	-34237.7891
68 COHO2	0.97		0.3249	-33891.6676
69 COHO3	13.64		0.0002	-33885.3330

STEP NUMBER 4 COHO3 IS ENTERED

LOG LIKELIHOOD = -33885.3330
IMPROVEMENT CHI-SQ (2*(LN(MPLR)) = 13.64 D.F.= 1 P = 0.0002
GLOBAL CHI-SQUARE = 657.14 D.F.= 4 P = 0.0000

VARIABLE	COEFFICIENT	STANDARD ERROR	COEFF./S.E.	EXP(COEFF.)
59 M59	-0.0057	0.0007	-7.6556	0.9943
67 NOJ	0.1493	0.0099	15.0449	1.1611
66 LFX	-0.0071	0.0003	-22.8015	0.9929
69 COHO3	0.1264	0.0340	3.7215	1.1348

STATISTICS TO ENTER OR REMOVE VARIABLES

VARIABLE NO. N A M E	APPROX. CHI-SQ. ENTER	APPROX. CHI-SQ. REMOVE	P-VALUE	LOG LIKELIHOOD
65 EDU	7.13		0.0076	-33881.7704
59 M59		62.99	0.0000	-33916.8291
67 NOJ		189.62	0.0000	-33980.1443
66 LFX		628.57	0.0000	-34199.6192
68 COHO2	9.46		0.0021	-33880.6030
69 COHO3		13.64	0.0002	-33892.1523

Table 5.2 continued

STEP NUMBER 5 COHO2 IS ENTERED

```
                  LOG LIKELIHOOD = -33880.6030
IMPROVEMENT CHI-SQ  ( 2*(LN(MPLR) ) =        9.46  D.F.=   1  P = 0.0021
                 GLOBAL CHI-SQUARE =      663.40  D.F.=   5  P = 0.0000
```

VARIABLE	COEFFICIENT	STANDARD ERROR	COEFF./S.E.	EXP(COEFF.)
59 M59	-0.0060	0.0008	-7.9313	0.9941
67 NOJ	0.1465	0.0100	14.7194	1.1578
66 LFX	-0.0070	0.0003	-22.3572	0.9930
68 COHO2	0.1075	0.0349	3.0753	1.1135
69 COHO3	0.1838	0.0390	4.7171	1.2017

STATISTICS TO ENTER OR REMOVE VARIABLES

VARIABLE NO. N A M E	APPROX. CHI-SQ. ENTER	APPROX. CHI-SQ. REMOVE	P-VALUE	LOG LIKELIHOOD
65 EDU	6.48		0.0109	-33877.3618
59 M59		67.82	0.0000	-33914.5117
67 NOJ		182.24	0.0000	-33971.7225
66 LFX		597.73	0.0000	-34179.4681
68 COHO2		9.46	0.0021	-33885.3330
69 COHO3		22.13	0.0000	-33891.6676

STEP NUMBER 6 EDU IS ENTERED

```
                  LOG LIKELIHOOD = -33877.3618
IMPROVEMENT CHI-SQ  ( 2*(LN(MPLR) ) =        6.48  D.F.=   1  P = 0.0109
                 GLOBAL CHI-SQUARE =      667.39  D.F.=   6  P = 0.0000
```

VARIABLE	COEFFICIENT	STANDARD ERROR	COEFF./S.E.	EXP(COEFF.)
65 EDU	0.0258	0.0100	2.5628	1.0261
59 M59	-0.0077	0.0010	-7.5078	0.9923
67 NOJ	0.1470	0.0100	14.7181	1.1584
66 LFX	-0.0070	0.0003	-22.2621	0.9930
68 COHO2	0.1038	0.0350	2.9690	1.1094
69 COHO3	0.1766	0.0391	4.5218	1.1932

Table 5.2 continued

STATISTICS TO ENTER OR REMOVE VARIABLES

NO.	VARIABLE NAME	APPROX. CHI-SQ. ENTER	APPROX. CHI-SQ. REMOVE	P-VALUE	LOG LIKELIHOOD
65	EDU		6.48	0.0109	-33880.6030
59	M59		58.80	0.0000	-33906.7628
67	NOJ		182.48	0.0000	-33968.6034
66	LFX		591.83	0.0000	-34173.2762
68	COHO2		8.82	0.0030	-33881.7704
69	COHO3		20.34	0.0000	-33887.5322

NO TERM PASSES THE REMOVE AND ENTER LIMITS (0.1500 0.1000) .

SUMMARY OF STEPWISE RESULTS

STEP NO	VARIABLE ENTERED	DF	VARIABLE REMOVED	LOG LIKELIHOOD	IMPROVEMENT CHI-SQUARE	IMPROVEMENT P-VALUE	GLOBAL CHI-SQUARE	GLOBAL P-VALUE
0				-34288.035				
1	66 LFX	1		-34017.481	541.109	0.000	462.240	0.000
2	67 NOJ	2		-33920.100	194.762	0.000	583.409	0.000
3	59 M59	3		-33892.152	55.894	0.000	643.018	0.000
4	69 COHO3	4		-33885.333	13.639	0.000	657.144	0.000
5	68 COHO2	5		-33880.603	9.460	0.002	663.397	0.000
6	65 EDU	6		-33877.362	6.482	0.011	667.389	0.000

After introducing the variable labor force experience into the model, the χ^2 values as well as their significance probabilities are once again calculated, assuming that in the next step only one of them will be introduced into the model. In so doing, the number of previously held jobs (NOJ) with a χ^2 value of 194.76 is the next best explanatory variable. Accordingly, this variable is introduced into the Cox model in STEP NUMBER 2. This, however, causes the $\hat{\beta}$ coefficient of the variable labor force experience to change from –0.0046 to –0.0074. This means that when controlling for the number of previously held occupations, which significantly affected the job shift rate ($\hat{\beta}_{NOJ} = 0.1520$), the negative effect of labor force experience is even stronger.

STEP NUMBER 3 then includes the variable prestige (M59) with an additional explanatory contribution of $\chi^2 = 55.89$. Interestingly, the β estimates and the χ^2 values of the variables previously included in the model are not significantly affected. Therefore, one must presume that through this new variable a basically new explanatory dimension, which is independent of $\hat{\beta}_{NOJ}$ and $\hat{\beta}_{LFX}$, is introduced into the regression equation.

After this third step, based upon the χ^2 values, it becomes evident that the youngest cohort (COHO3) differs the most (χ^2 value = 13.64), whereas the middle cohort (COHO2) is characterized by a χ^2 value = 0.97. Furthermore, the χ^2 value of COHO3 is larger than the χ^2 value of the education variable (8.32). Therefore, STEP NUMBER 4 includes the variable COHO3 in the Cox model.

In controlling the difference due to the youngest cohort, the dummy variable COHO2, which now only differentiates between the cohorts 1929-31 and 1939-41, also achieves a significant explanatory contribution (χ^2 value = 9.46). In STEP NUMBER 5 this variable is also included in the model.

Finally, in STEP NUMBER 6 the variable EDU (education) is introduced as the last covariate. It also produces a significant effect. One can, however, observe here that the $\hat{\beta}$ coefficients of the cohort dummies become somewhat smaller. A portion of the cohort differences may therefore be explained by the educational differences between the cohorts.

After this final step it can then be demonstrated that all of the variables have a significant effect and that the exclusion of one of them noticeably reduces the explanatory power of the Cox model (see the χ^2 values in the column APPROX. CHI-SQ. REMOVE).

Although the stepwise regression method leads to the same model as in the Program Example 5.1 (Table 5.1), the procedures imply different principles. The variables of Program Example 5.1 were introduced based on theoretical considerations, the stepwise regression method included them according to statistical criteria. The results of a stepwise regression may therefore not always be meaningful in substantive terms and may depend on the mechanism of selection. In particular, relevant variables may be omitted from the regression equation if they are highly correlated with variables that are hard to interpret, but are already included in the model. On the other hand, the explanatory contribution of a variable may first appear after other influences have been controlled, as was the case with the dummy variable COHO2. In general, one should apply the stepwise regression with great care. Above all, one should check whether the use of different methods (such as forward or backward elimination), as well as the variation of parameters that govern the Newton-Raphson algorithm, lead to different models.

5.4 Introducing Time-Dependent Covariates

The previous section discussed models that dealt with time-constant variables. However, the BMDP subprogram P2L can also be used to estimate the influence of time-dependent covariates. This is especially interesting because one may more realistically formulate the influence of the covariates on hazard rates and combine two or more parallel processes directly with one another. It is not uncommon that a duration dependency arises because the observed

Figure 5.2: Modeling the Influence of Marriage (a) on the Career Path (b) as a Time-Dependent Covariate, and (c) as a Time-Independent Covariate, Measured at the Beginning of Each Episode

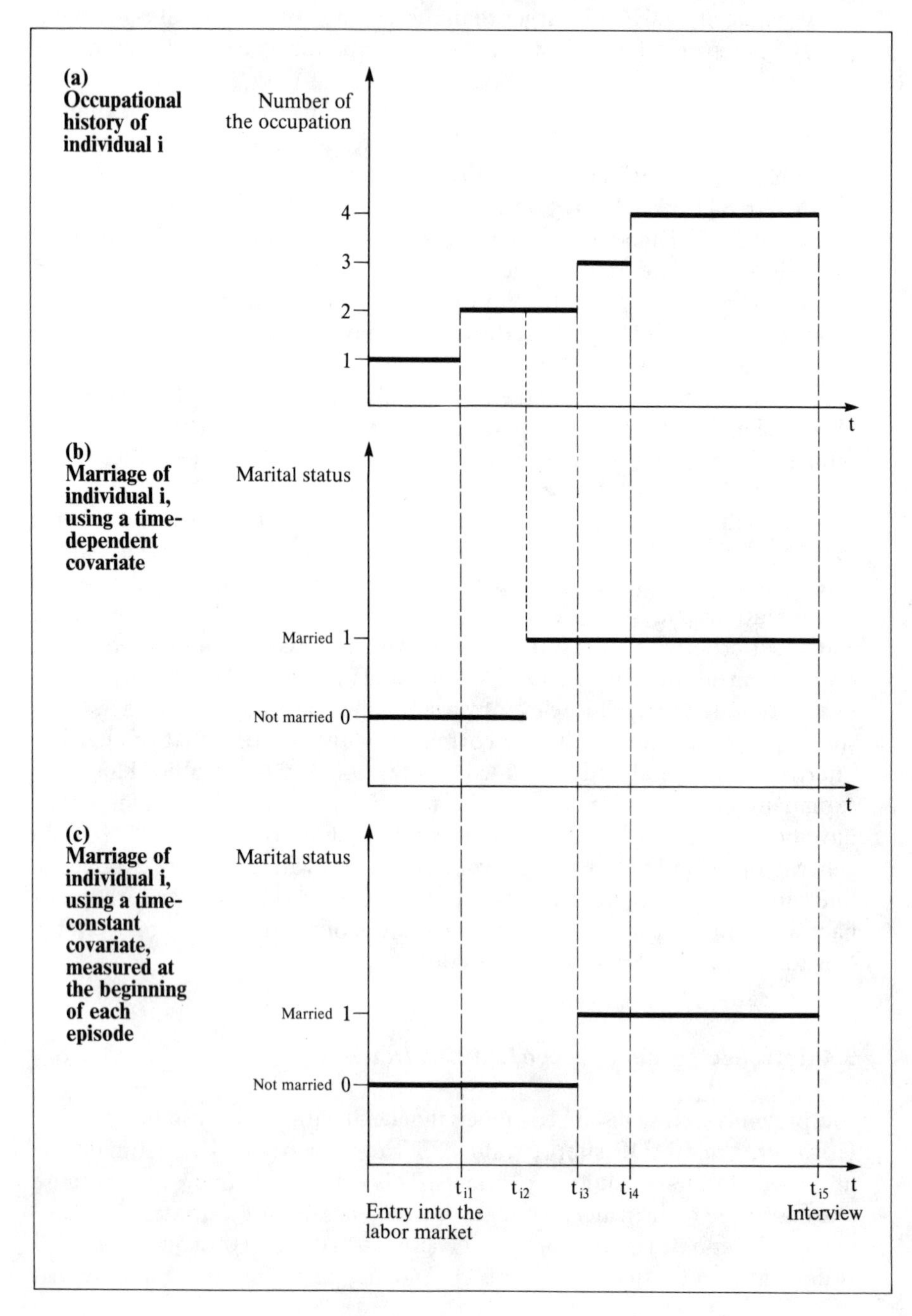

states are aggregates of unobserved states (see Section 3.9.1). This is due to the simple fact that time-dependent covariates are dealt with as time-independent covariates in the model.

Technically, one speaks of time-independent covariates when they are measured at the beginning of episode k and their values remain unchanged over the duration $v_k = t - t_{k-1}$ (see Figure 5.2(c)). Time-dependent covariates may, however, change their values within episode k. Given discrete time-dependent covariates, the values remain constant over certain subintervals v_{k_i}, $v_k = \sum_{i=1}^{s} v_{k_i}$ (see Figure 5.2(b)), whereas continuous time-dependent variables may change continously.

5.4.1 Discrete Time-Dependent Covariates

In general, economics and the social sciences deal with variables that do not continuously change over time. Such variables are characterized by a step function and influence duration by changing the rate within a given episode. For example, if one assumes that for men the event of marriage has a stabilizing effect on the process of job mobility (see Figure 5.2), then this relationship may be studied by introducing a time-dependent covariate.

If for an individual i, $t_{i,k-1}$ is designated as the beginning of an occupational episode k, and if v represents the duration of episode k, and t_i^H is the time of marriage of individual i, then the value of the time dependent dummy variable marriage $x_{ik}^H(v)$ is characterized by:

$$x_{ik}^H(v) = \begin{cases} 0 \text{ for } t_i^H - t_{i,k-1} \geq v \\ 1 \text{ for } t_i^H - t_{i,k-1} < v. \end{cases}$$

The covariate vector $\mathbf{x}_k$ of the Cox model is dependent on time, and the model is:

$$\lambda^k(v | \mathbf{x}_k(v)) = \lambda_0(v) \exp(\mathbf{x}_k'(v) \boldsymbol{\beta}) \qquad , k = 1, 2, \dots .$$

The following Program Example 5.4 demonstrates the realization of such a model applying BMDP by slightly modifying the program run given in Example 5.1. Unlike Program Example 5.1, where the variables were stratified according to sex, here the example concentrates only on men (USE = (M3 EQ 1)). For women the event of marriage will affect the employment process quite differently due to employment interruptions (family, child care, etc.).

Program Example 5.4:

```
/INPUT UNIT IS 30.
       CODE IS DATA.
/VARIABLE NAMES ARE (63)DUR,(64)CEN,(65)EDU,(66)LFX,
                    (67)NOJ,(68)COHO2,(69)COHO3.
```

```
         ADD IS 7.
/TRANSFORM USE = (M3 EQ 1).
          DUR = M51 - M50 + 1.
          CEN = 1.
          IF (M51 EQ M47) THEN CEN = 0.
          IF (M41 EQ 1 AND M42 EQ 1) THEN EDU = 9.
          IF (M41 EQ 1 AND (M42 EQ 2 OR M42 EQ 3)) THEN EDU = 11.
          IF (M41 EQ 2 AND M42 EQ 1) THEN EDU = 10.
          IF (M41 EQ 2 AND (M42 EQ 2 OR M42 EQ 3)) THEN EDU = 12.
          IF (M41 EQ 3 AND (M42 EQ 1 OR M42 EQ 2 OR M42 EQ 3))
          THEN EDU = 13.
          IF (M42 EQ 4) THEN EDU = 17.
          IF (M42 EQ 5) THEN EDU = 19.
          COHO2 = 0.
          COHO3 = 0.
          IF(M48 GE 468 AND M48 LE 504) THEN COHO2 = 1.
          IF(M48 GE 588 AND M48 LE 624) THEN COHO3 = 1.
          IF (M49 EQ 0) THEN M49 = 10000.
          M49 = M49 - M50.
          LFX = M50 - M43.
          NOJ = M5 - 1.
/FORM TIME IS DUR.
      STATUS IS CEN.
      RESPONSE IS 1.
/REGRESSION COVARIATES ARE EDU,M59,NOJ,LFX,
                           COHO2,COHO3.
           ADD IS MAR.
           AUXILIARY IS M49.
/FUNCTION MAR = 0.0.
         IF (TIME GT M49) THEN MAR = 1.0.
/PRINT CASES ARE 0.
/END
```

In the REGRESSION paragraph the covariates that are already known are introduced (see Appendix 1). The time-dependent covariate is specified through the dummy variable MAR which has also been introduced into the model with the ADD statement. The variable M49, which contains the time of marriage measured in months from the beginning of the century, is defined in the AUXILIARY statement.

The time-dependent variable MAR is generated in the FUNCTION paragraph. At first, we assign to the variable MAR the value "0." If the occupational duration (TIME) surpasses the time of marriage (M49), that is, as soon as the man has married, then the time-dependent variable MAR takes on the value "1."

In order to be in a position to do this comparison it is, however, necessary that the time of marriage in relation to duration (TIME) is standardized. Thus, in the TRANSFORM paragraph, the time of entrance into the *k*th occupational episode M50 is subtracted from the time of marriage (M49) (both are coded according to the number of months since the beginning of the century).

If a man is already married before entering an occupational episode, M49 will be negative, and the condition in the FUNCTION paragraph will always be true. The variable MAR has, therefore, the value "1" over the entire duration. On the other hand, all men who have not married over the entire observation period receive the value "0" for the variable time of marriage (M49). Thus, the TRANSFORM paragraph assigns a large value (10,000 months), so that the condition in the FUNCTION paragraph will never be true and the variable MAR maintains the value "0" over the entire process time.

The partial likelihood estimation of the influence of the time-dependent covariate MAR (marriage) is very time consuming. It requires intensive calculations since for each event and for all individuals who still remain exposed to the risk, the comparison is made in the FUNCTION paragraph. The results of the Program Example 5.4 are presented below in Table 5.3. Looking at the GLOBAL CHI-SQUARE value of 511.69 (given seven degrees of freedom) it once again becomes evident that the global hypothesis, that the introduced covariates explain nothing, must be discarded (P-VALUE = 0.0000).

A closer examination of the standardized coefficients in the column COEFF./S.E. illustrates, however, that not all of the variables have a significant influence. The absolute values of the standardized $\hat{\beta}$ coefficients of the variable education ($|\hat{\beta}_{EDU}/S.E.(\hat{\beta}_{EDU})| = 0.7756$) and the dummy variable COHO2 ($|\hat{\beta}_{COHO2}/S.E.(\hat{\beta}_{COHO2})| = 1.2379$) are, for instance, smaller than the value 1.96 and therefore not significantly different from 0 for an error probability of 0.05. Differences in the level of education among men thus have no significant effect on the rate of job shifts. Furthermore, men belonging to the oldest cohort and the middle cohort show no major difference with respect to job shifts. Finally, the variables prestige (M59), the number of previously held jobs (NOJ), labor force experience (LFX) and COHO3, all have the same effect as in Table 5.1 and may be interpreted accordingly.

The most interesting covariate in the above program example is of course the time-dependent covariate MAR. Its $\hat{\beta}$ coefficient is significant and negative as expected, which means that the rate to change occupations after marriage is greatly reduced. Compared with the mobility rate of unmarried men, the mobility rate of married men decreases by 27.44 percent [$(0.7256 - 1) \cdot 100\% = -27.44\%$].

Another example of coupling parallel processes by introducing discrete time-dependent covariates in a Cox model is presented in Mayer and Wagner

Table 5.3: Results of the Cox Model From Program Example 5.4

```
  LOG LIKELIHOOD = -19012.0115
GLOBAL CHI-SQUARE =     511.69  D.F.=    7  P-VALUE =0.0000
```

VARIABLE	COEFFICIENT	STANDARD ERROR	COEFF./S.E.	EXP(COEFF.)
65 EDU	0.0113	0.0145	0.7756	1.0113
59 M59	-0.0045	0.0014	-3.2215	0.9955
67 NOJ	0.1609	0.0121	13.2832	1.1745
66 LFX	-0.0076	0.0005	-15.6909	0.9925
68 COHO2	0.0581	0.0469	1.2379	1.0598
69 COHO3	0.1888	0.0533	3.5452	1.2078
70 MAR	-0.3208	0.0511	-6.2807	0.7256

(1988). Based on the German Life History Study (GLHS), they analyze time synchronous educational and occupational events as well as the effects of fertility behavior upon moving out of the parental household.

They define duration as the time period from the age of 15 until the age at which one moves out of the parental household (one-episode case). Their analysis models two different types of events (see the following section on the construction of multistate models): leaving the parental household and not marrying at the same time (NS = nonsynchronized) and moving out at the time of marriage (S = synchronized). The time of moving out and getting married were considered to be synchronous when marriage occurred not more than two months before or after leaving the parental household. For each of these types of events separate regression equations were estimated (see Table 5.4) for the three birth cohorts (those born in 1929–31, 1939–41, and 1949–51). In these estimates, along with various time-constant covariates (which we do not discuss here) four time-dependent dummy variables were introduced: (1) the variable labor force experience (JOBERF), which illustrates whether or not an individual was employed prior to moving out of the parental household; (2) the variable education (AUSB), which indicates whether or not (within the postulated time period of plus/minus two months) moving out of the parental household occurred simultaneously with enrollment in a vocational or professional training program; (3) the variable employment activity (JOB), which illustrates whether a new employment activity was simultaneously started or not upon leaving the parental household (within plus/minus two months); and (4) the variable birth (GEB), which indicates whether or not, in a time period of two months before up until five months after the act of moving out, a child was born or not. The results of the partial likelihood estimations are presented in Table 5.4. Table 5.4 shows that the rate to leave the parental household significantly increases for all birth cohorts given the beginning of

Table 5.4: Partial Likelihood Estimations of the Effect of Time-Dependent and Time-Constant Covariates on the Rate of Moving Out of the Parental Household[1]

	Year of Birth					
	1929–31		1939–41		1949–51	
	S[2]	NS[2]	S[2]	NS[2]	S[2]	NS[2]
SEX	0.54*	-0.01	0.55*	0.56*	1.14*	0.48*
MS	-0.15	0.01	0.05	0.09	0.13	0.11
ABI	0.49	0.22	-0.89*	-0.40	-0.75	0.45*
AUSB	2.61*	–[3]	1.92*	4.48*	1.91*	2.82*
JOBERF	0.67*	-0.26	0.74*	0.14	1.07*	-0.05
JOB	3.80*	4.19*	2.73*	3.98*	1.94*	3.01*
GEB	1.61*	1.45*	2.12*	1.18*	2.43*	0.72*
BILDV	-0.02	0.09*	0.03	-0.01	0.00	0.00
BILDM	0.12*	-0.01	-0.06	-0.04	-0.07	0.02
ERWERBM	-0.38*	0.07	-0.19	-0.03	0.10	0.10
UNFAM	0.32	-0.12	0.14	0.16	0.15	0.04
LANDWVA	-0.57*	-0.10	0.14	0.07	-0.40	0.45*
SELSTVA	0.22	-0.14	-0.36	-0.06	0.10	0.16
BEAMTVA	-0.26	0.14	-0.11	-0.30	0.11	-0.06
ANGESTVA	-0.53*	0.03	-0.45*	0.24	-0.13	0.00
BILDDIF	0.06*	0.00	0.02	0.05*	-0.03	0.00
ALTDIF	-0.05*	0.00	0.00	0.00	0.01	-0.01
GEZIFF	0.04	0.00	0.05	0.04	0.09*	0.03
GEBLAND	-0.49*	1.04*	-0.12	0.39*	-0.24	-0.18
ANZMIG15	-0.03	-0.04	0.04	0.04	-0.16*	0.04
ORT15	-0.05	0.07	-0.32*	0.24	-0.16	0.13
WOART15	-0.19	-0.25	-0.29*	-0.22	-0.36*	-0.40*
BELDI15	-0.19	0.16	-0.38*	-0.20	0.04	-0.06
χ^2	800.34	2902.96	733.32	4068.64	911.96	2824.90
d.f.	23	23	23	23	23	23
Number of Events	234	283	283	280	270	334
Number of Persons	607	607	649	649	701	701

[1] Values with a * are statistically significant at 0.05 level

[2] S: Leaving the parent's household at time of marriage. NS: Leaving the parents's household without simultaneous marrying.

[3] Model NS for cohort 1929–31 was only possible to estimate, leaving out variable AUSB.

an occupational activity (variable JOB). If leaving the parental household is not associated with marriage (see columns NS), the autarchy and financial independence which begin with employment, have an even greater effect on the act of leaving the parental household.

Labor force experience occurring before one moves out of the parental household (variable JOBERF) has a similar influence. This effect is, however, only significant if the individual is simultaneously in the process of establishing his own family (see columns S).

Moving out of the parental household is also expedited when an individual begins a new educational phase (variable AUSB). This effect is greater for those individuals who do not get married at the same time (see columns NS).

Fertility behavior (variable GEB) also influences the time in which an individual moves out of the parental household. If a child is born, then the likelihood of an individual leaving the parental household increases significantly. This effect is especially great when an individual simultaneously gets married (see columns S).

5.4.2 Continuous Time-Dependent Covariates

The examples just presented, which showed the inclusion of discrete time-dependent covariates, show that realistic models can be built with the methods of event history analysis. This increases the appeal of collecting and analyzing event history data in economics and the social sciences.

The same is true when continuous time-dependent covariates are modeled. In comparison to the possibilities of introducing continuous time covariates which are presented here for parametric standard models (see Section 6.3), the Cox model offers the advantage that one can directly use duration information (in the BMDP program P2L through the variable TIME). In the partial likelihood estimation only those probabilities are considered for which, under a given risk set, an individual experiences an event. The covariate vector in the FUNCTION paragraph is then updated for each of these events and for all individuals who are still exposed to the risk.

If one knows the functional form of the continuous time-dependent covariate, or if it may be approximated by a polynomial or a step function, then this process may be coupled with the variable TIME without great difficulty. With regard to the career example, this means that changes in social structures and economic cycles may be included in the model. Examples are changes in the unemployment rate, the growth of the social product, the proportion of individuals gainfully employed in tertiary sector, and so on (see Blossfeld, 1986). With the aid of these macro-data that are usually accessible from official statistical offices as time series, interesting hypotheses may be studied. For example, the hypothesis that with increasing unemployment the rate of job shifts decreases, or with rising per capita income and corresponding economic independency a job change becomes more probable. Because the micro-data (of the individually recorded occupational paths) and the macro-data (the aggregates of measured social structures and economic indicators) are synchronizable over time or may be linked by lag functions, event history

Figure 5.3: Modeling the Influence of the Labor Force Experience (a) on the Career Path, (b) as a Time-Dependent Covariate, and (c) as a Time-Independent Covariate, Measured at the Beginning of Each Episode

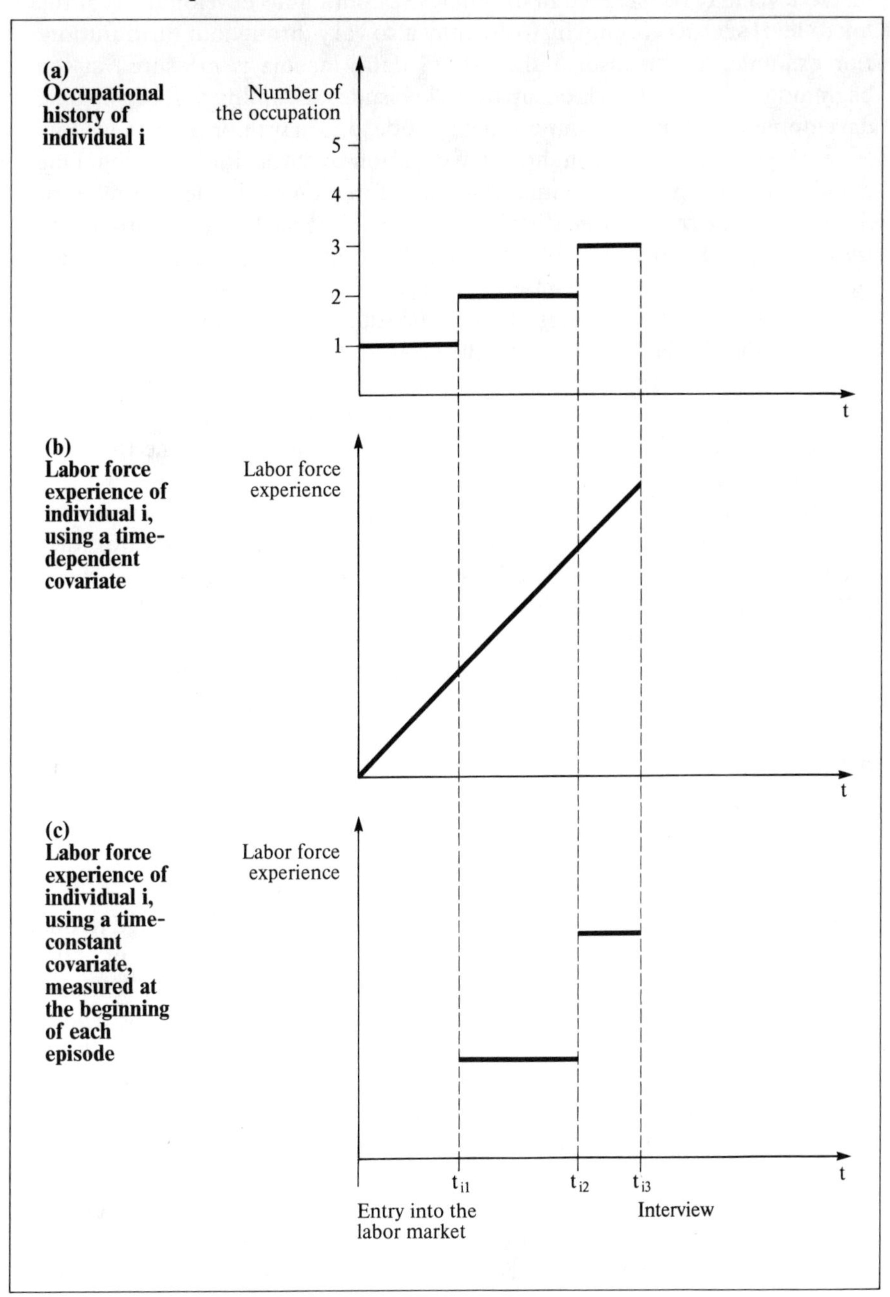

analysis offers the possibility of causally modeling macro-data on individual data.

Not only period effects, cyclical fluctuations and long-term trends at the macro level may be included in the analysis. Continuous developments at the micro level such as earning histories may also vary throughout the duration. For example, in the case of the GLHS data, income is measured at the beginning and end of each occupational episode. Assuming a linear income development within the occupational episode, one may approximate income for each point of time throughout the duration by linear interpolation. One could also attempt to estimate a nonlinear function of the development of lifetime income over consecutive job episodes. This could then be applied to the path of each individual job episode dependent upon an individual's age. For both methods of approximation, it is possible to say that the approximation is generally better, the shorter the occupational episodes are and generally worse for the longer occupational episodes.

It is usually advantageous when modeling a continuous time-dependent covariate to approximate its path in a first step by a parameter sparse linear, nonlinear, or step function and then in a second step to combine this with TIME in the FUNCTION paragraph.

Unfortunately, with the partial likelihood method it is not possible to estimate time-dependent covariates for which the dependency of the duration is introduced in the Cox model in the form $x(v) = x^* + g(v)$. For example, if one includes the variable labor force experience (LFX) as a time-dependent covariate, approximated by a linear function (see Figure 5.3(b); $x_{LFX}(t) = x_{LFX}(t_{i,k-1}) + v$, with $v = t - t_{i,k-1}$), then as equation (3.8.6) illustrates, v is canceled out for each event and the labor force experience is estimated in fact as a time-constant covariate ($x_{LFX}(t) = x_{LFX}(t_{i,k-1})$. This type of influence may be estimated only by applying parametric methods (see Sections 6.3.2 and 6.5).

5.5 Modeling Multistate Models

In economics and the social sciences one is usually not only interested in one type of event (such as the event "change of job" or the event "moving out of the parental household"), but one also wants to know how specific covariates affect the transition rate of various competing end states (such as the transition from a worker to an employee or from a worker to a self-employed; or, with regard to moving out of the parental household with or without simultaneously getting married). This section demonstrates how multistate or competing risk models can be estimated with the partial likelihood method of Cox.

As has already been presented in Sections 3.4 and 4.3, when observing a specific type of event (or a specific destination state), the realization of multistate models is achieved by regarding the competing events as censored.

In order to demonstrate the procedure for modeling a multistate model in BMDP, we once again use the event oriented employment activity data set (see Appendix 1).

To obtain several starting and end states, the occupational activities of the employment histories are classified into 12 occupational groups (see Table 4.4). Then for each employment episode, the corresponding occupational group code (variable M61) (starting state) and the occupational group code of the following employment episode (variable M62) (end state) were stored. If the last employment episode of an employment history is followed by no further episode, then the end state is coded by 0 (censored).

The following BMDP run shows a somewhat modified version of the Program Example 5.1. It illustrates how the specific transition from an unskilled manual occupation (code 2) to a skilled manual occupation (code 3) may be estimated in a Cox model:

$$\lambda_{23}^{k}(v|\mathbf{x}) = \lambda_{023}(v) \exp(\mathbf{x}_k \beta) \qquad , k = 1, 2, \ldots .$$

Program Example 5.5:

```
/INPUT UNIT IS 30.
      CODE IS DATA.
/VARIABLE NAMES ARE (63)DUR,(64)CEN,(65)EDU,(66)LFX,
                    (67)NOJ,(68)COHO2,(69)COHO3.
         ADD IS 7.
/TRANSFORM IF (M3 NE 1 OR M61 NE 2) THEN USE = -1.
           DUR = M51 - M50 + 1.
           CEN = 1.
           IF (M51 EQ M47) THEN CEN = 0.
           IF (M62 NE 3) THEN CEN = 0.
           IF (M41 EQ 1 AND M42 EQ 1) THEN EDU = 9.
           IF (M41 EQ 1 AND (M42 EQ 2 OR M42 EQ 3)) THEN EDU = 11.
           IF (M41 EQ 2 AND M42 EQ 1) THEN EDU = 10.
           IF (M41 EQ 2 AND (M42 EQ 2 OR M42 EQ 3)) THEN EDU = 12.
           IF (M41 EQ 3 AND (M42 EQ 1 OR M42 EQ 2 OR M42 EQ 3))
           THEN EDU = 13.
           IF (M42 EQ 4) THEN EDU = 17.
           IF (M42 EQ 5) THEN EDU = 19.
           COHO2 = 0.
           COHO3 = 0.
           IF(M48 GE 468 AND M48 LE 504) THEN COHO2 = 1.
           IF(M48 GE 588 AND M48 LE 624) THEN COHO3 = 1.
           LFX = M50 - M43.
           NOJ = M5 - 1.
```

```
/FORM TIME IS DUR.
      STATUS IS CEN.
      RESPONSE IS 1.
/REGRESSION COVARIATES ARE EDU,M59,NOJ,LFX,
                           COHO2,COHO3.
/PLOT TYPE = LOG.
/PRINT CASES ARE 0.
/END
```

In contrast to Program Example 5.1, we once again select men (M3) and use only those episodes in which individuals are in an unskilled manual occupation (M61). The risk set thus consists of the group of men who are employed in an unskilled manual occupation.

The program regards all episodes that do not coincide with the end state (M62) "skilled manual jobs" (with the code 3) as being censored. Therefore, in the TRANSFORM paragraph for all other events (M62 NE 3), the censoring variable CEN is replaced by a 0. This means that men in an unskilled manual occupation are regarded as being exposed to the "risk" of advancing into a skilled manual occupation until a transition into one of the competing events (changing to one of the other occupational groups) has occurred. As already described in Section 4.3, these censored episodes are included in the partial likelihood estimation as long as they affect the risk set of the successive occurring events.

The other commands in the FORM and REGRESSION paragraph lead to the estimation of a Cox model with the known covariates education (EDU), prestige (M59), number of previously performed occupations (NOJ), labor force experience (LFX), and the cohort dummies COHO2 and COHO3.

The results of the estimation of this specific job transition are presented in Table 5.5. On the basis of the GLOBAL CHI-SQUARE value of 26.11 with six degrees of freedom and a significance probability of 0.0002, the hypothesis must be rejected that none of the introduced covariates explains the rate of transition from an unskilled manual occupation to a skilled manual occupation ($\alpha = 0.05$).

In the column headed COEFF./S.E., representing the standardized $\hat{\beta}$ coefficients, it is observed, however, that only the variable education (EDU) is significant. All other remaining standardized $\hat{\beta}$ coefficients are absolutely smaller than the value 1.96 and thus not significant given an error probability of 0.05. The rate of transition from an unskilled manual occupation to a skilled manual occupation therefore only depends on education and increases with each additional year of training by 61.91 percent $[(1.6191 - 1) \cdot 100\% = 61.91\%]$.

Although this result is quite plausible, it must be noted that the results of the significance tests depend on the number of actual events and therefore

Table 5.5: Results of the Cox Model From Program Example 5.5

```
    LOG LIKELIHOOD =    -404.0954
GLOBAL CHI-SQUARE =        26.11  D.F.=   6  P-VALUE =0.0002
```

VARIABLE	COEFFICIENT	STANDARD ERROR	COEFF./S.E.	EXP(COEFF.)
65 EDU	0.4819	0.1080	4.4617	1.6191
59 M59	-0.0273	0.0144	-1.8920	0.9731
67 NOJ	0.0863	0.0761	1.1340	1.0902
66 LFX	-0.0046	0.0025	-1.8230	0.9955
68 COHO2	-0.1413	0.2825	-0.5002	0.8682
69 COHO3	0.2323	0.3264	0.7116	1.2615

indirectly on the size of the sample. For example, the number of episodes is reduced by restricting the analysis to men and the starting state of unskilled manual occupations from 6,732 in the total data file to 705. Because very few individuals are in a position to advance to a skilled manual occupation (the number of events is reduced to 72) raising the censorings to a share of 89.79 percent. Thus, it is possible that smaller effects may be significant when the sample set is larger and therefore contains a larger number of events. The size of the sample, therefore sets narrow boundaries for the number of starting and end states.

Similar to the previous example for multistate models, one could also estimate all other occupational transitions and compare the coefficients of the Cox regressions with one another.

Instead of doing this, we proceed with a second example for applying multistate models, developed by Carroll and Mayer (1986). Their study also analyzed career processes based on the GLHS. The aim of the study was to show that upward, downward, or horizontal mobility, as well as mobility within and between firms, is dependent on the size of the organization, provided various individual and structural characteristics are controlled. The hypothesis tested was that the size of an organization should have a negative effect on the rate of job change because employees in large organizations who expect to be promoted are less inclined to leave their present position. On the other hand, employees in smaller organizations perceive the possibility of improving their current status only through a change of firms.

Compared to this, the expected differences in mobility patterns between larger and smaller firms should prove to be paradoxical, if one assumes that the bureaucratic career ladders are "rationalized." In this context "rationalization" means that universal standards are applied to restrict the rate and level of promotion. Too many promotions occurring too rapidly or promotions which imply large increases in gratification threaten the legitimacy of established hierarchy systems. Due to this fact large organizations offer bureaucratic career ladders with many steps and small relative increases. Therefore, one can assume that the size of an organization does have a positive effect upon job change within a company, but has a negative effect on job change between firms. These job movements within firms must, however, have more of a horizontal than ascending nature.

As defined by Caroll and Mayer, an upward movement occurs if the initial earnings of a successive job is higher than 15 percent points of the final earnings obtained in the previous position. An occupational downward move occurs if earnings decrease. Horizontal occupational mobility is characterized by a change in earnings ranging between 0 and 15 percent.

As already illustrated in the previous example, in estimating multistate models those events that are not of interest are censored. For example, in examining the event "upward mobility," all other occupational transitions that are not characterized by an increase in earnings of more than 15 percent points are also censored during the estimation (see the column UPWARD MOVES). Accordingly, all other transitions presented in Table 5.6 are estimated in a similar manner.

The first model of Table 5.6 relates to all job shifts (ANY MOVE). From this equation it is observed that for each younger cohort (C2 $\hat{=}$ cohort 1939–41 and C3 $\hat{=}$ cohort 1949–51), for women (SEX) and for different levels of general education (GED) a significantly higher rate of job change is estimated. Significantly lower rates are obtained, on the other hand, for the first occupation (JN1), the status variable (STATUS), earnings (LNWAGE), and for each higher level of vocational training (OED). The variable size of organization (LNSIZE), which is actually the focal point of the analysis, has, as expected, a negative effect on the rate of job change. This supports the hypothesis that due to expectations of promotion, employees of large firms are less inclined to leave their present job than individuals working in smaller organizations.

The other estimations presented in Table 5.6 support the "rationalization hypothesis." Organizational size has its largest negative effect on upward mobility (UPWARD MOVES) and is less important for horizontal mobility (LATERAL MOVES) as well as for downward movements (DOWNWARD MOVES). More important, however, is the fact that the $\hat{\beta}$ coefficients of the variable organizational size differ depending on whether or not one examines job changes within a firm (WITHIN FIRM) or between firms (ACROSS FIRMS). The opportunity for an occupational change within the same firm is greater the larger the firm is, whereas changes across firms increase with decreasing organizational size.

The remaining four columns in Table 5.6 give a detailed insight into the relationship of occupational mobility and firm size. Although the variable organizational size (LNSIZE) has a positive effect upon both types of job changes within a firm (UPWARD WITHIN FIRM and LATERAL WITHIN FIRM), it is only significant for horizontal occupational movements (LAT-

Table 5.6: Partial Likelihood Estimations of the Effects of Organizational Size on the Rate of Job Change (Standard Error of the β Estimates in Brackets)

	Any Move	Upward Moves	Laternal Moves	Downward Moves	Within Firm	Across Firms	Upward Within Firm	Laternal Within Firm	Upward Across Firms	Laternal Across Firms
C2	.140 (.077)	.188 (.046)	.071 (.088)	.241 (.114)	.120 (.114)	.192 (.054)	.138 (.211)	.049 (.202)	.147 (.084)	.060 (.099)
C3	.178 (.092)	.309 (.053)	.226 (.101)	.331 (.130)	.467 (.132)	.203 (.064)	.488 (.253)	.565 (.224)	.146 (.100)	.124 (.116)
SEX	.128 (.072)	-.111 (.040)	-.063 (.083)	.853 (.100)	-.563 (.109)	-.089 (.049)	-.527 (.210)	-.609 (.205)	-.074 (.077)	.080 (.093)
LFX	-.006 (.001)	-.005 (.000)	-.005 (.001)	-.005 (.001)	-.001 (.001)	-.006 (.000)	-.001 (.002)	-.002 (.001)	-.007 (.001)	-.006 (.001)
JN1	-.126 (.083)	-.285 (.048)	-.334 (.094)	-.701 (.126)	.149 (.120)	-.420 (.057)	.142 (.228)	.041 (.209)	-.157 (.089)	-.483 (.109)
STATUS	-.008 (.002)	-.005 (.001)	-.004 (.002)	-.007 (.003)	-.007 (.003)	-.004 (.002)	-.012 (.006)	-.005 (.005)	-.007 (.003)	-.004 (.003)
GED	.293 (.069)	.169 (.039)	-.115 (.075)	-.009 (.095)	.323 (.095)	.136 (.048)	.569 (.177)	.409 (.153)	.247 (.076)	-.009 (.089)
OED	-.048 (.054)	-.026 (.029)	-.022 (.056)	-.223 (.074)	.180 (.069)	-.108 (.036)	.089 (.133)	.134 (.111)	-.076 (.059)	.010 (.066)
LNSIZE	-.088 (.015)	-.045 (.008)	-.025 (.016)	-.028 (.020)	.074 (.020)	-.082 (.010)	.063 (.039)	.146 (.034)	-.116 (.017)	-.078 (.018)
LNWAGE	-.486 (.039)	-.189 (.025)	-.084 (.052)	.367 (.073)	-.382 (.061)	-.178 (.030)	-.665 (.107)	-.301 (.121)	-.452 (.042)	-.040 (.059)
χ^2	644.1	570.9	111.9	206.3	142.4	446.6	76.6	72.8	597.3	113.7
d.f.	10	10	10	10	10	10	10	10	10	10

ERAL WITHIN FIRM). Thus, within large firms only moderate promotions actually occur in accordance with the "rationalization hypothesis." On the other side, organizational size has a negative effect on the upward movements (UPWARD ACROSS FIRMS) as well as on the horizontal movements (LATERAL ACROSS FIRMS), under the condition that a job change is combined with a change of firm. This means, the smaller a firm is, the greater is the probability of advancing one's status by changing firms.

In comparing the various models presented in Table 5.6 we, finally, draw attention to the fact that each of the models is based on a different number of events. If, as in this example, the variable organizational size (LNSIZE) proves not to be significant in the model UPWARD WITHIN FIRM, then this may be related to the fact that given the sample size, comparatively few events were available in estimating this transition and thus only big effects were significant.

Although applying a multistate model increases the number of parameters to be estimated and the number of initial and end states as well as the data required for estimation, the example just presented clearly shows that with the aid of event history analysis and appropriate choice of multiple states, one may analyze extremely differentiated lines of argument (see Galler, 1988).

In the following chapter on parametric methods we will not deal with the multistate case because it is easy to transfer the examples given in the present chapter to parametric methodology.

Chapter 6: Parametric Regression Models

The advantage of a Cox model is that the researcher can determine the influence of covariates without having to make further assumptions about the form of the hazard-rate path. A disadvantage, however, is that part of the regression model remains unknown and unspecified. This does not present a problem as long as the researcher does not have specific hypotheses regarding the hazard-rate path or when changes in event risks over the duration are unknown or occur so unsystematically that they may not be approximated by a parametric model. However, whenever fundamental hypotheses are to be tested or the researcher has clear ideas concerning the duration dependency, one should—as far as possible—rely on parametric models.

In this chapter, several examples are given of the application of the most commonly used parametric models. These models are presented in detail in Sections 3.2.2, 3.3.2, 3.6.3, and 3.9.2. After a graphic examination of the distribution assumptions (Section 6.1), there is a detailed discussion of the exponential model and its interpretation including advice about examining the residuals (Section 6.2). In Section 6.3 we describe how time-dependent covariates can be included in parametric models and in Section 6.4 we look at models in which the duration is divided into periods. Special duration distributions are presented in Section 6.5. Detailed examples are given for the Gompertz-(Makeham) (Section 6.5.1), the Weibull (Section 6.5.2), the log-logistic (Section 6.5.3), and for the lognormal distribution models (Section 6.5.4). In the final section (Section 6.6) we consider how to deal with unobserved population heterogeneity.

Unfortunately, no generally accessible and user-friendly program package exists to date to estimate the parametric models mentioned above.[1] Programs such as GLIM or P3R of BMDP make use of macros and subprograms to

[1]With the LIFEREG procedure of SAS, for example, it is indeed possible to estimate the exponential, Weibull, lognormal, gamma, and log-logistic models; however, the Gompertz model may not be estimated.

calculate maximum likelihood estimates but these programs require a reasonable knowledge of mathematics and statistics, as well as programming experience.

The goal of this chapter is to acquaint the reader with the software available for estimating parametric models. We estimate the exponential model with the FORTRAN subprogram P3RFUN (written by Trond Petersen, 1985), RATE (developed by Nancy Tuma, 1979), and SAS; the Gompertz model with P3RFUN and RATE; parametric models with periodic durations with RATE; the Weibull and the log-logistic models with P3RFUN, GLIM (see the listing of the applied macros in Appendix 3 written by Roger and Peacock, 1983), and SAS; and the lognormal model with SAS. Models with unobserved heterogeneity (given a gamma distribution) are estimated with RATE and SAS.

6.1 Checking the Parametric Distribution Assumptions Graphically

Since regressions of parametric models specify duration dependency, the question of the suitability of the assumed distribution model is extremely important. Below, as we present the various parametric models we will consider how to analyze residuals after a model has been estimated and how to verify a model by estimation of a more general distribution model containing other distributions as special cases (e.g., the Weibull model contains the exponential model as a special case). But, first, we look at the possibility of checking the type of distribution using a parameter-free method of estimating the survivor function (see Section 3.7.1).

In this instance, the survivor function must be transformed in such a way that an assessment can easily be made. This is simple if the relationship can be expressed in a linear form ($y = a + bx$) and deviations from the distribution assumption are reflected in deviations from this line.

According to (3.2.17), the survivor function of the *exponential model* is

$$S(t) = \exp(-\lambda t).$$

Taking the logarithm, the linear form

$$\ln S(t) = -\lambda t,$$

is obtained with $y = \ln S(t)$, $a = 0$, $b = -\lambda$, $x = t$.

In order to fulfill the distribution assumption, a plot of the estimated path of $\hat{S}(t)$ with respect to t should result in a line going approximately through the origin. Assuming that the model is valid, one can compute least squares estimates for the parameters a and b based upon the path of the estimated and transformed survivor functions. It should thus be the case that $\hat{a} \sim 0$ and $-\hat{b} \sim \hat{\lambda}$.

According to (3.3.20), the survivor function of the *Weibull model* is

$$S(t) = \exp(-(\lambda t)^{\alpha}).$$

The logarithm of this expression is

$$\ln S(t) = (-\lambda t)^{\alpha}.$$

Taking the logarithm again results in the linear form

$$\ln(-\ln S(t)) = \alpha \ln\lambda + \alpha \ln t,$$

with $y = \ln(-\ln S(t))$, $a = \alpha\ln\lambda$, $b = \alpha$, $x = \ln t$.

Once again, if the log-minus-log survivor function $\ln(-\ln\hat{S}(t))$ is plotted against the logarithm of time $\ln t$, one should approximately obtain a line if the distribution assumption is valid. Under the assumption that the model is valid and based upon the path of the estimated and transformed survivor function, least squares estimates may be computed for the parameters a and b for which $\hat{a} \sim \hat{\alpha}\ln\hat{\lambda}$ and $\hat{b} \sim \hat{\alpha}$.

For the *Gompertz model* the survivor function takes the form

$$S(t) = \exp(-\frac{\lambda_0}{\gamma_0} (\exp(\gamma_0 t)-1)).$$

The logarithm of this expression is

$$\ln S(t) = -\frac{\lambda_0}{\gamma_0} (\exp(\gamma_0 t)-1).$$

For time t+1 we obtain

$$\ln S(t+1) = -\frac{\lambda_0}{\gamma_0} (\exp(\gamma_0(t+1))-1).$$

Taking the difference it follows that

$$\ln \frac{S(t)}{S(t+1)} = -\frac{\lambda_0}{\gamma_0} \exp(\gamma_0 t)\,(1 - \exp(\gamma_0)).$$

Taking the logarithm again leads to the linear form

$$\ln(\ln \frac{S(t)}{S(t+1)}) = \ln(-\frac{\lambda_0}{\gamma_0}(1 - \exp(\gamma_0))) + \gamma_0 t,$$

with $y = \ln(\ln \frac{S(t)}{S(t+1)})$, $a = \ln(-\frac{\lambda_0}{\gamma_0}(1 - \exp(\gamma_0)))$, $b = \gamma_0$, $x = t$.

For a Gompertz distribution, the plot of $\ln(\ln \frac{\hat{S}(t)}{\hat{S}(t+1)})$ against t should also approximate a line. Assuming that the model is valid, least squares estimates for the parameters a and b of this line can again be computed on the basis of the path of the estimated and transformed survivor function. As such, $\hat{a} \sim \ln(-\frac{\hat{\lambda}_0}{\hat{\gamma}_0}(1 - \exp(\hat{\gamma}_0)))$ and $\hat{b} \sim \hat{\gamma}_0$.

Finally, according to (3.2.29) the survivor function of the *log-logistic model* is

$$S(t) = \frac{1}{1+(\lambda t)^{\alpha}}.$$

Consequently,

$$1\text{–}S(t) = \frac{(\lambda t)^{\alpha}}{1+(\lambda t)^{\alpha}}$$

and therefore

$$\frac{1\text{–}S(t)}{S(t)} = (\lambda t)^{\alpha}.$$

Taking the logarithm one obtains the line

$$\ln\frac{1\text{–}S(t)}{S(t)} = \alpha\ln\lambda + \alpha\ln t,$$

with $y = \ln\frac{1\text{–}S(t)}{S(t)}$, $a = \alpha\ln\lambda$, $b = \alpha$, $x = \ln t$.

If the assumed distribution is correct, then the transformed survivor function should approximate a line. Assuming that the model is valid, least squares estimates for the parameters a and b of this line can again be computed based on the path of the estimated and transformed survivor function, in which $\hat{a} \sim \hat{\alpha}\ln\hat{\lambda}$ and $\hat{b} \sim \hat{\alpha}$.

In order to demonstrate the application of this graphic method of checking distribution assumptions, we examine below whether or not the risk of changing jobs among men remains constant over the duration or whether a Weibull, a Gompertz, or a log-logistic distribution is more appropriate. In order to do this, one can, for example, estimate the survivor function $\hat{S}(t)$ with the software package SPSS, creating a life table in raw data form (OPTIONS 8).[2] Thereby, the interval range is chosen to be the smallest possible (in this case one month), in order to obtain a most accurate estimation.

[2]The SPSS program used to create a life table in raw data form:

```
SPACE              100000
GET FILE           DATA
SELECT IF          (M3 EQ 1)
COMPUTE            DUR = M51 - M50 + 1
COMPUTE            CEN = 1
IF                 (M51 EQ M47) CEN = 0
RAW OUTPUT UNIT    15
SURVIVAL           TABLES = DUR/
                   INTERVALS = THRU 300 BY 1/
                   STATUS = CEN (1)/
                   PLOTS (LOGSURV)
OPTIONS            8
FINISH
```

An exact description of this format and the variables is given in Hull and Nie (1981).

In a second step, SPSS is applied to transform the raw data (according to the transformations described above) and to obtain a plot with the command SCATTERGRAM:

Program Example 6.1:

```
VARIABLE LIST    TYPE,TABNR,BOINT,NEIT,WITHDR,RISK,EVENTS,
                 TERMI,PROBSURV,SURV,DENS,HAZ,SE1,SE2,SE3
INPUT FORMAT     FIXED (F2.0,F5.0,F6.2,4F8.2/7X,5F8.6,3F7.4)
INPUT MEDIUM     DISK
N OF CASES       300
COMPUTE          TIME = BOINT
SELECT IF        (TIME LE 160)
COMPUTE          LOGTIME = LN(BOINT)
COMPUTE          EXPOSURV = LN(SURV)
COMPUTE          WEIBSURV = LN(-LN(SURV))
LAG              GOSURV = SURV
IF               (GOSURV GT SURV)
                 GOMPSURV = LN(LN(GOSURV/SURV))
COMPUTE          LOGLOG =LN((1-SURV)/SURV)
SCATTERGRAM      EXPOSURV(-4,0) WITH BOINT (0,160)
OPTIONS          4
SCATTERGRAM      WEIBSURV(-3,2) WITH LOGTIME (0,6)
OPTIONS          4
SCATTERGRAM      GOMPSURV(-20,20) WITH TIME (0,160)
OPTIONS          4
SCATTERGRAM      LOGLOG(-3,5) WITH LOGTIME(0,6)
OPTIONS          4
FINISH
```

The result of this procedure is illustrated in Figures 6.1 through 6.4. Visual inspection of these curves can be simplified by estimating a line for each of the individual figures with the aid of the least squares estimation. The fit of the data points to this line can be assessed using the R^2 values:

Program Example 6.2:

```
VARIABLE LIST    TYPE,TABNR,BOINT,NEIT,WITHDR,RISK,EVENTS,
                 TERMI,PROBSURV,SURV,DENS,HAZ,SE1,SE2,SE3
INPUT FORMAT     FIXED (F2.0,F5.0,F6.2,4F8.2/7X,5F8.6,3F7.4)
INPUT MEDIUM     DISK
N OF CASES       300
COMPUTE          TIME = BOINT
SELECT IF        (TIME LE 160)
```

```
COMPUTE         LOGTIME = LN(BOINT)
COMPUTE         EXPOSURV = LN(SURV)
COMPUTE         WEIBSURV = LN(-LN(SURV))
LAG             GOSURV = SURV
IF              (GOSURV GT SURV)
                GOMPSURV = LN(LN(GOSURV/SURV))
COMPUTE         LOGLOG =LN((1-SURV)/SURV)
REGRESSION      VARIABLES = EXPOSURV,WEIBSURV,GOMPSURV,
                            GOMPSURV,LOGLOG,
                            TIME,LOGTIME/
                REGRESSION = EXPOSURV WITH TIME (2)/
                REGRESSION = WEIBSURV WITH LOGTIME (2)/
                REGRESSION = GOMPSURV WITH TIME (2)/
                REGRESSION = LOGLOG WITH LOGTIME (2)/
STATISTICS      ALL
FINISH
```

This procedure results in the following estimates for the different types of distributions as shown below in Table 6.1.

Table 6.1: Least Squares Estimates for Evaluating the Distribution Assumptions

Model	$\hat{a}$	$\hat{b}$	R^2
Exponential distribution	–0.2029	–0.0106	0.9642
Weibull distribution	–3.6950	0.8621	0.8534
Gompertz distribution	–3.9092	–0.0095	0.4114
Log-logistic distribution	–4.2100	1.1313	0.9068

For an exponential distribution, an estimated λ coefficient of 0.0106 [$\hat{b} = -\hat{\lambda} = -0.0106$] is obtained in Table 6.1. The path of the curve is graphed in Figure 6.1. It is easy to see that the curve bends slightly downward and that, based upon the regression estimation, the line does not go directly through the origin, but intercepts the y axis at a value of $\hat{a} = -0.2029$. Despite these deviations, the exponential distribution still possesses the "best data fit" ($R^2 = 0.9642$) of all the distribution types tested.

For the Weibull distribution the estimated values $\hat{\alpha} = 0.8621$ [$\hat{b} = \hat{\alpha} = 0.8621$] and $\hat{\lambda} = 0.0138$ [$-3.6950 = 0.8621 \ln\hat{\lambda}$] are obtained. Since $\hat{\alpha} < 1$, one should expect the hazard rate to monotonically decrease over the entire duration (see Figure 3.5). However, as is evident from Figure 6.2 the linear "fit" is not quite as good as for the exponential model ($R^2 = 0.8534$).

Figure 6.1: Exponential Distribution Graphic Test

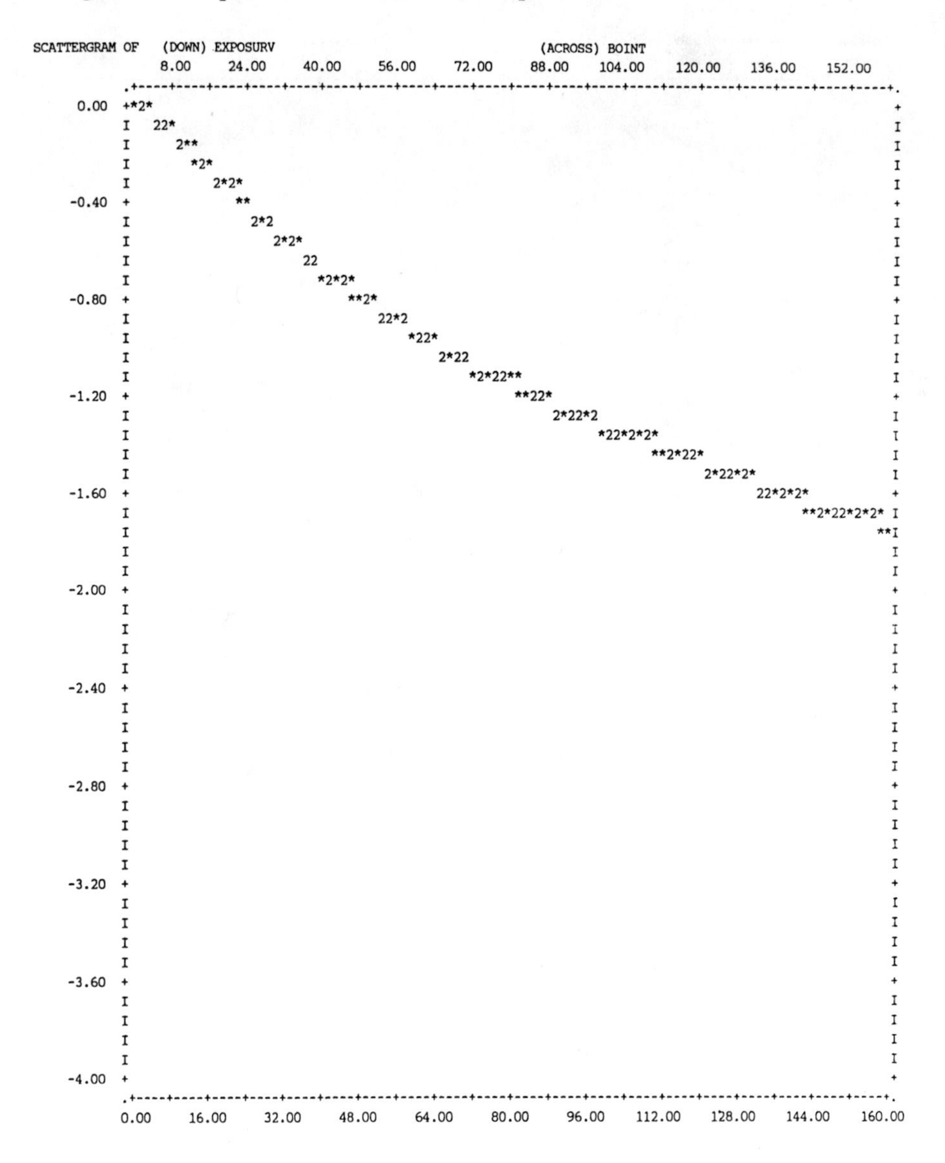

For the Gompertz model the estimate for the γ_0 coefficient is a value of –0.0095 [$\hat{b} = \hat{\gamma}_0 = -0.0095$] and for the λ_0 coefficient a value of 0.00202 [$-3.9092 = \ln(-\frac{\hat{\lambda}_0}{-0.0095}(1-\exp(-0.0095)))$]. Since $\hat{\gamma}_0 < 0$, the hazard rate is

Figure 6.2: Weibull Distribution Graphic Test

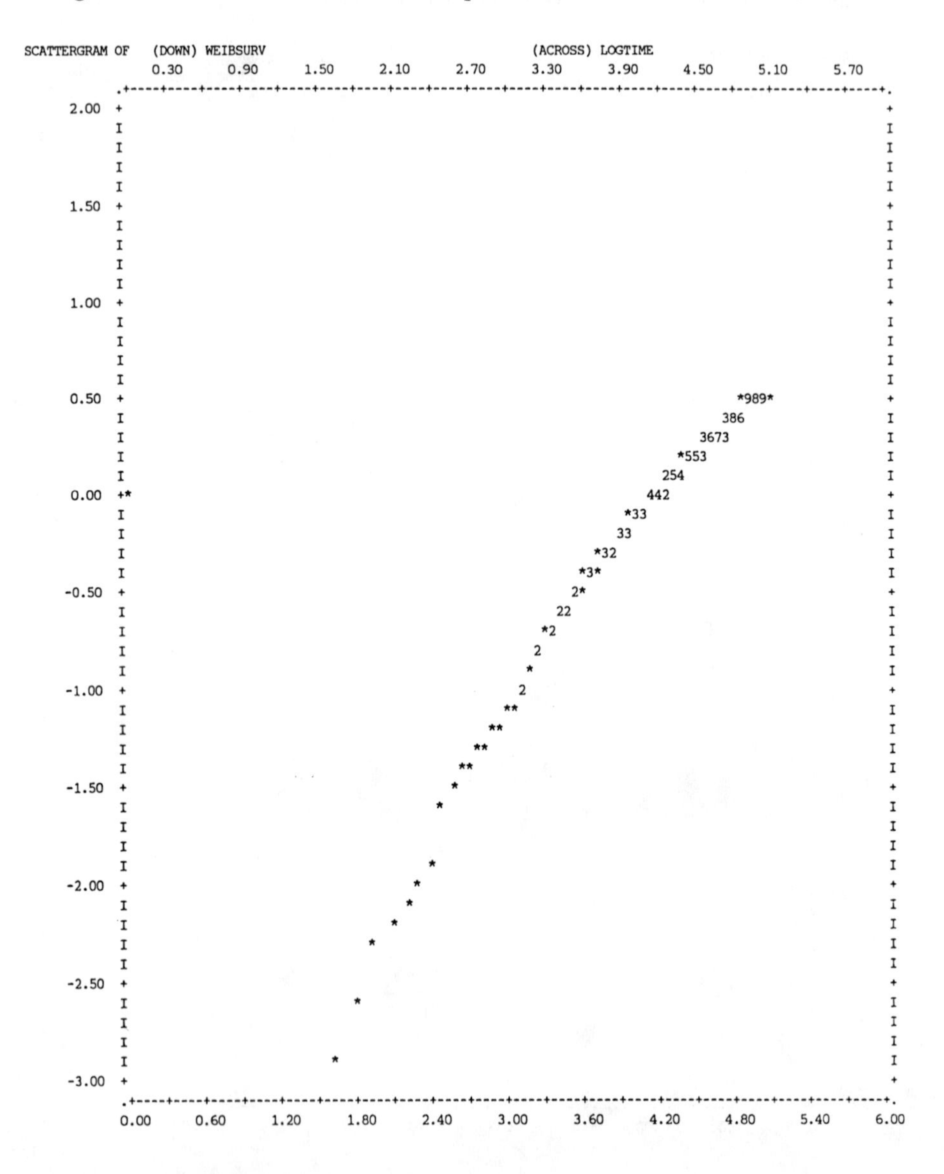

once again expected to decrease monotonically over the duration (see Figure 3.8). The linear "fit" of the data points in Figure 6.3, with a R^2 of 0.4114, is, however, even worse than for the Weibull distribution.

Finally, one obtains an estimated value of 1.1313 [$\hat{b} = \hat{\alpha} = 1.1313$] for the α coefficient of the log-logistic distribution and 0.024 [$-4.2100 = 1.1313 \ln\hat{\lambda}$] for

Figure 6.3: Gompertz Distribution Graphic Test

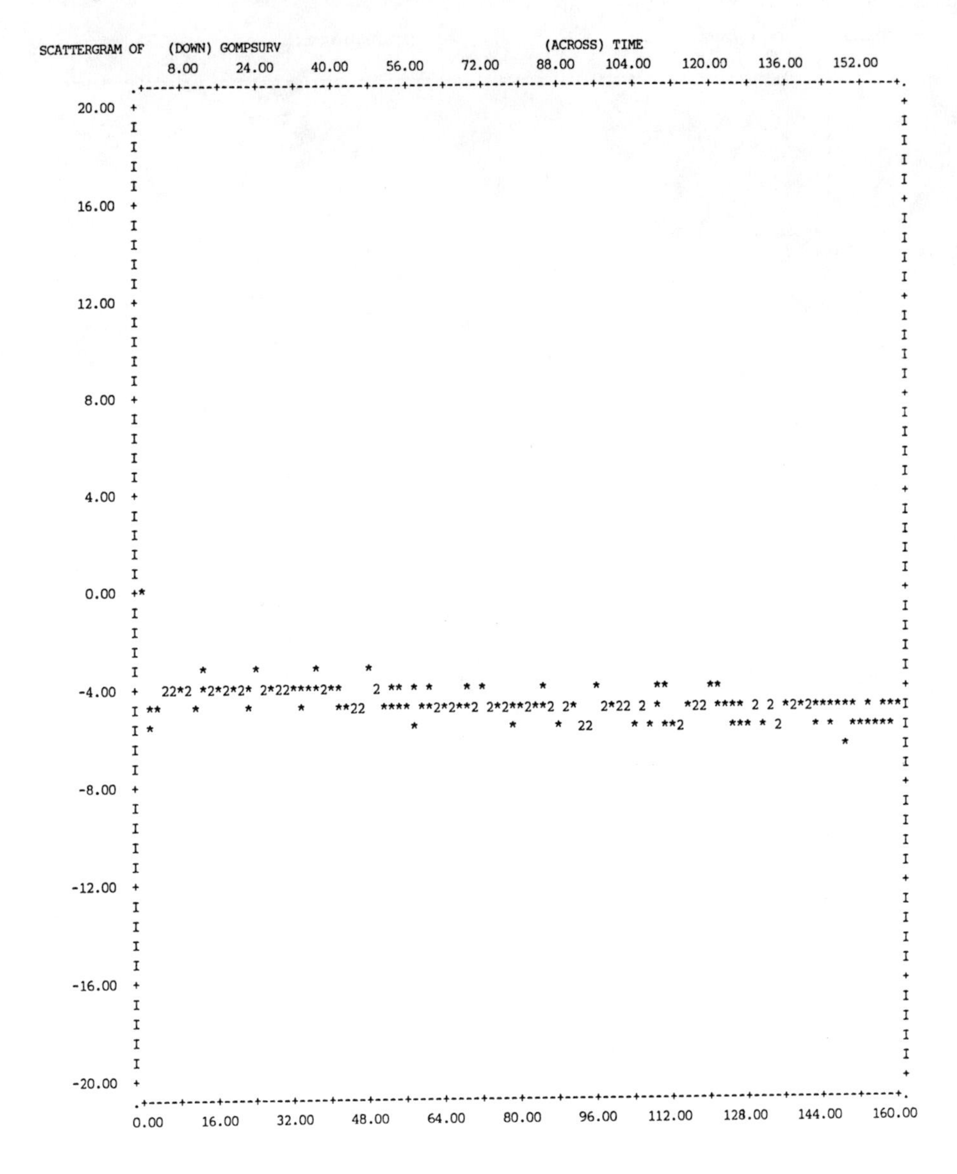

the λ coefficient. According to these estimates, the hazard rate over the duration is nonmonotonic (see Figure 3.7): The risk of changing jobs increases for men at the beginning of each new job, reaches a climax after about seven months $[(\hat{\alpha} - 1)^{1/\hat{\alpha}}/\hat{\lambda} = (1.1313 - 1)^{1/1.1313}/0.024 = 6.92]$ and decreases with increased duration. A comparison of the data points "fit" of the estimated line

Figure 6.4: Log-Logistic Distribution Graphic Test

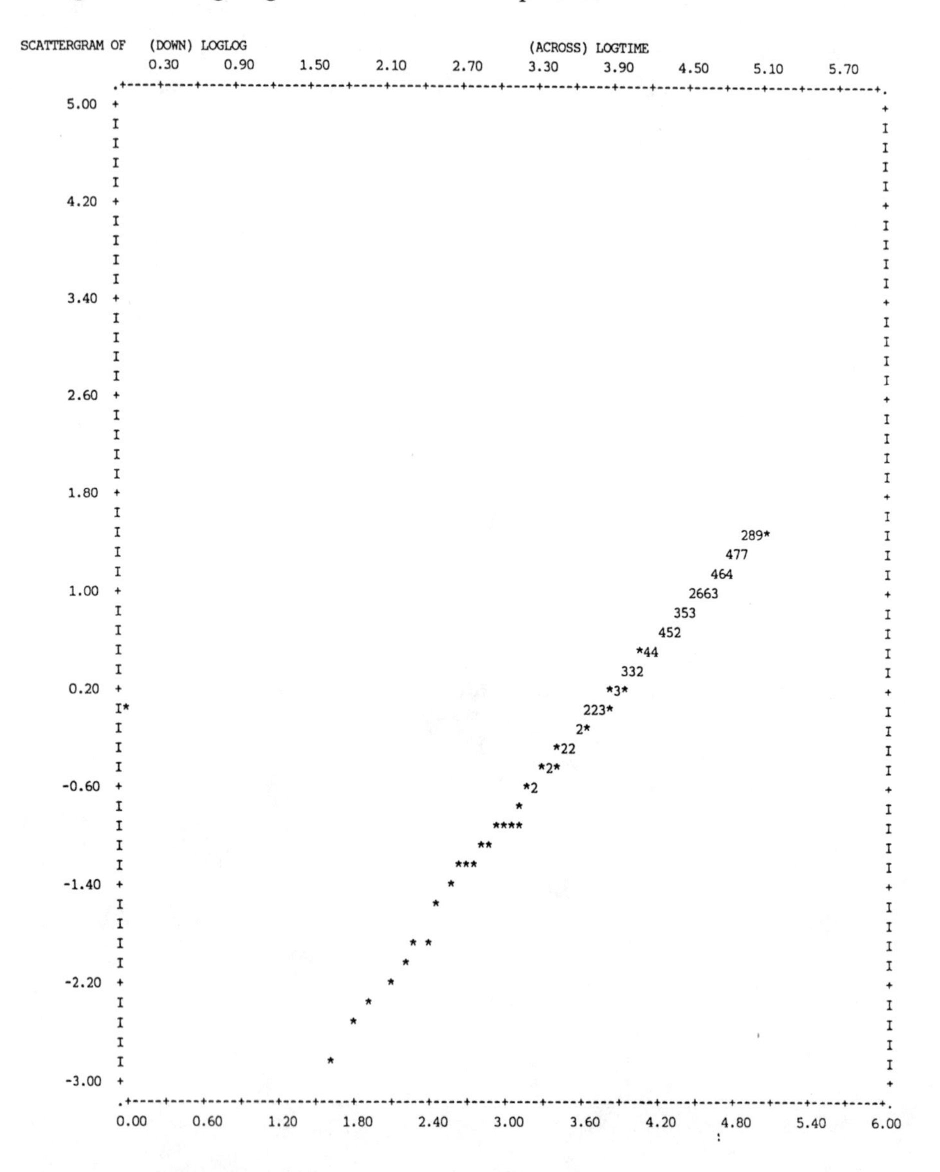

in Figure 6.4 with the "fits" of the Weibull and the Gompertz model, indicates that the log-logistic model is relatively good ($R^2 = 0.9068$).

Based upon a graphic examination of the distribution assumptions, a rather contradictory picture emerges in regard to the risk path of job change for men: (1) the exponential model resulted in the "best fit" (although the assumption $a \sim 0$ was not satisfied and a slightly falling path was observed) and led to a model with a constant rate; (2) the Weibull and Gompertz distributions revealed a "poorer fit," and both suggested a monotonically decreasing risk path; (3) finally, the log-logistic distribution, led to a "relatively good data fit," and to the assumption of a nonmonotonic risk.

Good explanations can be developed for each of these results. For this reason the method of graphic analysis of the distribution assumptions cannot be used as a final test in the process of finding the right model structure. Graphic analysis should, therefore, only be used to obtain preliminary information with regard to duration dependencies.

One also should not attempt to evaluate the different distribution models simply on the basis of the R^2 values obtained, since the least squares estimation assumption of homoscedasticity is not fulfilled and the residuals are correlated. Further, it is important to remember that we are dealing with an examination without controlling for covariates and that the implicit aggregation of the heterogeneous subgroup rates, even when these are constant over time, may result in apparent duration dependency (see Section 3.9.1).

When examining various distribution assumptions with the aid of graphic tests, it would be more appropriate to take the different values of the independent criteria, to estimate the survivor function in a nonstratified Cox model, and to transform and plot these estimates as shown above. One can indeed apply the subprogram P2L of BMDP to each constellation of covariates and plot the estimated survivor functions $S(t|\mathbf{x})$ using the pattern command of the PLOT paragraph. It is, however, not possible to use these estimates directly to perform the transformations needed for the graphic analysis.

Furthermore, even when GLIM is used to estimate and transform the survivor functions for various subgroups (requiring some lengthy calculations and a good working knowledge of GLIM), the result is a large number of curves which do not necessarily simplify the decision regarding the suitability of some specific parametric model. In this vein, it is better not to depend too much upon this type of graphic distribution test. Instead, graphic analysis should be followed-up with comparisons of means and medians, residual tests, and comparisons of different models (e.g., between parametric models and the Cox model, or in case of an exponential distribution between a Weibull model and a model with unobserved heterogeneity).

6.2 Models Without Duration Dependency of the Hazard Rate: The Exponential Model

The exponential model rests on the assumption of a constant event risk and thus implies the assumption of proportional risks over the duration. This model lends itself to easy interpretation and is commonly applied as the basic or reference model to which estimates of more complex distribution models are compared.

6.2.1 The Exponential Model Without Covariates

To illustrate the interpretation of the exponential model we will study the average risk of job change for men. A model without covariates, that is, a model with only one regression constant β_0, is formulated and estimated:

$$\lambda^k(v) = \exp(\beta_0), \qquad k = 1, 2, \ldots .$$

The maximum likelihood estimation is calculated with the aid of the P3RFUN subprogram written by Trond Petersen (1985). The source program is found in Appendix 2. It may easily be implemented in the subprogram P3R of BMDP. Above all, this program has the advantage that a series of parametric models (the exponential, Weibull, Gompertz, and log-logistic models) can be estimated using the widely distributed software package BMDP.

Program Example 6.3:

```
/INPUT UNIT IS 30.
       CODE IS DATA.
/VARIABLE NAMES ARE (63)DUR,(64)CEN,(65)EDU,(66)LFX,
                    (67)NOJ,(68)COHO2,(69)COHO3,(70)X1,(71)DP.
          ADD IS 9.
/TRANSFORM USE = (M3 EQ 1).
           DUR = M51 - M50 + 1.
           CEN = 1.
           IF (M51 EQ M47) THEN CEN = 0.
           IF (M41 EQ 1 AND M42 EQ 1) THEN EDU = 9.
           IF (M41 EQ 1 AND (M42 EQ 2 OR M42 EQ 3)) THEN EDU = 11.
           IF (M41 EQ 2 AND M42 EQ 1) THEN EDU = 10.
           IF (M41 EQ 2 AND (M42 EQ 2 OR M42 EQ 3)) THEN EDU = 12.
           IF (M41 EQ 3 AND (M42 EQ 1 OR M42 EQ 2 OR M42 EQ 3))
           THEN EDU = 13.
           IF (M42 EQ 4) THEN EDU = 17.
           IF (M42 EQ 5) THEN EDU = 19.
           COHO2 = 0.
           COHO3 = 0.
           IF(M48 GE 468 AND M48 LE 504) THEN COHO2 = 1.
           IF(M48 GE 588 AND M48 LE 624) THEN COHO3 = 1.
           LFX = M50 - M43.
           NOJ = M5 - 1.
           DP = 0.
           X1 = 0.
```

```
/REGRESS DEPENDENT IS DP.
         PARAMETERS ARE 1.
         PRINT IS 0.
         MEANSQUARE IS 1.0.
         ITERATIONS ARE 100.
         LOSS.
/PARAMETER INITIAL ARE -3.8.
           NAMES ARE CONST.
/END.
/COMMENT'
M 1
T 70 63
D 71
C 64
I  0
 '.
```

The first step, as demonstrated in Program Example 6.3 and as was the case for the examples of the Cox model, is to input the BMDP system file. In addition to listing the known variables, it is also necessary to define the variable X1 (the initial value of the duration in episode k) and DP (**dep**endent variable). Both of these variables are set to a value of 0 in the TRANSFORM paragraph.

In the REGRESSION paragraph for DEPENDENT IS one must state the name of the dependent variable, DP. This program estimates one parameter (namely, β_0) (PARAMETERS ARE 1). The command PRINT IS 0 suppresses a printout of the residuals and the observed and estimated values of the dependent variable. Implementing MEAN SQUARE IS 1.0 causes the value of the "residual mean square," for the estimation of the asymptotic standard deviation, to be replaced by the value 1. At the most, 100 iterations should be carried out (ITERATIONS ARE 100). If no convergence occurs after 100 iterations, the algorithm is stopped. In the maximum likelihood estimation (see Section 3.6.3) for the convergence calculations, the LOSS function ($\triangleq$ –(log likelihood)) is implemented. The Gauß-Newton algorithm of the program P3R thus attempts to find that set of parameters that maximizes the log likelihood function or minimizes the LOSS function. Remaining decisions about the iteration algorithm are discussed by Dixon et al. (1983).

After the END statement of the BMDP program, parameters must still be provided for the subprogram P3RFUN. These are read in as comment statements in the form /COMMENT '....' after the BMDP command. The first statement specifies the exponential **m**odel, which carries the number 1 (M1) in the program P3RFUN. The T statement (**T**ime) designates the BMDP internal numbers of the variables in which the initial and terminal values of the duration are stated. These are the number 70 for the variable X1 (duration begin) and the number 63 for the variable DUR (end of duration). Thereafter,

P3RFUN must also be provided with the BMDP internal number 71 of the **dep**endent variable DP. This is done using the D statement (**d**ependent). Censoring information contained in the variable CEN must be introduced in the C statement (**c**ensored) by the BMDP internal number 64. Since the model is to be estimated with only one regression constant β_0, the I statement (**i**ndependent) includes the number 0, no covariates.

The estimation results are presented in Table 6.2. The value of the log likelihood function ($\triangleq$ –LOSS) is –14434.35, and $\hat{\beta}_0 = -4.582$. The estimated average rate of changing jobs is then $\hat{\lambda} = \exp(-4.582) = 0.0102$. Comparison of this maximum likelihood estimate with the estimate gained using the graphic approach of Section 6.1 ($\hat{\lambda} = 0.0106$), reveals that both estimates are very similar.

Using the relationship (Section 3.2.2)

$$E(T) = \frac{1}{\lambda}$$

an estimate of the average duration of job employment for men is

$$\frac{1}{\hat{\lambda}} = \frac{1}{0.0102} = 98.04 \text{ months.}$$

That is, on the average somewhat more than eight years pass before a male changes his job.

Furthermore, with the aid of the relationship (3.2.17)

$$S(t) = \exp(-\lambda t),$$

the median duration M* with

$$S(M^*) = 0.5,$$

may be estimated as

$$S(\hat{M}^*) = \exp(-0.0102\ \hat{M}^*)$$
$$\hat{M}^* = 67.96 \text{ months.}$$

The median job duration is shorter than the mean of the duration on the job, since the waiting times for an exponential distribution are to the right skewed. The median equals 69.34 percent of the mean $[0.5 = \exp(\frac{1}{E(T)} M^*)]$.[3]

The probability that a man is still employed at the same job after an eight-year period is $\hat{S}(96) = \exp(-0.0102 \cdot 96) = 0.3756$, which is 37.56 percent.

---

[3]Since for an exponentially distributed duration, the number N of events during a specific time period is characterized by a Poisson distribution with an expected value

$$E(N) = \lambda t,$$

we forecast an annual average $\hat{\lambda} \cdot v = 0.0102 \cdot 12 = 0.1224$ of job changes for men. Or, in other words, within an eight-year time period we expect an average of one job change for each man.

Table 6.2: Results of the Exponential Model From Program Example 6.3

```
PARAMETER          ESTIMATE          ASYMPTOTIC
                               STANDARD DEVIATION


CONST              -4.581673           0.015316
```

On the other hand, the probability that he has left his job within the same time span is 62.44 percent. Naturally, the validity of the interpretation just discussed, is based upon the assumption that the risk is constant across the men in the sample. As has already been demonstrated in using the Cox model in Section 5.2, the covariates have a significant influence and should also be included in an exponential model.

6.2.2 The Exponential Model With Time-Constant Covariates

If, in estimating the model, additional covariates are considered, not only do the possibilities for interpretation improve, but the models also become more realistic. For this reason, in the following, the risk of job change among men is estimated as being dependent on the known variables education (EDU), prestige (M59), number of previously held jobs (NOJ), labor force experience at the beginning of each job (LFX), as well as the cohort dummies COHO2 and COHO3 (see Appendix 1):

$$\lambda^k(v|\mathbf{x}_k) = \exp(\mathbf{x}_k'\boldsymbol{\beta}), \qquad k = 1, 2, \ldots, .$$

The maximum likelihood estimation is once again calculated using the BMDP program P3R and Trond Petersen's integrated subprogram P3RFUN:[4]

---

[4]The following program example shows the estimation of this exponential model with the procedure LIFEREG of SAS. SAS estimates the exponential model as a "regression model" for the log of the duration ($\log v = -\boldsymbol{\beta}'\mathbf{x} + \omega$), where $\boldsymbol{\beta}'\mathbf{x} = \log\lambda$ and the disturbance ω has an extreme value distribution (see Kalbfleisch and Prentice, 1980, p. 21 and Section 3.2.2). The exponential distribution is treated as a Weibull distribution with the SCALE parameter restricted to the value 1:

```
LIBNAME SAS'[MAIR.MAIR.SAS]';
DATA EXP;
 SET SAS.SYS;
 IF M3 = 1;
 DUR = M51 - M50 + 1;
 CEN = 1;
 IF (M51 EQ M47) THEN CEN = 0;
 IF (M41 = 1 AND M42 = 1) THEN EDU = 9;
 IF (M41 = 1 AND (M42 = 2 OR M42 = 3)) THEN EDU = 11;
 IF (M41 = 2 AND M42 = 1) THEN EDU = 10;
 IF (M41 = 2 AND (M42 = 2 OR M42 = 3)) THEN EDU = 12;
```

Program Example 6.4:

```
/INPUT UNIT IS 30.
       CODE IS DATA.
/VARIABLE NAMES ARE (63)DUR,(64)CEN,(65)EDU,(66)LFX,
                    (67)NOJ,(68)COHO2,(69)COHO3,(70)X1,(71)DP.
          ADD IS 9.
/TRANSFORM USE = (M3 EQ 1).
           DUR = M51 - M50 + 1.
           CEN = 1.
           IF (M51 EQ M47) THEN CEN = 0.
           IF (M41 EQ 1 AND M42 EQ 1) THEN EDU = 9.
           IF (M41 EQ 1 AND (M42 EQ 2 OR M42 EQ 3)) THEN EDU = 11.
           IF (M41 EQ 2 AND M42 EQ 1) THEN EDU = 10.
           IF (M41 EQ 2 AND (M42 EQ 2 OR M42 EQ 3)) THEN EDU = 12.
           IF (M41 EQ 3 AND (M42 EQ 1 OR M42 EQ 2 OR M42 EQ 3))
           THEN EDU = 13.
           IF (M42 EQ 4) THEN EDU = 17.
           IF (M42 EQ 5) THEN EDU = 19.
           COHO2 = 0.
           COHO3 = 0.
           IF(M48 GE 468 AND M48 LE 504) THEN COHO2 = 1.
           IF(M48 GE 588 AND M48 LE 624) THEN COHO3 = 1.
           LFX = M50 - M43.
           NOJ = M5 - 1.
           DP = 0.
           X1 = 0.
/REGRESS DEPENDENT IS DP.
         PARAMETERS ARE 7.
         PRINT IS 0.
         MEANSQUARE IS 1.0.
         ITERATIONS ARE 100.
         LOSS.
/PARAMETER INITIAL ARE -3.8,0.04,-0.01,0.15,-0.01,0.1,0.17.
           NAMES ARE CONST,EDU,M59,NOJ,LFX,COHO2,COHO3.
```

Footnote 4 continued

```
 IF (M41 = 3 AND (M42 = 1 OR M42 = 2 OR M42 = 3)) THEN EDU = 13;
 IF (M42 = 4) THEN EDU = 17;
 IF (M42 = 5) THEN EDU = 19;
 COHO2 = 0;
 COHO3 = 0;
 IF (M48 GE 468 AND M48 LE 504) THEN COHO2 = 1;
 IF (M48 GE 588 AND M48 LE 624) THEN COHO3 = 1;
 LFX = M50 - M43;
 NOJ = M5 - 1;
PROC LIFEREG DATA=EXP;
 MODEL DUR*CEN(0)=EDU M59 NOJ LFX COHO2 COHO3/D=EXPONENTIAL;
  SET ALL;
```

```
/END.
/COMMENT'
M 1
T 70 63
D 71
C 64
I  6 65 59 67 66 68 69
 '.
```

The preparation of the variables in the TRANSFORM paragraph is the same as that described in detail in Program Examples 5.1 and 6.3. The commands included in Program Example 6.4 once again document that program runs to estimate the Cox model need only slight modifications for P3R.

In comparison to Program Example 6.3, however, 7 β coefficients are now estimated (PARAMETERS ARE 7). These are the regression constant β_0 and the 6 β coefficients of the independent variables. For each of the β coefficients an initial value is given in the paragraph PARAMETER INITIAL ARE. Here, these are based upon the Cox estimations presented in Chapter 5.

Compared to Program Example 6.3, the parameters of the subprogram P3RFUN remain unchanged except for the independent variables command (I). Here the number of covariates is included, in this case 6, followed by a list of the covariates according to their BMDP assigned numbers: 65 ($\triangleq$ EDU), 59 ($\triangleq$ M59, prestige), 67 ($\triangleq$ NOJ), 66 ($\triangleq$ LFX), 68 ($\triangleq$ COHO2), and 69 ($\triangleq$ COHO3). The results of this program run are presented in Table 6.3.

First of all, Table 6.3 provides a value of the log likelihood function ($\triangleq$ –LOSS) equal to –14081.40. If one compares this model with the model without covariates from Section 6.2.1, then based upon the likelihood ratio test (see Section 3.7.3) a χ^2 value of

$$Lq = 2(-14081.40 - (-14434.35)) = 705.90,$$

Table 6.3: Results of the Exponential Model From Program Example 6.4

PARAMETER	ESTIMATE	ASYMPTOTIC STANDARD DEVIATION
CONST	-4.337501	0.103741
EDU	0.012903	0.011682
M59	-0.005213	0.001116
NOJ	0.171426	0.008598
LFX	-0.008856	0.000369
COHO2	0.179615	0.035365
COHO3	0.486224	0.041957

with six degrees of freedom is obtained. The included covariates thus additionally contribute to explaining the risk of men changing jobs, and the null hypothesis which states that none of the additionally introduced β coefficients is different from zero must be rejected.

A significance test may be implemented for the individual regression parameters by dividing the coefficients $\hat{\beta}_i$ by their estimated asymptotic standard errors $s(\hat{\beta}_i)$. Assuming the hypothesis $H_0 : \beta_i = 0$, these test values are approximately characterized by a standard normal distribution (see Section 3.7.3). If one uses a 0.05 significance level and a two-sided test, then the covariates have a significant effect if the absolute value of their standardized coefficients satisfy

$$\left| \frac{\hat{\beta}_i}{s(\hat{\beta}_i)} \right| > 1.96.$$

This is the case for the constant β_0 (CONST) as well as for the variables M59 (prestige), NOJ, LFX, COHO2, and COHO3.[5] Only education (EDU) does not have a significant influence upon the rate of job change among men.

---

[5]Using the procedure LIFEREG of SAS, the following result is obtained:

```
                                   SAS

                       L I F E R E G   P R O C E D U R E

DATA SET           =WORK.EXP
DEPENDENT VARIABLE=DUR
CENSORING VARIABLE=CEN
CENSORING VALUE(S)=     0
NONCENSORED VALUES=2586  CENSORED VALUES= 930
OBSERVATIONS WITH MISSING VALUES= 330

LOGLIKELIHOOD FOR WEIBULL         -5483.11

VARIABLE   DF   ESTIMATE STD ERR CHISQUARE  PR>CHI LABEL/VALUE

INTERCPT    1     4.33789 0.128009  1148.35  0.0001 INTERCEPT
EDU         1 -0.0128712 .0146136 0.775754  0.3784
M59         1 0.00520771 .0014008  13.8207  0.0002
NOJ         1  -0.171315 .0123017  193.939  0.0001
LFX         1 0.00885129  4.7E-04  357.037  0.0001
COHO2       1  -0.179925 .0465685  14.9278  0.0001
COHO3       1  -0.486628 .0520417   87.436  0.0001
SCALE       0          1        0                  EXTREME VALUE SCALE PARAMETER
LAGRANGE MULTIPLIER CHI-SQUARE FOR SCALE       127.583 PR>CHI .0001
```

Given the parametrization in SAS [$\lambda = \exp(-\beta_0^* - \mathbf{x}'\boldsymbol{\beta}^*)$] (see (3.3.4)), the coefficients of the covariates (including the regression coefficient β_0^*) must be transformed as follows in order to obtain the parameters in Table 6.3:

$$\hat{\beta}_0 = \hat{\beta}_0^* \cdot (-1),$$
$$\hat{\boldsymbol{\beta}} = \hat{\boldsymbol{\beta}}^* \cdot (-1).$$

For example:

$$\hat{\beta}_0 = 4.33789 \cdot (-1) = -4.33789,$$
$$\hat{\beta}_1 = -0.0128712 \cdot (-1) = 0.0128712.$$

As in the Cox model (see Section 5.2), one can also easily interpret the influence of a covariate x_i, given the constancy of the remaining covariates $\mathbf{x}'\boldsymbol{\beta}$, by demonstrating the percentage change of the rate given an increase in the covariate x_i by a specific value Δx_i. For example, an increase in the number of previously held jobs (NOJ) of one unit results in an increase of the rate equivalent to 18.70 percent [$(\exp(0.171426) - 1) \cdot 100\% = 18.70\%$]. On the other hand, an increase in prestige (M59) of 20 units leads to a decline in the inclination to change jobs of about 9.9 percent [$(\exp(-0.00521)^{20} - 1) \cdot 100\% = -9.9\%$]. A simultaneous change of NOJ by one unit and prestige (M59) by 20 units, which would represent a career advance, raises the rate only by 6.95 percent [$(\exp(0.171426)^1 \cdot \exp(-0.00521)^{20} - 1) \cdot 100\% = 6.95\%$] and not of the magnitude of 8.80 percent [$18.70\% - 9.9\% = 8.80\%$].

Applying the relationship

$$E(T|\mathbf{x}) = \frac{1}{\lambda(\mathbf{x})} = \frac{1}{\exp(\mathbf{x}'\boldsymbol{\beta})},$$

given an exponential distribution, one may explicitly state, how, when all other covariates are constant, the average duration $E(T|\mathbf{x})$ changes if one increases the value of the independent variable x_i by the amount Δx_i:

$$\delta_{\Delta x_i} = \frac{\dfrac{1}{\exp(\mathbf{x}'\boldsymbol{\beta} + \beta_i (x_i + \Delta x_i))} - \dfrac{1}{\exp(\mathbf{x}'\boldsymbol{\beta} + \beta_i x_i)}}{\dfrac{1}{\exp(\mathbf{x}'\boldsymbol{\beta} + \beta_i x_i)}} \cdot 100\%$$

$$\delta_{\Delta x_i} = \left(\frac{1}{\exp(\beta_i)^{\Delta x_i}} - 1\right) \cdot 100\%.$$

Increasing the number of previously held jobs (NOJ) by one accordingly results in a decrease in average employment duration by 15.75 percent [$(1/\exp(0.171426) - 1) \cdot 100\% = -15.75\%$]. On the other hand, an increase in prestige (M59) by 20 units results in an increase in employment duration by 10.98 percent [$(1/\exp(-0.00521)^{20} - 1) \cdot 100\% = 10.98\%$]. A simultaneous change of NOJ by one unit and prestige (M59) by 20 units, which again would represent a career jump, lowers the average duration by about 6.5 percent [$(1/(\exp(0.171426)^1 \exp(-0.00521)^{20}) - 1) \cdot 100\% = -6.5\%$] rather than by 4.77 percent [$10.98\% - 15.75\% = -4.77\%$].

For selected subgroups, prognoses may also be made concerning the average duration, the median of the duration, the average number of events occurring in a given time period, and the probability of remaining in the same state at a given point in time.

For example, if one observes a man from the cohort 1939–1941 (COHO2 = 1 and COHO3 = 0) who is employed in an occupation valued at the 50-point prestige level (M59 = 50) and who previously worked in 10 jobs (NOJ = 10) as well as having collected 100 months of labor force experience (LFX = 100), the following forecast equation for the rate of job change is derived:[6]

$$\hat{\lambda} = \exp(-4.338 - 0.005 \cdot 50 + 0.171 \cdot 10 - 0.009 \cdot 100 + 0.180 \cdot 1)$$
$$= 0.0274.$$

Consequently, this person is characterized by an average duration of 36.5 months $[\frac{1}{\hat{\lambda}} = 1/0.0274 = 36.496]$, which is well below the average of 98.04 months. Furthermore, one can forecast the median duration of employment as $\hat{M}^* = 0.6934 \cdot 36.5 = 25.31$ months and in one year expect an average of $\hat{\lambda}v = 0.0274 \cdot 12 = 0.3288$ job changes. Finally, the probability that this individual is still employed in the same occupation after 8 years is 7.2 percent $[\hat{S}(96) = \exp(-0.0274 \cdot 96) = 0.072]$, whereas the total average for all men is 37.56 percent.

By means of further prognoses for additional subgroups, one obtains a well differentiated picture of job changing behavior among men as well as an indication of the significance of varying factors of influence.

6.2.3 Examination of the Residuals in the Exponential Model

The interpretation of the exponential models presented in Sections 6.2.1 and 6.2.2 is valid only if a constant risk of job change over the duration actually does exist. The fact that this supposition is justifiable has already been seen by the graphic analysis of the distribution assumptions discussed in Section 6.2.1. In that section, it was, however, also noted that it is necessary to implement additional criteria for evaluating the model. One further method of evaluating the model is, as in normal regression, to examine the residuals from the model.

As discussed in detail in Section 3.7.1, given the validity of the distribution assumptions, one can, in general, regard the cumulative hazard rate $\Lambda(t|x_i)$ as a residual r_i that is characterized by a standard exponential distribution (with $\lambda = 1$):

$$r_i = \Lambda(t|x_i) = -\ln S(t|x_i).$$

The survivor function of this exponentially distributed stochastic variable R is then

$$S(r|x) = \exp(-r)$$

and given the validity of the distributive assumption, a logarithmic transformation (see Section 4.4.2.1) results in a line with

$$y = \ln S(r|x), \ a = 0, \ b = -1, \ x = r.$$

---

[6]The educational variable (EDU) does not have to be included in the prognosis since its $\hat{\beta}$ coefficient is not significantly different from zero.

For cases with multiepisodes, one obtains an estimation of the residuals r_{ik} as follows:

$$\hat{r}_{ik} = \hat{\Lambda}\,(v_{ik}|x_{ik}) = \exp(x'_{ik}\hat{\boldsymbol{\beta}}) \cdot v_{ik}.$$

Based upon these calculated residuals and given censored data, the survivor function may be estimated with the aid of the product-limit method and, after taking logarithms, plotted against r. If the assumption of constant risks over the entire duration is valid, one should then obtain a line with the slope of –1.

The following two program runs with the subprogram P1L of BMDP demonstrate this residual test for the exponential model with and without covariates:

Program Example 6.5:

```
/INPUT UNIT IS 30.
       CODE IS DATA.
/VARIABLE NAMES ARE (63)DUR,(64)CEN,(65)EDU,(66)LFX,
                    (67)NOJ,(68)COHO2,(69)COHO3,(70)RESID1.
          ADD IS 8.
/TRANSFORM USE = (M3 EQ 1).
           DUR = M51 - M50 + 1.
           CEN = 1.
           IF (M51 EQ M47) THEN CEN = 0.
           IF (M41 EQ 1 AND M42 EQ 1) THEN EDU = 9.
           IF (M41 EQ 1 AND (M42 EQ 2 OR M42 EQ 3)) THEN EDU = 11.
           IF (M41 EQ 2 AND M42 EQ 1) THEN EDU = 10.
           IF (M41 EQ 2 AND (M42 EQ 2 OR M42 EQ 3)) THEN EDU = 12.
           IF (M41 EQ 3 AND (M42 EQ 1 OR M42 EQ 2 OR M42 EQ 3))
           THEN EDU = 13.
           IF (M42 EQ 4) THEN EDU = 17.
           IF (M42 EQ 5) THEN EDU = 19.
           COHO2 = 0.
           COHO3 = 0.
           IF(M48 GE 468 AND M48 LE 504) THEN COHO2 = 1.
           IF(M48 GE 588 AND M48 LE 624) THEN COHO3 = 1.
           LFX = M50 - M43.
           NOJ = M5 - 1.
           RESID1 = DUR * EXP(-4.581673).
/FORM TIME IS RESID1.
      STATUS IS CEN.
      RESPONSE IS 1.
/ESTIMATE METHOD IS PROD.
          PLOTS ARE LOG.
/PRINT CASES ARE 0.
/END
```

Program Example 6.6:

```
/INPUT UNIT IS 30.
       CODE IS DATA.
/VARIABLE NAMES ARE (63)DUR,(64)CEN,(65)EDU,(66)LFX,
                    (67)NOJ,(68)COHO2,(69)COHO3,(70)RESID2.
          ADD IS 8.
/TRANSFORM USE = (M3 EQ 1).
           DUR = M51 - M50 + 1.
           CEN = 1.
           IF (M51 EQ M47) THEN CEN = 0.
           IF (M41 EQ 1 AND M42 EQ 1) THEN EDU = 9.
           IF (M41 EQ 1 AND (M42 EQ 2 OR M42 EQ 3)) THEN EDU = 11.
           IF (M41 EQ 2 AND M42 EQ 1) THEN EDU = 10.
           IF (M41 EQ 2 AND (M42 EQ 2 OR M42 EQ 3)) THEN EDU = 12.
           IF (M41 EQ 3 AND (M42 EQ 1 OR M42 EQ 2 OR M42 EQ 3))
           THEN EDU = 13.
           IF (M42 EQ 4) THEN EDU = 17.
           IF (M42 EQ 5) THEN EDU = 19.
           COHO2 = 0.
           COHO3 = 0.
           IF(M48 GE 468 AND M48 LE 504) THEN COHO2 = 1.
           IF(M48 GE 588 AND M48 LE 624) THEN COHO3 = 1.
           LFX = M50 - M43.
           NOJ = M5 - 1.
           RESID2 = DUR * EXP(-4.337501+0.012903*EDU-0.00521*M59
                     +0.171426*NOJ-0.008856*LFX+0.179615*COHO2
                     +0.486224*COHO3).
/FORM TIME IS RESID2.
      STATUS IS CEN.
      RESPONSE IS 1.
/ESTIMATE METHOD IS PROD.
          PLOTS ARE LOG.
/PRINT CASES ARE 0.
/END
```

First of all, in the TRANSFORM paragraphs in the above program examples residuals (RESID1 or RESID2) are estimated based upon the β estimates of the exponential model from Tables 6.2 and 6.3. In the FORM paragraphs these are then introduced along with the censoring variables (STATUS IS CEN) as durations (TIME IS RESID1 or TIME IS RESID2). In estimating the survivor function, the product-limit method is applied each time (METHOD IS PROD), followed finally by the plotting of the logarithmic survivor functions (PLOTS ARE LOG).

Figure 6.5: Residual Plot for the Exponential Model Without Covariates (Only Regression Constant β_0)

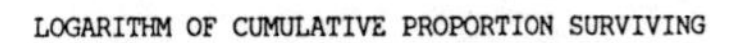

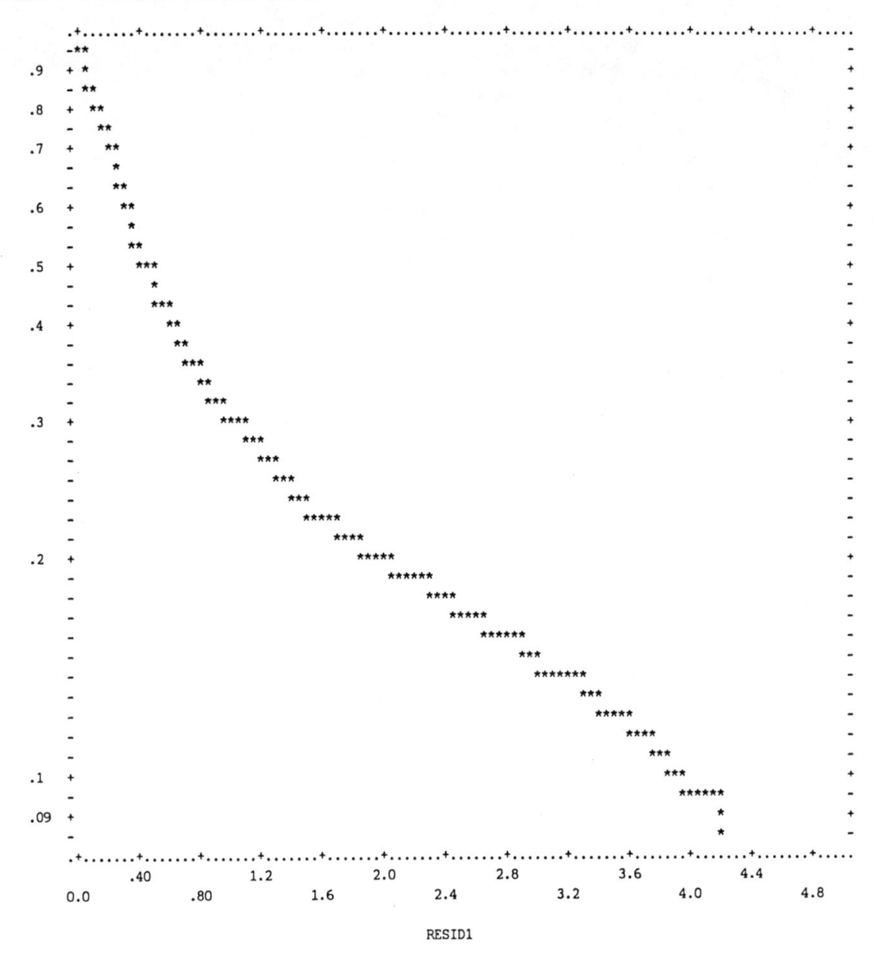

The results of these runs are presented in Figures 6.5 (for the model without covariates from Section 6.2.1) and 6.6 (for the model with covariates from Section 6.2.2).

Observing Figure 6.5, one first finds that the path of the transformed survivor function $\ln\hat{S}(r)$ clearly deviates from being linear. Especially for small residuals, the curve possesses a strongly downward curvature, which contradicts the assumption of an exponentially distributed duration with the rate $\exp(\beta_0)$.

Figure 6.6: Residual Plot for the Exponential Model With Covariates

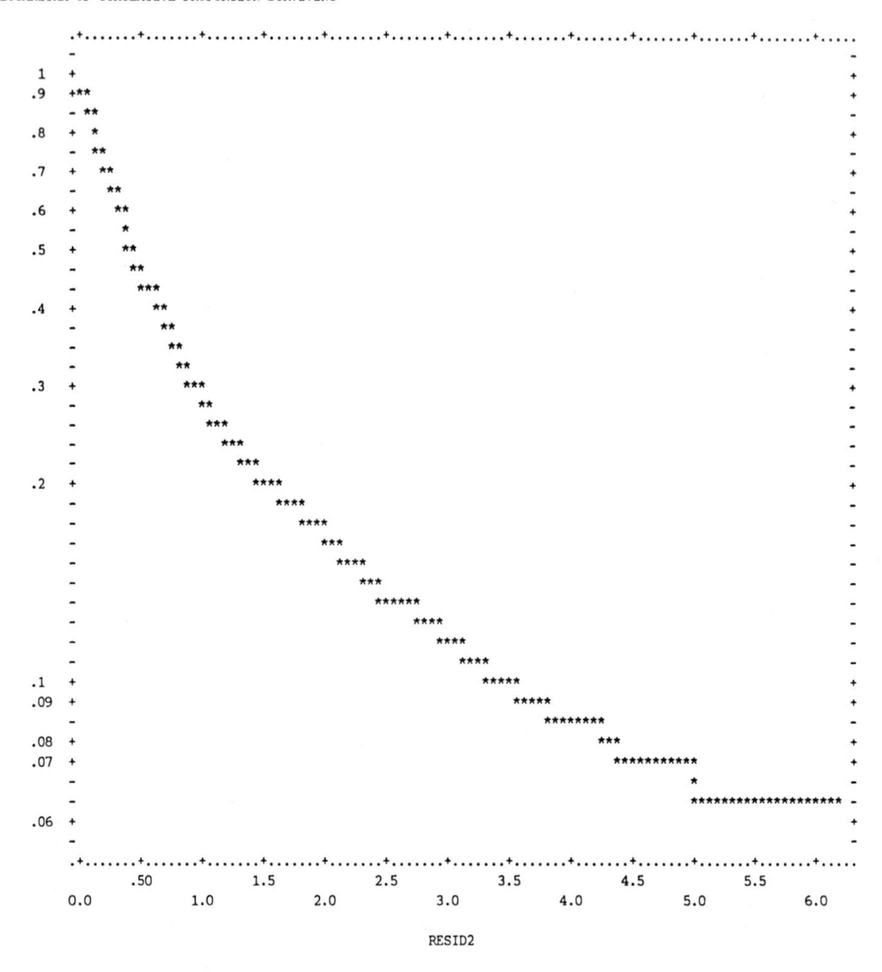

Also, for the exponential model with covariates the fit with a line of slope of –1 is not much better (Figure 6.6). The path of the transformed survivor function also clearly curves downward contradicting the notion of a constant rate over the duration.

Although, based upon these findings, one might be inclined not to accept the assumption of an exponentially distributed duration for jobs, one should consider that the above defined residuals are neither independent nor do they possess an identical distribution (see Section 3.7.1). Thus, the procedure just described should only be regarded as an approximation of a strict test. Further information regarding the question of whether or not the assumption of constant risks for job changes among men is right, may be obtained by an estimation of the Weibull model which contains the exponential distribution as a special case. However, before additional methods with specific duration dependencies are presented we will describe how time-dependent covariates may be included in parametric models.

6.3 Inclusion of Time-Dependent Covariates in Parametric Models

The importance of time-dependent covariates for event history analysis in economics and the social sciences has already been discussed in Section 5.4. Time-dependent covariates enable a more realistic formulation of the influence of covariates upon the hazard rate and also allow two or more parallel processes to be connected in the model.

Compared to partial likelihood models estimated with P2L of BMDP, parametric methods do not allow for the inclusion of time-dependent covariates with the help of a TIME variable (see Sections 5.4.1 and 5.4.2). In these cases, it is neccessary to find other solutions.

6.3.1 Method of Episode Splitting Given Discrete Time-Dependent Covariates

Including discrete time-dependent covariates within a maximum likelihood estimation procedure is relatively easy. These follow a step function over time and are constant over subepisodes (see Figure 5.2(b)). Designating $t_0 < t_1 < \ldots < t_s$ as the time points of change of the covariate vector within the duration interval $[0,t)$ and given $t_{s+1} = t$, then according to the discussion in Section 3.8 the cumulative hazard rate may be decomposed into an integral sum. The probability that no event occurs until the time point t is obtained from the product of the survivor functions of the subepisodes in which the covariate vector remains unchanged:

$$S(t|\mathbf{x}(t)) = \prod_{r=1}^{s+1} S(t_r|t_{r-1}, \mathbf{x}(t_{r-1})).$$

The maximum likelihood estimation is gained by splitting the observed duration t_i based upon the s_i time points of change into s_i+1 independent subepisodes. The hazard rate is estimated as in the case of time-constant covariates. Thus, for each of these subepisodes in which the covariate vector remains unchanged, the newly created data record contains the following information:

1) The values of the covariates at the beginning of each subepisode;
2) the duration at the beginning and end of a subepisode (the duration as such is sufficient only for the exponential model);
3) censoring information indicating whether the subepisode ends with an event (CEN = 1) or not (CEN = 0).

In order to illustrate the application of the method of episode splitting given discrete time-dependent covariates, we will once again refer to the marriage example presented in Section 5.4.1. Now, with the aid of the exponential model, we examine whether or not the event "marriage" has a stabilizing effect upon men's employment. This is estimated with the program RATE. In the RATE program, data transformations and data recordings are complicated,[7] so we recommend that the data first be prepared, for example, using the SPSS software package, and printed as a raw data file:

Program Example 6.7:

```
GET FILE
SELECT IF        (M3 EQ 1)
COMPUTE          DUR = M51 - M50 + 1
COMPUTE          CEN = 1
IF               (M51 EQ M47) CEN = 0
IF               (M41 EQ 1 AND M42 EQ 1) EDU = 9
IF               (M41 EQ 1 AND (M42 EQ 2 OR M42 EQ 3)) EDU = 11
IF               (M41 EQ 2 AND M42 EQ 1) EDU = 10
IF               (M41 EQ 2 AND (M42 EQ 2 OR M42 EQ 3)) EDU = 12
IF               (M41 EQ 3 AND (M42 EQ 1 OR M42 EQ 2 OR M42 EQ 3))
                 EDU = 13
IF               (M42 EQ 4) EDU = 17
IF               (M42 EQ 5) EDU = 19
COMPUTE          COHO2 = 0
COMPUTE          COHO3 = 0
IF               (M48 GE 468 AND M48 LE 504) COHO2 = 1
IF               (M48 GE 588 AND M48 LE 624) COHO3 = 1
COMPUTE          LFX = M50 - M43
COMPUTE          NOJ = M5 - 1
COMPUTE          TS = M50
COMPUTE          TF = M51 + 1
COMPUTE          PRES = M59
COMPUTE          JOBN = M61
COMPUTE          JOBN1 = M62
COMPUTE          TMAR = M49
WRITE CASES      (13F5.0) TS,TF,CEN,EDU,PRES,NOJ,LFX,
                 COHO2,COHO3,DUR,JOBN,JOBN1,TMAR
FINISH
```

[7]This no longer applies to the new version of RATE.

The output of the above SPSS program is an event oriented data file (Table 6.4) containing the variables TS (initial time point of the episode, measured in months from the beginning of the century), TF (the terminal time point of the episode, measured in months from the beginning of the century), CEN (censoring variable), EDU (education level, measured in number of school years), PRES (Wegener's prestige score), NOJ (number of previously held jobs), LFX (labor force experience at the time of episode entry, measured in months), COH2 (dummy variable for the cohort 1939–41), COH3 (dummy variable for the cohort 1949–51), DUR (duration in months), JOBN (occupational group of the job in the episode), JOBN1 (occupational group of the subsequent employment), and TMAR (time of marriage, measured in months from the beginning of the century) (see Appendix 1).

The first record of the file in Table 6.4 may, for example, be interpreted as follows: It states the first job episode (NOJ = 0, LFX = 0) of a man in the cohort 1929–31 (COH2 = 0, COH3 = 0) who after obtaining a lower secondary school degree with vocational training (EDU = 11) was employed in a skilled commercial or administrative job (JOBN = 11) (with a prestige score of 66 points, i.e., PRES = 66). The date of job entrance was March 1946 (TS = 555 months), and the man was employed in this position up to the time of the interview (CEN = 0, JOBN1 = 0) in November 1981 (TF = 983 months) without any interruptions (DUR = 428). Finally, the man married (TMAR = 679) in July 1956.

If one is interested in taking the family status (married, not married) into consideration as a time-dependent covariate to estimate the risk of job change for men in an exponential model, then the job episodes of the file presented in Table 6.4 can be split up according to the time of marriage. The new event oriented file (see Table 6.5) is prepared so that for each time interval within a given job episode in which the covariate family status remains unchanged, a

Table 6.4: Example of an Event Oriented Data File

TS	TF	CEN	EDU	PRES	NOJ	LFX	COH2	COH3	DUR	JOBN	JOBN1	TMAR
555	983	0	11	66	0	0	0	0	428	11	0	679
583	651	1	11	50	0	0	0	0	68	3	3	701
651	788	1	11	50	1	68	0	0	137	3	3	701
788	983	0	11	50	2	205	0	0	195	3	0	701
691	717	1	9	39	0	0	1	0	26	6	3	781
728	754	1	11	56	1	37	1	0	26	3	3	781
771	847	1	11	56	2	80	1	0	76	3	0	781
.	.	.	.	.	.	.	.	.	.	.	.	.
.	.	.	.	.	.	.	.	.	.	.	.	.
.	.	.	.	.	.	.	.	.	.	.	.	.

Table 6.5: Example of an Event Oriented Data Set Where the Episodes Have Been Split Into Subepisodes After the Time of Marriage

TS	TF	CEN	EDU	PRES	NOJ	LFX	COH2	COH3	DUR	JOBN	JOBN1	TMA	MAR
555	679	0	11	66	0	0	0	0	124	11	0	679	0
679	983	0	11	66	0	0	0	0	304	11	0	679	1
583	651	1	11	50	0	0	0	0	68	3	3	701	0
651	701	0	11	50	1	68	0	0	50	3	3	701	0
701	788	1	11	50	1	68	0	0	87	3	3	701	1
788	983	0	11	50	2	205	0	0	195	3	0	701	1
691	717	1	9	39	0	0	1	0	26	6	3	781	0
728	754	1	11	56	1	37	1	0	26	3	3	781	0
771	781	0	11	56	2	80	1	0	10	3	0	781	0
781	847	1	11	56	2	80	1	0	66	3	0	781	1
.	.	.	.	.	.	.	.	.	.	.	.	.	.
.	.	.	.	.	.	.	.	.	.	.	.	.	.
.	.	.	.	.	.	.	.	.	.	.	.	.	.

separate data record is produced. In these, the covariate values at the beginning of each subepisode (including the new dummy variable MAR) are stored, along with the actualized beginning (TS) and end time point (TF), the resulting duration (DUR)[8] and the censoring information indicating whether the subepisode ended with the event "job change" (CEN = 1) or not (CEN = 0).

Unfortunately, because it is not possible to carry out episode splitting with the file oriented software packages such as SPSS or BMDP, the solution to this problem is generally a data bank system (such as SIR or SAS) or software such as GLIM or even self-programming. In the example under consideration, the new event oriented file (see Table 6.5) was compiled with the aid of a FORTRAN program.[9]

In the new event history file (Table 6.5) the first episode from the file in Table 6.4 has been split into two subepisodes. The first subepisode begins with the entrance into the employment system in March 1946 (TS = 555) and lasts until the time of marriage in July 1956 (TF = 679, TMA = 679, MAR = 0, and CEN = 0). The second subepisode begins with the marriage (TS = 679, TMA = 679,

[8]Given an exponential model with a time-constant rate, it suffices to state the duration DUR, whereas for the Weibull, Gompertz, and log-logistic models, the initial value as well as the end value of the duration of a subepisode is needed. In the program P3RFUN, for example, the beginning of the duration may be included for estimation with the variable X1 and the end of the duration with the variable DUR (see Program Example 6.4). Also, the program RATE permits one to specify the initial and end time points of subepisodes using the T AND S command.

[9]This program may be found in Appendix 3.

MAR = 1) and terminates in November 1981 (TF = 983), the time of the interview (CEN = 0). All other job episodes have been dealt with accordingly. It should be noted that this episode splitting process changes nothing with regard to the duration of the specific state itself, although the number of records increases from 3516 in the file presented in Table 6.4 to 4268 in the file of Table 6.5.

The newly created event history data set which has split episodes according to the time of marriage (see Table 6.5), may now be treated in the same way as described for time-constant covariates. The data set is accessed by the program RATE to estimate the exponential model with the time-dependent marriage variable

$$\lambda^k(v|\mathbf{x}_k(v)) = \exp(\mathbf{x}_k'(v)\boldsymbol{\beta}), \qquad k = 1, 2, \dots .$$

Program Example 6.8:

```
RUN NAME        EXPONENTIAL-MODEL
N OF CASES      UNKNOWN
VARIABLES       14
TS              1
TF              2
ICEN            3
EDU             4
PRES            5
NOJ             6
ILFX            7
COHO2           8
COHO3           9
IDUR            10
JOBN            11
JOBN1           12
TMAR            13
MAR             14
READ DATA
(14F5.0)
T AND S         10 3
MODEL           (1) A=1
VECTOR          (1) 4 5 6 7 8 9 14
SOLVE
FINISH
```

In the RATE program example just presented (see Appendix 1), the RATE assigned number of the duration variables IDUR (≙ 10) and the censoring variable ICEN (≙ 3) are introduced on the T AND S command, (1) after stating the variables' names which are numbered successively and (2) after following the READ DATA and the FORMAT statement. Given the exponential model, we are then dealing with a type one RATE model in which the letter A is modeled in a log linear fashion with the covariate vector **x** (A = 1) : A = exp($\mathbf{x}'\boldsymbol{\beta}$). The variables specified on the VECTOR command with their RATE assigned numbers are included in the covariate vector **x**: 4 ≙ education (EDU), 5 ≙ prestige (PRES), 6 ≙ number of previously held jobs (NOJ), 7 ≙ labor force experience at the beginning of the episode (ILFX), 8 ≙ cohort dummy for those born in 1939–41 (COHO2), 9 ≙ cohort dummy for those born in 1949–51 (COHO3), and 14 ≙ family status (MAR). Finally, the SOLVE command starts the calculation of maximum likelihood estimates of the β coefficients (including the regression constant β_0). The results of this run are presented in Table 6.6 below.

Table 6.6: Results of the Exponential Model for Program Example 6.8

DESTINATION	UNWEIGHTED FREQUENCY	WEIGHTED FREQUENCY	PROPORTION OF ALL	PROPORTION OF ORIGIN	ESTIMATED RATE	LOG OF RATE	MAX(LOG OF L)
1	2586	2586.0	0.60449	0.60449	1.0237D-02	-4.5817D+00	-1.44343D+04
NO CHANGE	1692	1692.0	0.39551	0.39551			

DESTINATION	UNWEIGHTED FREQUENCY	WEIGHTED FREQUENCY	MAX(LOG OF L) NULL HYPOTHESIS	MAX(LOG OF L) ALTERNATIVE HYPOTHESIS	PSEUDO R-SQUARED	CHI-SQUARED	DF	PROBABILITY LEVEL
1	2586	2586.0	-1.443432D+04	-1.394939D+04	0.0336	969.87	7	0.00D+00

DESTINATION 1 LETTER A LOG-LINEAR TIME-INDEPENDENT VECTOR

INTERNAL NUMBER	VARIABLE NUMBER	VARIABLE NAME	VECTOR 1 PARAMETER	PARAMETER STANDARD ERROR	PARAMETER F RATIO	ANTILOG OF THE PARAMETER	ANTILOG STANDARD ERROR	ANTILOG F RATIO
1		(CONSTANT)	-4.283D+00	1.245D-01	1183.109	1.380D-02		
2	4	EDU	2.524D-02	1.441D-02	3.069	1.026D+00	1.478D-02	2.993
3	5	PRES	-4.088D-03	1.408D-03	8.426	9.959D-01	1.403D-03	8.461
4	6	NOJ	1.731D-01	1.211D-02	204.442	1.189D+00	1.440D-02	172.374
5	7	ILFX	-6.593D-03	4.766D-04	191.393	9.934D-01	4.734D-04	192.659
6	8	COHO2	1.558D-01	4.663D-02	11.163	1.169D+00	5.449D-02	9.572
7	9	COHO3	4.150D-01	5.231D-02	62.937	1.514D+00	7.921D-02	42.161
8	14	MAR	-7.141D-01	4.459D-02	256.438	4.897D-01	2.183D-02	546.344

In the first part of the RATE output in Table 6.6 we find that 2586 of the inserted subepisodes from the file presented in Table 6.5 ended with a job change and 1692 subepisodes were censored. This is equivalent to a censoring share of 39.55 percent.

For the RATE model without covariates the estimated rate $\hat{\lambda}$ (ESTIMATED RATE = 0.010237) and the logarithmic value, that is, the estimation for β_0 (LOG OF RATE = –4.5817) are given. This estimation is in accordance with the results obtained in Table 6.2 as is the log likelihood value of –14434.3. Through episode splitting, then, the model estimates have not been affected.

In a second step, the RATE program calculates for the model with covariates, the log likelihood value which is shown in the column MAX (LOG OF L) ALTERNATIVE HYPOTHESIS. It amounts to –13949.39. In comparing the model without covariates with the model possessing covariates, based upon a likelihood ratio test, one obtains—with seven degrees of freedom—a χ^2 value of 969.87 [Lq = 2(–13949.39 – (–14434.32)) = 969.87]. Thus, at least one of the covariates included makes a significant contribution.

In addition, RATE also calculates and prints the value of a PSEUDO-R^2 which is computed as:

$$\text{PSEUDO R-SQUARED} = 1 - \frac{\ln L(\text{present model})}{\ln L(\text{model without covariates})}$$

In the above example, the value of PSEUDO R-SQUARED = 1 – (–13943.39)/(–14434.32) = 0.0336. Unfortunately, PSEUDO-R^2 can not be interpreted in the same way as R^2 in a normal regression (i.e., as an index of explained variance). Rather PSEUDO-R^2 expresses the relative decrease of the log likelihood function of the present model as opposed to the model without covariates.

Finally, in the third part of the RATE output, the estimate $\hat{\beta}_i$ and its asymptotic standard error $s(\hat{\beta}_i)$ are given for each covariate and, in order to examine the level of significance, the resulting F values (with $d_1 = 1$ and $d_2 = 1$ degrees of freedom) are printed out as well. Since the root of an F-distributed stochastic variable (with $d_1 = 1$ and $d_2 = 1$) yields a t-distributed stochastic variable (with one degree of freedom) and this is (with a large sample size) approximately normally distributed, then in the above example the $\hat{\beta}$ coefficients, allowing for an error probability of 0.05, are significant if the corresponding value is larger than $1.96^2 = 3.84$. One observes that all of the covariates, with the exception of the variable education (EDU), are significantly different from zero.

Since the influence of these covariates upon the risk of job change among men has already been estimated with a Cox model (see Table 5.3), a comparison of the estimations of the above exponential model may be made.

First of all, the $\hat{\beta}$ coefficients in both models have the same sign as well as the same level of significance. The absolute size of the $\hat{\beta}$ coefficients is also approximately the same. Larger differences are found solely for the variables education (EDU), the dummy variable COHO3, and the variable MAR, whose $\hat{\beta}$ coefficients in the above exponential model are approximately double the size of those in the Cox model (Table 5.3). For the time-dependent covariate MAR, this implies that according to the exponential model, the rate of mobility of married men would decrease by some 51.03 percent [(0.4897 – 1) · 100% = –51.03%], whereas in the Cox model a decrease of only 27.44 percent would occur. The influence of the variable marriage (MAR) on the inclination to change jobs is thus significantly higher in the exponential model. Since both models presuppose proportional risks and the Cox model is the more comprehensive (the baseline hazard rate of the Cox model is constant given the validity of the exponential model), one would assume that actually constant rates are not given, but rather that the risk of job change in the above exponential model is falsely specified. In a following step, therefore, a continuous time-dependent covariate is introduced into the exponential model.

6.3.2 Method of Episode Splitting Given Continuous Time-Dependent Covariates

Whereas discrete time-dependent covariates can very simply be included in parametric rate models by splitting the original durations into subepisodes characterized by constancy, this simple way does not exist given continuous time-dependent covariates. If the continuous time-dependent covariates are assumed to be a function of the duration then a Weibull, a Gompertz-(Makeham), a log-logistic, a lognormal, or a gamma model may be used to do the estimation. Such methods will be discussed in detail in Sections 6.5.1 to 6.5.5.

If these parametric models are, however, not appropriate, then a solution may be derived by approximating the path of the cumulative hazard rate (see Section 3.8), by considering the continuous covariate to be constant over segments of the subepisodes, and proceeding as in the case of discrete time-dependent covariates (Blossfeld, 1986). The values of the continuous time-dependent covariates are calculated for some specifically stated points of time and the episodes are then artificially split at these time points. As shown in Figure 6.7, such an approximation is illustrated by a step function. The approximation is better the smaller the chosen interval and the more often the continuous time-dependent covariates are remeasured.

Labor force experience is used as an example for an approximation of the path of a continuous time-dependent covariate. Up until now labor force experience was considered to be constant over duration in the exponential model and was measured only at the beginning of each new job episode. If one assumes, as one has until now, that labor force experience rises linearly over the time a person spends in the employment system, then one may approximate labor force experience by the number of months spent in the employment system at the point of entry into each subepisode (see Figure 6.7(c)).

Table 6.7 demonstrates an event oriented data file in which the original job duration from the data in Table 6.4 has been artifically split into a maximum of 60 month-long subepisodes. Labor force experience is measured at the beginning of each subepisode.[10]

For example, in the event oriented file presented in Table 6.7 the first episode of the file presented in Table 6.4 has been split into eight subepisodes, in which the variable occupational experience (LFX) has been updated at each respective start.

It should be pointed out that through episode splitting the duration, in a state itself, does not change.

[10]The FORTRAN program used to prepare the data is presented in Appendix 4.

Figure 6.7: Modeling the Continuous Variable Labor Force Experience (a) With Regard to its Influence Upon the Job Path, (b) as a Time-Dependent Covariate, and (c) as an Approximation With the Aid of a Step Function

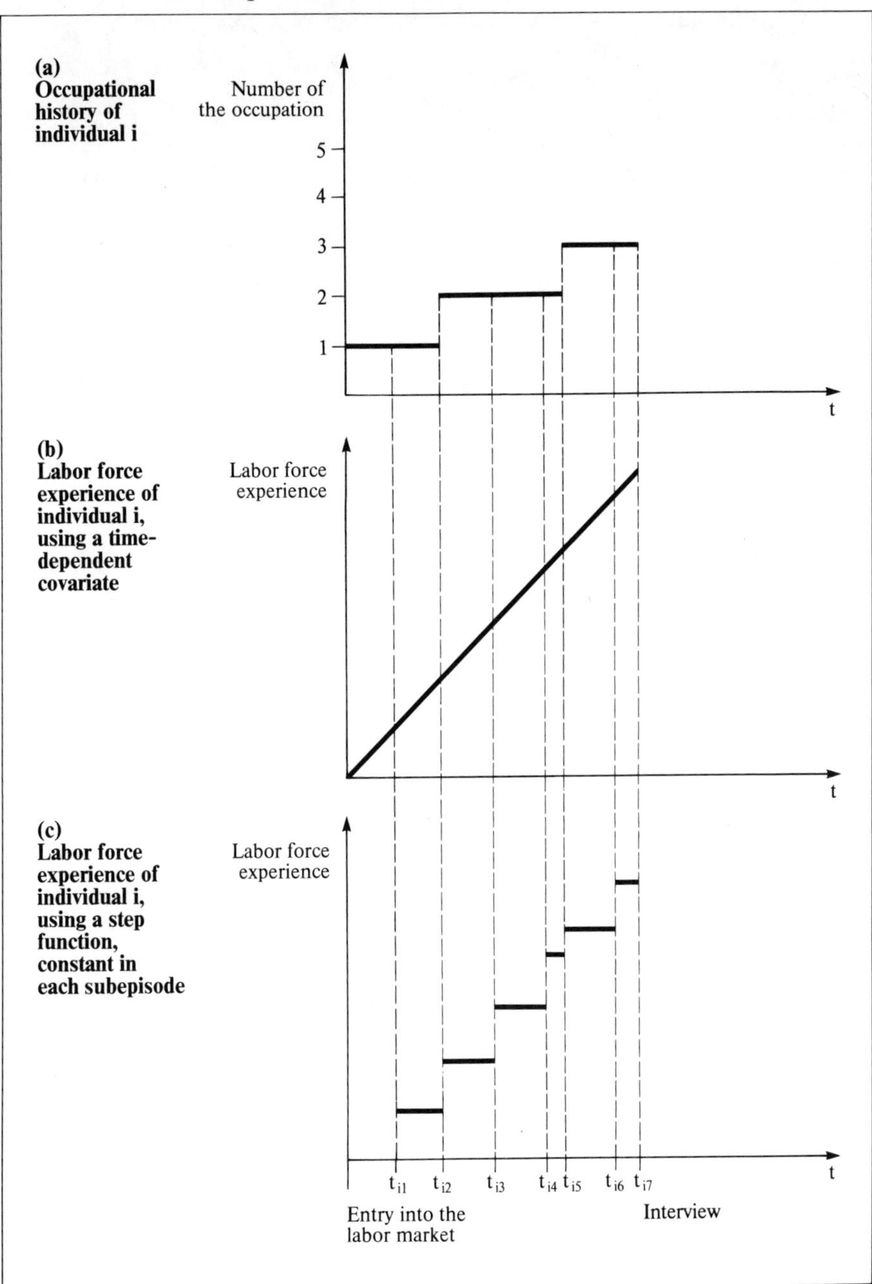

Table 6.7: Example of an Event Oriented Data File Where the Episodes Have Been Split Into a Maximum of 60 Month-Long Subepisodes

TS	TF	CEN	EDU	PRES	NOJ	LFX	COH2	COH3	DUR	JOBN	JOBN1	TMAR
555	615	0	11	66	0	0	0	0	60	11	0	679
615	675	0	11	66	0	60	0	0	60	11	0	679
675	735	0	11	66	0	120	0	0	60	11	0	679
735	795	0	11	66	0	180	0	0	60	11	0	679
795	855	0	11	66	0	240	0	0	60	11	0	679
855	915	0	11	66	0	300	0	0	60	11	0	679
915	975	0	11	66	0	360	0	0	60	11	0	679
975	983	0	11	66	0	420	0	0	8	11	0	679
583	643	0	11	50	0	0	0	0	60	3	3	701
643	651	1	11	50	0	60	0	0	8	3	3	701
651	711	0	11	50	1	68	0	0	60	3	3	701
711	771	0	11	50	1	128	0	0	60	3	3	701
771	788	1	11	50	1	188	0	0	17	3	3	701
788	848	0	11	50	2	205	0	0	60	3	0	701
848	908	0	11	50	2	265	0	0	60	3	0	701
908	968	0	11	50	2	325	0	0	60	3	0	701
968	983	0	11	50	2	385	0	0	15	3	0	701
691	717	1	9	39	0	0	1	0	26	6	3	781
728	754	1	11	56	1	37	1	0	26	3	3	781
771	831	0	11	56	2	80	1	0	60	3	0	781
.	.	.	.	.	.	.	.	.	.	.	.	.
.	.	.	.	.	.	.	.	.	.	.	.	.

In order to demonstrate how estimates behave when the maximal interval span of the subepisodes is reduced and the approximation of the time-dependent covariate labor force experience is improved, two further event oriented data files are created in which the episodes are divided into a maximal interval span of 24 months and 12 months. With the following RATE program (Program Example 6.9) an exponential model is estimated for each of these three event oriented input files with the variables education (EDU), prestige (PRES), number of previously held jobs (NOJ), labor force experience at the beginning of each subepisode (ILFX), as well as the cohort dummies COHO2 and COHO3:

$$\lambda^k(v \mid \mathbf{x}_k(v)) = \exp(\mathbf{x}_k'(v)\boldsymbol{\beta}), \qquad k = 1, 2, \dots .$$

Program Example 6.9:

```
RUN NAME        EXPONENTIAL-MODEL
N OF CASES      UNKNOWN
VARIABLES       13
TS              1
TF              2
ICEN            3
EDU             4
PRES            5
NOJ             6
ILFX            7
COHO2           8
COHO3           9
IDUR            10
JOBN            11
JOBN1           12
TMAR            13
READ DATA
(13F5.0)
T AND S         10 3
MODEL           (1) A=1
VECTOR          (1) 4 5 6 7 8 9
SOLVE
FINISH
```

The above RATE program is identical with the Program Example 6.8 with the exception that the variable MAR was not stated in the Program Example 6.9 nor included in the covariate vector. The explanations of the various commands in Program Example 6.8 thus also apply to Program Example 6.9. Results for the three RATE estimations for the maximal subepisode lengths of 60 months, 24 months, and 12 months are presented in Table 6.8.

First, one can compare the exponential model in which the continuous time-dependent covariate labor force experience is approximated based on the subepisodes with a maximal interval range of 60 months (see Table 6.8(a)), with the exponential model in which occupational experience is regarded as constant over the duration (see Table 6.3). In so doing it is clear that, compared to the model without covariates, a marked improvement in the χ^2 value from 705.9—given time-independent labor force experience—to a value of 1164.16—given time-dependent labor force experience—occurs. The dynamization of labor force experience can thus significantly increase the explanatory power of the model.

Table 6.8: Results of the Exponential Model From Program Example 6.9 (Occupational Episodes Split Into Maximal Interval Spans of 60, 24 and 12 Months)

(a) Maximal Interval Span of 60 Months

DESTINATION	UNWEIGHTED FREQUENCY	WEIGHTED FREQUENCY	MAX(LOG OF L) NULL HYPOTHESIS	MAX(LOG OF L) ALTERNATIVE HYPOTHESIS	PSEUDO R-SQUARED	CHI-SQUARED	DF	PROBABILITY LEVEL
1	2586	2586.0	-1.443432D+04	-1.385225D+04	0.0403	1164.16	6	0.00D+00

DESTINATION 1 LETTER A LOG-LINEAR TIME-INDEPENDENT VECTOR

INTERNAL NUMBER	VARIABLE NUMBER	VARIABLE NAME	VECTOR 1 PARAMETER	PARAMETER STANDARD ERROR	PARAMETER F RATIO	ANTILOG OF THE PARAMETER	ANTILOG STANDARD ERROR	ANTILOG F RATIO
1		(CONSTANT)	-3.813D+00	1.280D-01	887.078	2.209D-02		
2	4	EDU	4.505D-03	1.452D-02	0.096	1.005D+00	1.458D-02	0.096
3	5	PRES	-5.172D-03	1.376D-03	14.118	9.948D-01	1.369D-03	14.191
4	6	NOJ	1.503D-01	1.052D-02	204.019	1.162D+00	1.223D-02	175.883
5	7	ILFX	-8.367D-03	3.101D-04	727.855	9.917D-01	3.076D-04	733.975
6	8	COHO2	4.051D-02	4.661D-02	0.755	1.041D+00	4.854D-02	0.725
7	9	COHO3	1.763D-01	5.280D-02	11.146	1.193D+00	6.298D-02	9.369

(b) Maximal Interval Span of 24 Months

DESTINATION	UNWEIGHTED FREQUENCY	WEIGHTED FREQUENCY	MAX(LOG OF L) NULL HYPOTHESIS	MAX(LOG OF L) ALTERNATIVE HYPOTHESIS	PSEUDO R-SQUARED	CHI-SQUARED	DF	PROBABILITY LEVEL
1	2586	2586.0	-1.443432D+04	-1.383443D+04	0.0416	1199.79	6	0.00D+00

DESTINATION 1 LETTER A LOG-LINEAR TIME-INDEPENDENT VECTOR

INTERNAL NUMBER	VARIABLE NUMBER	VARIABLE NAME	VECTOR 1 PARAMETER	PARAMETER STANDARD ERROR	PARAMETER F RATIO	ANTILOG OF THE PARAMETER	ANTILOG STANDARD ERROR	ANTILOG F RATIO
1		(CONSTANT)	-3.698D+00	1.284D-01	828.997	2.477D-02		
2	4	EDU	3.255D-03	1.450D-02	0.050	1.003D+00	1.455D-02	0.050
3	5	PRES	-5.073D-03	1.374D-03	13.632	9.949D-01	1.367D-03	13.701
4	6	NOJ	1.472D-01	1.040D-02	200.415	1.159D+00	1.204D-02	173.300
5	7	ILFX	-8.227D-03	2.963D-04	770.761	9.918D-01	2.939D-04	777.133
6	8	COHO2	3.607D-02	4.662D-02	0.599	1.037D+00	4.833D-02	0.578
7	9	COHO3	1.583D-01	5.289D-02	8.957	1.172D+00	6.196D-02	7.662

(c) Maximal Interval Span of 12 Months

DESTINATION	UNWEIGHTED FREQUENCY	WEIGHTED FREQUENCY	MAX(LOG OF L) NULL HYPOTHESIS	MAX(LOG OF L) ALTERNATIVE HYPOTHESIS	PSEUDO R-SQUARED	CHI-SQUARED	DF	PROBABILITY LEVEL
1	2586	2586.0	-1.443432D+04	-1.383425D+04	0.0416	1200.16	6	0.00D+00

DESTINATION 1 LETTER A LOG-LINEAR TIME-INDEPENDENT VECTOR

INTERNAL NUMBER	VARIABLE NUMBER	VARIABLE NAME	VECTOR 1 PARAMETER	PARAMETER STANDARD ERROR	PARAMETER F RATIO	ANTILOG OF THE PARAMETER	ANTILOG STANDARD ERROR	ANTILOG F RATIO
1		(CONSTANT)	-3.656D+00	1.287D-01	807.335	2.583D-02		
2	4	EDU	3.057D-03	1.449D-02	0.045	1.003D+00	1.454D-02	0.044
3	5	PRES	-5.050D-03	1.374D-03	13.517	9.950D-01	1.367D-03	13.585
4	6	NOJ	1.461D-01	1.038D-02	198.144	1.157D+00	1.201D-02	171.512
5	7	ILFX	-8.179D-03	2.940D-04	773.943	9.919D-01	2.916D-04	780.303
6	8	COHO2	3.587D-02	4.662D-02	0.592	1.037D+00	4.832D-02	0.571
7	9	COHO3	1.573D-01	5.290D-02	8.844	1.170D+00	6.192D-02	7.572

Not only is the model as a whole significantly better, but the conclusions that can be drawn about the model change as well. In particular the significant effect related to the cohort variable COHO2 in Table 6.3 disappears. This means, that the difference between the 1929–31 and 1939–41 cohorts in Table

6.3 may have been due to differences in the length of labor force experience within jobs. The same is true for the difference between the 1929–31 and 1949–51 cohorts (see the $\hat{\beta}$ coefficients for COHO3) which, as observed in Table 6.8(a), is reduced substantially, but still remains significant. The remaining $\hat{\beta}$ coefficients, with the exception of the nonsignificant one for the variable education, remain largely unchanged.

In a second step one can then examine how the model behaves if one reduces the maximal interval span from 60 months to 24 months and finally to 12 months. First, the approximation of the continuous time-dependent covariate labor force experience steadily improves. Further, compared to the model without covariates and a maximal interval span of 60 months, the χ^2 value of 1164.16 improves only slightly to 1199.79 when the maximal interval span is 24 months. Comparing the individual $\hat{\beta}$ coefficients and F values, it is evident that the estimations are only slightly modified. Practically the same estimates are obtained when one compares the 24 month maximal interval span model with the 12 month maximal interval span model. Thus, we can conclude that given a maximal interval span of 24 months, one achieves relatively good approximations of the continuous time-dependent covariate labor force experience.

6.4 Models With Periodic Durations

In the examples presented in Section 6.3 it was assumed that the values of the covariates could vary over duration, but that the covariates themselves and their β coefficients remained unchanged. Such a formulation of risk of job change is not always appropriate, as the following example demonstrates. As presented in Section 3.3, an alternative method of modeling time-dependent hazard rates is to divide the duration into periods and to estimate the hazard rates with covariates and/or β coefficients varying from period to period.

In the case of examining men's job change behavior, we could, for example, argue along the lines of filter or signaling theory (see Arrow, 1973; Spence, 1973, 1974) that certain characteristics such as educational qualification, number of previously held jobs, and present labor force experience serve as signals for the employer. These characteristics allow an employer to infer differences in real productivity among applicants. When an employee is hired, then the employer gains more and more direct information on the employee and has less and less need to rely on indirect signals of competence and productivity. So the importance of such signals for job decisions should decline with increasing duration. On the other hand, it is also possible that the behavior of an employee, especially at the beginning of each new job, depends strongly upon the image (or prestige) connected with the respective job and that this image slowly adjusts to the actual situation at hand. Both arguments lead to the supposition that the effect of certain fixed variables in the form of their β coefficients over the duration is not constant.

In the present example it is therefore sensible to divide the job duration into periods and to let the β coefficients of the covariates vary over these periods. Here, the problem of deciding how many periods to be modeled and what length these periods should be presents itself. On the one hand, as has already been noted in Section 3.3, the period model leads with a moderate number of periods to a large number of parameters to be estimated which accordingly require large data sets. It is, therefore, important not to choose too many periods when constructing models. On the other hand the choice of periods should be realistic and not lead to erroneous specifications. Thus, in estimating job change risk we decided to choose three periods: a period from entry into each new job up to a duration of two years (24 months), a second period from two years to five years (60 months), and a third period beginning after five years. It is to be expected that the risk of a job change is especially high in the first period due to the varying adjustments of employee and employer expectations. The "filter variables" should also be very important in this period. In the second period, the influence of these variables should decline as both sides become increasingly oriented to actual relationships. Finally, in the third period these variables should be expected to no longer play an important role for the risk of job change.

The estimation of this period specific exponential model

$$\lambda_p^k(v|x_k) = \exp(x_k'\beta_p) \qquad \text{with } p = 1, 2, 3, k = 1, 2, \ldots$$

is done with the following RATE program run:

Program Example 6.10:

```
RUN NAME         MODEL WITH PERIODIC DURATION
N OF CASES       3516
VARIABLES        12
TS               1
TF               2
CEN              3
EDU              4
PRES             5
NOJ              6
LFX              7
COHO2            8
COHO3            9
DUR              10
JOBN             11
JOBN1            12
READ DATA
(12F5.0)
TIME INTERVALS   0.0 24.0 60.0
```

```
T AND S          10 3
MODEL            (4) A=2 B=0 C=0
VECTOR           (1) 4 5 6 7 8 9
SOLVE            (2)=30
FINISH
```

As was the case in Program Examples 6.8 and 6.9, the RATE run given above first lists the variables with their RATE assigned numbers and inserts the stated format. The TIME INTERVALS command then specifies the interval length for the periods, namely 0, 24, and 60 months. On the MODEL statement the model is specified with the number (4). The letters B and C are set to zero and the letter A is related period specifically and log-linearly with the covariate vectors (A = 2) : $A = \exp(\mathbf{x}_k'\boldsymbol{\beta}_p)$.[11] Since only one vector (1) is specified on the VECTOR command, in all three periods the same variables are automatically included in the period specific covariate vectors. The variables are the same as those already discussed in Program Examples 6.8 and 6.9. The SOLVE command, which initiates the maximum likelihood estimation of the period specific coefficients, contains the command (2) = 30 which sets the maximal number of iterations through to 30. The results obtained for this RATE program are found in Table 6.9.

Based on the likelihood ratio test for the period specific exponential model in Table 6.9 (compared to the exponential model without covariates and only one regression constant β_0), a χ^2 value of 1163.79 is obtained. Now, however, there are 20 degrees of freedom since for each of these three periods 7 β parameters have been estimated. The model is significant (PROBABILITY LEVEL = 0.00), and the null hypothesis that none of the 20 additionally introduced regression parameters explain something about the risk of job change for men must be rejected.

The results of the first period (0 to 24 months) show (Table 6.9(a)), that the variables education (EDU) and COHO2 do not have any significant influence (both F test statistics are smaller than the value 3.84 ($\triangleq$ significance level 0.05)). Contrary to our original hypothesis, level of education is not relevant in explaining job change risk even in the first phase of each new job activity. The situation is entirely different for the variables prestige (PRES), number of previously held jobs (NOJ), and labor force experience (LFX). A comparison of their $\hat{\beta}$ coefficients with the corresponding $\hat{\beta}$ coefficients of the simple exponential model presented in Table 6.3 clearly shows, that at the beginning of each new employment phase the influence of these variables is stronger than the average: prestige (–0.0068 compared to –0.0052), number of previously held jobs (0.1950 compared to 0.1714), and labor force experience (–0.0091

[11] With an additional linear and period specific specification of the letter C (C = –2) : $C = \mathbf{x}_k'\boldsymbol{\beta}_p$ one could, however, also model a period specific Gompertz model using RATE.

Table 6.9: Results of the Exponential Model With Periodic Duration From Program Example 6.10

DESTINATION	UNWEIGHTED FREQUENCY	WEIGHTED FREQUENCY	MAX(LOG OF L) NULL HYPOTHESIS	MAX(LOG OF L) ALTERNATIVE HYPOTHESIS	PSEUDO R-SQUARED	CHI-SQUARED	DF	PROBABILITY LEVEL
1	2586	2586.0	-1.443432D+04	-1.385243D+04	0.0403	1163.79	20	0.00D+00

(a) Period 0–24 Months

DESTINATION 1 LETTER A LOG-LINEAR TIME-DEPENDENT VECTOR, PERIOD 1

INTERNAL NUMBER	VARIABLE NUMBER	VARIABLE NAME	VECTOR 1 PARAMETER	PARAMETER STANDARD ERROR	PARAMETER F RATIO	ANTILOG OF THE PARAMETER	ANTILOG STANDARD ERROR	ANTILOG F RATIO
1		(CONSTANT)	-3.845D+00	1.877D-01	419.460	2.140D-02		
2	4	EDU	1.742D-02	2.133D-02	0.667	1.018D+00	2.170D-02	0.656
3	5	PRES	-6.779D-03	2.090D-03	10.524	9.932D-01	2.076D-03	10.596
4	6	NOJ	1.950D-01	1.582D-02	152.043	1.215D+00	1.922D-02	125.501
5	7	LFX	-9.060D-03	6.972D-04	168.866	9.910D-01	6.909D-04	170.404
6	8	COHO2	6.601D-02	7.384D-02	0.799	1.068D+00	7.888D-02	0.748
7	9	COHO3	2.160D-01	7.675D-02	7.919	1.241D+00	9.525D-02	6.406

(b) Period 24–60 Months

DESTINATION 1 LETTER A LOG-LINEAR TIME-DEPENDENT VECTOR, PERIOD 2

INTERNAL NUMBER	VARIABLE NUMBER	VARIABLE NAME	VECTOR 2 PARAMETER	PARAMETER STANDARD ERROR	PARAMETER F RATIO	ANTILOG OF THE PARAMETER	ANTILOG STANDARD ERROR	ANTILOG F RATIO
8		(CONSTANT)	-4.061D+00	2.229D-01	331.862	1.723D-02		
9	4	EDU	1.130D-02	2.569D-02	0.194	1.011D+00	2.598D-02	0.191
10	5	PRES	-3.793D-03	2.428D-03	2.440	9.962D-01	2.419D-03	2.450
11	6	NOJ	1.105D-01	2.526D-02	19.139	1.117D+00	2.821D-02	17.154
12	7	LFX	-8.332D-03	8.340D-04	99.814	9.917D-01	8.271D-04	100.650
13	8	COHO2	1.591D-01	8.535D-02	3.473	1.172D+00	1.001D-01	2.968
14	9	COHO3	2.931D-01	9.305D-02	9.918	1.341D+00	1.247D-01	7.452

(c) Period More Than 60 Months

DESTINATION 1 LETTER A LOG-LINEAR TIME-DEPENDENT VECTOR, PERIOD 3

INTERNAL NUMBER	VARIABLE NUMBER	VARIABLE NAME	VECTOR 3 PARAMETER	PARAMETER STANDARD ERROR	PARAMETER F RATIO	ANTILOG OF THE PARAMETER	ANTILOG STANDARD ERROR	ANTILOG F RATIO
15		(CONSTANT)	-4.494D+00	2.757D-01	265.719	1.117D-02		
16	4	EDU	-2.447D-02	3.123D-02	0.614	9.758D-01	3.047D-02	0.629
17	5	PRES	-2.972D-03	2.831D-03	1.102	9.970D-01	2.822D-03	1.106
18	6	NOJ	8.782D-02	3.119D-02	7.927	1.092D+00	3.405D-02	7.265
19	7	LFX	-6.876D-03	9.082D-04	57.309	9.931D-01	9.020D-04	57.705
20	8	COHO2	4.801D-02	8.671D-02	0.307	1.049D+00	9.097D-02	0.292
21	9	COHO3	1.687D-01	1.326D-01	1.620	1.184D+00	1.569D-01	1.372

compared to –0.0084). This speaks in defense of the hypothesis that these variables, especially at the beginning of each new job episode, are very important, thus supporting the "filter" or "signal" hypothesis.

This result becomes even more definitive if one compares the absolute and standardized $\hat{\beta}$ coefficients of these variables over the three periods. For example, after the first period, the variable prestige no longer has a significant effect and the effect of the significant variables NOJ and LFX decreases from period to period: $\hat{\beta}^{P1}_{NOJ} = 0.1950$, $\hat{\beta}^{P2}_{NOJ} = 0.1105$ and $\hat{\beta}^{P3}_{NOJ} = 0.0878$; furthermore, $\hat{\beta}^{P1}_{LFX} = -0.0091$, $\hat{\beta}^{P2}_{LFX} = -0.0083$ and $\hat{\beta}^{P3}_{LFX} = -0.0069$. This also supports the hypothesis that, given increasing duration in a job, behavior becomes more and more oriented to the actual situation, and due to this, the explanatory power of the "filter" variables of job change behavior decreases.

In a second step, all covariates in Table 6.9 which, in the three periods, were not found to be significant should be eliminated from the model. This can easily be realized using RATE,[12] although we do not present a further detailed discussion at this point.

In general, the possibility of permitting a period-specific varying hazard rate offers an additional instrument with which stochastic processes in economics and the social sciences may be modeled more realistically. The hazard rate may be analyzed with a period specific rate model, where the standard methods of duration dependency may fail. The disadvantage of this type of model is, however, that with a rising number of periods and covariates the number of parameters needed to be estimated rapidly increases. Even for a moderate number of periods this requires not only a very large data set, but also means that the complexity of results increases. If at all possible, one should therefore attempt to apply a parametric duration model and to explain the process with a limited number of parameters.

6.5 Models With Duration Dependency of the Hazard Rate: The Gompertz-(Makeham), Weibull, Log-Logistic, and Lognormal Model

In modeling the duration dependency of parametric models an important question arises: How should this time dependency be interpreted? For the example of occupational experience, one should be aware that given the exponential model in Table 6.8, not only the continuous time-dependent covariate labor force experience, but a special Gompertz model was approximated at the same time, using a step function. If one assumes, as in this example, that labor force experience increases linearly over the time which a person i spends in the employment system,

$$x_{LFX}(t) = x_{LFX}(t_{i,k-1}) + v \qquad \text{, with } v = t - t_{i,k-1} \; (v \geq 0),$$

and includes this dynamic labor force experience variable $x_{LFX}(t)$ in an exponential model, then the model may be reformulated as follows:

$$\begin{aligned} \lambda^k(v \mid \mathbf{x}_k(t)) &= \exp(\mathbf{x}_k'(t)\boldsymbol{\beta}) \\ &= \exp(\mathbf{x}_k'\boldsymbol{\beta} + \beta_{LFX}\, x_{LFX}(t)) \\ &= \exp(\mathbf{x}_k'\boldsymbol{\beta} + \beta_{LFX}\, (x_{LFX}(t_{k-1}) + v)) \\ &= \exp(\mathbf{x}_k'\boldsymbol{\beta} + \beta_{LFX}\, x_{LFX}(t_{k-1}))\, \exp(\beta_{LFX}\, v). \end{aligned}$$

---

[12]For the above example one only includes on the VECTOR card the respective significant variables: VECTOR (1) 5 6 7 9 (2) 6 7 9 (3) 6 7.

Here $x_{LFX}(t_{k-1})$ designates the labor force experience at the beginning of each new job episode k and v represents the duration in episode k. This is, however, according to (3.3.15) a Gompertz model with $\lambda_0(\mathbf{x}_k) = \exp(\mathbf{x}_k'\boldsymbol{\beta} + \beta_{LFX} x_{LFX}(t_{k-1}))$ and $\gamma_0 = \beta_{LFX}$ (see Section 6.5.1):

$$\lambda^k(v) = \lambda_0(\mathbf{x}_k) \exp(\gamma_0 \cdot v).$$

Accordingly, a Weibull model would have been obtained if one had had good reason to assume that labor force experience of an individual i would not increase linearly, but rather that it would increase logarithmically over the time t which the individual spends in the employment system:

$$x_{LFX}(t) = x_{LFX}(t_{i,k-1}) + \ln v \qquad \text{, with } v = t - t_{i,k-1} \ (v \geq 0).$$

The exponential model may then be written as follows:

$$\lambda^k(v|\mathbf{x}_k(t)) = \exp(\mathbf{x}_k'\boldsymbol{\beta} + \beta_{LFX} x_{LFX}(t_{k-1}))\, v^{\beta_{LFX}}.$$

According to (3.2.19) with $\lambda^*(\mathbf{x}_k) = \exp(\mathbf{x}_k'\boldsymbol{\beta} + \beta_{LFX} x_{LFX}(t_{k-1}))$ as well as $\gamma^* = \beta_{LFX}$ one obtains the following special Weibull model (see Section 6.5.2):

$$\lambda^k(v) = \lambda^*(\mathbf{x}_k)\, v^{\gamma^*}.$$

In both cases, based upon substantive hypotheses and with the help of a continuous time-dependent covariate, one would directly arrive at a substantive interpretation of the time dependency.

Unfortunately, we are not always in the position of being able to measure exactly the change of continuous time-dependent covariates. It is more often the case that one has to use proxy variables in modeling the effects of these variables.[13]

One of these proxy variables may be the duration in a state. As a countinuous time-dependent covariate it may affect the hazard rate through specific distribution models. Depending on substantive considerations one has, therefore, to choose a specific distribution model. In principle, any arbitrary probability distribution can be used if it is appropriate to characterize the duration dependency. Based upon collective research experience from the areas of medicine, biology, demography, technology, psychology, as well as economics and the social sciences, it has been found that it is possible to limit oneself to a few distribution models. In order to model the covariates, which lead to a monotonically decreasing or increasing path of the hazard rate, the Gompertz-(Makeham) (see Figure 3.8) or the Weibull distribution (see Figure 3.5) may be applied. These functions are the ones which are probably most frequently modeled. For those influence factors which first cause a monotonically increasing and then later after a specific point lead to a monotonically

---

[13]In fact, in the above examples labor force experience has been introduced into the analysis through the proxy variable "number of years in the employment system" (see also Tuma, 1985, p. 340).

decreasing path of the hazard rate, the log-logistic or the lognormal model is recommended (see Figure 3.7). The following examples and interpretations are thus restricted to these distribution types.

6.5.1 Gompertz-(Makeham) Model

As already described in Section 3.2, the Gompertz-(Makeham) distribution has been widely used in demographic and insurance studies. It occurs, for example, if many causes of death exist (all independent from each other) and of these, the one most likely to occur actually ends the life.

The Gompertz-Makeham law, which states that the hazard rate decreases monotonically with increasing duration, has been effectively applied especially in studying the lifetime of organizations (see Carroll and Delacroix, 1982; Freeman, Carroll, and Hannan, 1983) and in analyzing job mobility processes (Sørensen and Tuma, 1981; Sørensen, 1979; Blossfeld and Mayer, 1988).

Gompertz Model Without Covariates

In order to demonstrate the interpretation of the Gompertz distribution a model without covariates is estimated, once again based on the multiepisode case for the average job change risk of men. In so doing the λ_0 coefficient of the Gompertz distribution is related to the regression constant β_0 log-linearly and γ_0 is estimated directly:

$$\lambda^k(v) = \exp(\beta_0) \cdot \exp(\gamma_0 v) \quad , k = 1, 2, \ldots .$$

In this example, the duration v is regarded as a proxy variable for the newly acquired job specific abilities in each new job. This implies that the job specific experience begins—as the duration itself—with each new job at zero and grows linearly with the time spent on the job. Since the acquisition of these job specific knowledge and abilities is understood to be an investment which incurs costs that are largely lost if a job change occurs, we hypothesize that with increasing job specific experience the hazard rate sinks monotonically. This suggests that given a Gompertz distribution, validity of the theory requires a significant $\hat{\gamma}_0$ with a negative sign (see Figure 3.8).

The estimation of this model is calculated again with the BMDP program P3R and the subprogram P3RFUN:

Program Example 6.11:

```
/INPUT UNIT IS 30.
       CODE IS DATA.
/VARIABLE NAMES ARE (63)DUR,(64)CEN,(65)EDU,(66)LFX,
                    (67)NOJ,(68)COHO2,(69)COHO3,(70)X1,(71)DP.
          ADD IS 9.
```

```
/TRANSFORM USE = (M3 EQ 1).
           DUR = M51 - M50 + 1.
           CEN = 1.
           IF (M51 EQ M47) THEN CEN = 0.
           IF (M41 EQ 1 AND M42 EQ 1) THEN EDU = 9.
           IF (M41 EQ 1 AND (M42 EQ 2 OR M42 EQ 3)) THEN EDU = 11.
           IF (M41 EQ 2 AND M42 EQ 1) THEN EDU = 10.
           IF (M41 EQ 2 AND (M42 EQ 2 OR M42 EQ 3)) THEN EDU = 12.
           IF (M41 EQ 3 AND (M42 EQ 1 OR M42 EQ 2 OR M42 EQ 3))
           THEN EDU = 13.
           IF (M42 EQ 4) THEN EDU = 17.
           IF (M42 EQ 5) THEN EDU = 19.
           COHO2 = 0.
           COHO3 = 0.
           IF(M48 GE 468 AND M48 LE 504) THEN COHO2 = 1.
           IF(M48 GE 588 AND M48 LE 624) THEN COHO3 = 1.
           LFX = M50 - M43.
           NOJ = M5 - 1.
           DP = 0.
           X1 = 0.
/REGRESS DEPENDENT IS DP.
         PARAMETERS ARE 2.
         PRINT IS 0.
         MEANSQUARE IS 1.0.
         ITERATIONS ARE 100.
         LOSS.
/PARAMETER INITIAL ARE -3.8,-0.01.
           NAMES ARE CONST,TDEP.
/END.
/COMMENT'
M 2
T 70 63
D 71
C 64
I 0
 '.
```

The BMDP program above differs from Program Example 6.3 only in that now for PARAMETERS ARE the number 2 is inserted for the two parameters to be estimated and for the INITIAL ARE command a starting value is given (TDEP) for the γ_0 parameter (see Appendix 1). Also, the control commands for the subprogram P3RFUN, which come after the END statement, change only slightly by modifying the model card M in which the number 2 for the

Gompertz model is now inserted. The results of the estimation are presented below in Table 6.10.

Table 6.10: Results of the Gompertz Model From Program Example 6.11

PARAMETER	ESTIMATE	ASYMPTOTIC STANDARD DEVIATION
CONST	-4.080915	0.026859
TDEP	-0.008132	0.000393

According to Table 6.10, given a value of the log likelihood function ($\triangleq$-LOSS) of -14147.10 the estimations $\hat{\beta}_0 = -4.0809$ and $\hat{\gamma}_0 = -0.0081$ result. A comparison of this Gompertz model with the exponential model without covariates in Table 6.2 and based upon a likelihood ratio test results in a χ^2 value of

$$Lq = 2(-14147.10 - (-14434.35)) = 574.5,$$

with one degree of freedom. Inclusion of the proxy variable duration, which, in this model, is representative of the acquisition of job specific knowledge, improves the estimation significantly.

A comparison of the above maximum likelihood estimates $\hat{\lambda}_0 = \exp(-4.0809) = 0.0169$ and $\hat{\gamma}_0 = -0.0081$ with estimates obtained from Section 6.1 for the Gompertz model ($\hat{\lambda}_0 = 0.0202$ and $\hat{\gamma}_0 = -0.0095$) illustrates once again that the estimates based upon the graphic method were relatively good and contained only slight deviations.

Especially important from a substantive point of view is the fact that the estimated γ_0 coefficient actually does possess the expected sign, meaning that the inclination to change jobs decreases monotonically with increasing job specific knowledge. If one compares, for example, a man who has just started a new job [$\lambda^k(0) = \exp(-4.0809) = 0.0169$], with a man who has already been working 10 years at the same job [$\lambda^k(120) = \exp(-4.0809) \cdot \exp(-0.0081 \cdot 120) = 0.0064$], then the inclination of the second man to change his job has been reduced by the accumulation of job specific knowledge by 62.13 percent [$((0.0064 - 0.0169)/0.0169) \cdot 100\% = -62.13\%$].

With the aid of the relationship

$$S(t) = \exp(-\frac{\lambda_0}{\gamma_0} (\exp(\gamma_0 t) - 1)),$$

the median of the duration for men in a job M*, with

$$S(M^*) = 0.5,$$

may be estimated

$$S(\hat{M}^*) = \exp(-\frac{0.0169}{-0.008132} (\exp(-0.008132 \cdot \hat{M}^*) - 1))$$

$$\hat{M}^* = 49.90 \text{ months.}$$

According to the Gompertz model, the median amounts to somewhat more than four years and thus is much smaller than the median of the exponential distribution in Section 6.2.1.

In addition to this, one can, for example, compute the probability that a man after a time period of eight years still works in the same job. The probability postulating a Gompertz distribution is $\hat{S}(96) = \exp(-\frac{0.0169}{-0.008132}(\exp(-0.008132 \cdot 96) - 1)) = 0{,}32426$, or some 32.43 percent. In comparison to this the exponential model in Section 6.2.1 gives a probability of 37.56 percent.

The interpretations made above are only valid under the presumption that these men comprise a homogeneous population. That this is indeed not the case has often been demonstrated in this book. But the argument of heterogeneous subgroups in the present example is especially important, since given even constant rates in the subgroups, one obtains an apparent monotonically decreasing duration dependency as may be expressed in a significant γ_0 coefficient. The interpretation that the hazard rate falls due to the accumulation of job specific knowledge and abilities would then be a false interpretation. Thus, in the next step we estimate the Gompertz model from the standpoint of observed heterogeneity.

The Gompertz Model With Covariates in the λ_0 Term

Covariates may first of all be included in the Gompertz model in such a way that one relates the λ_0 parameter log-linearly with the covariate vector (see Section 3.3.2): $\lambda_0(\mathbf{x}) = \exp(\mathbf{x}'\boldsymbol{\beta})$.

In the job change example this means that if job specific knowledge (as measured by the proxy variable duration) is equal to zero at the time of entry into the new job, a number of time-independent covariates affect the inclination of changing jobs (as was the case in the exponential model). Given an increasing duration on the job and the corresponding increasing job specific knowledge, our hypothesis is that the hazard rate should fall monotonically. This is the case if the estimation of γ_0 is significantly different from zero and has a negative sign. This version of the Gompertz model may then be stated as follows:

$$\lambda^k(v|\mathbf{x}_k) = \exp(\mathbf{x}_k'\boldsymbol{\beta}) \cdot \exp(\gamma_0 v) \qquad , k = 1, 2, \ldots .$$

The covariate vector $\mathbf{x}$, once again contains education (EDU), prestige (M59), number of previously held jobs (NOJ), labor force experience at the time of job entry (LFX), as well as the cohort dummies COHO2 and COHO3 (see Appendix 1). The actual job specific experience at time t ($x_{LFX}(t)$) in this model is thus divided into two components; first, the general labor force experience ($x_{LFX}(t_{k-1})$) that an individual possesses at the time of job entry, and second, the job specific labor force experience ($x_{LFX}(v)$) that an individual has accumulated in the job:[14]

$$x_{LFX}(t) = \beta_{LFX} x_{LFX}(t_{k-1}) + \gamma_0 x_{LFX}(v).$$

As before, the maximum likelihood estimates of the parameters are calculated with the BMDP program P3R and the integrated subprogram P3RFUN written by Trond Petersen.

Program Example 6.12:

```
/INPUT UNIT IS 30.
       CODE IS DATA.
/VARIABLE NAMES ARE (63)DUR,(64)CEN,(65)EDU,(66)LFX,
                    (67)NOJ,(68)COHO2,(69)COHO3,(70)X1,(71)DP.
          ADD IS 9.
/TRANSFORM USE = (M3 EQ 1).
           DUR = M51 - M50 + 1.
           CEN = 1.
           IF (M51 EQ M47) THEN CEN = 0.
           IF (M41 EQ 1 AND M42 EQ 1) THEN EDU = 9.
           IF (M41 EQ 1 AND (M42 EQ 2 OR M42 EQ 3)) THEN EDU = 11.
           IF (M41 EQ 2 AND M42 EQ 1) THEN EDU = 10.
           IF (M41 EQ 2 AND (M42 EQ 2 OR M42 EQ 3)) THEN EDU = 12.
           IF (M41 EQ 3 AND (M42 EQ 1 OR M42 EQ 2 OR M42 EQ 3))
           THEN EDU = 13.
           IF (M42 EQ 4) THEN EDU = 17.
           IF (M42 EQ 5) THEN EDU = 19.
           COHO2 = 0.
           COHO3 = 0.
           IF(M48 GE 468 AND M48 LE 504) THEN COHO2 = 1.
           IF(M48 GE 588 AND M48 LE 624) THEN COHO3 = 1.
           LFX = M50 - M43.
           NOJ = M5 - 1.
           DP = 0.
           X1 = 0.
/REGRESS DEPENDENT IS DP.
         PARAMETERS ARE 8.
         PRINT IS 0.
         MEANSQUARE IS 1.0.
         ITERATIONS ARE 100.
         LOSS.
```

[14]Both types of labor force experience may only be distinguished in the multiepisode case in which the variable "labor force experience at the beginning of each job" varies.

```
/PARAMETER INITIAL ARE -3.8,-0.01,0.04,-0.01,0.15,-0.01,0.1,0.17.
           NAMES ARE CONST,TDEP,EDU,M59,NOJ,LFX,COHO2,COHO3.
/END.
/COMMENT'
M 2
T 70 63
D 71
C 64
I  6 65 59 67 66 68 69
 '.
```

Preparation of the variables in the TRANSFORM paragraph is done the same way as shown for Program Examples 5.1 and 6.3. However, in comparison to Program Example 6.11, now 8 coefficients (PARAMETERS ARE 8) are estimated, namely, the regression constant β_0, 6 β coefficients for the independent variables and the γ_0 coefficient expressing the duration dependency (TDEP). For each of these coefficients a starting value is given in the paragraph PARAMETER. These values are based on those discussed in the Cox estimations of Chapter 5.

Also the commands of the subprogram P3RFUN remain unchanged from Program Example 6.11 with the exception of the I command, which lists the BMDP assigned numbers of the independent variables. First, the number of covariates is stated, in this case 6, and then the covariates follow with their BMDP assigned numbers: 65 (≙ EDU), 59 (≙ M59, prestige), 67 (≙ NOJ), 66 (≙ LFX), 68 (≙ COHO2), and 69 (≙ COHO3). The results of this estimation are presented in Table 6.11.

As can be seen in Table 6.11 for this model the value of the log likelihood function (≙ –LOSS) is –13854.60. If one compares this model with the

Table 6.11: Results of the Gompertz Model From Program Example 6.12

PARAMETER	ESTIMATE	ASYMPTOTIC STANDARD DEVIATION
CONST	-3.651509	0.127242
TDEP	-0.007335	0.000397
EDU	0.003205	0.013920
M59	-0.004916	0.001314
NOJ	0.155978	0.009915
LFX	-0.008679	0.000416
COHO2	0.037889	0.046693
COHO3	0.167833	0.052601

Gompertz model without covariates, the likelihood ratio test (see Section 3.7) has a χ^2 value of

$$Lq = 2(-13854.60 - (-14147.10)) = 585.00$$

with six degrees of freedom. The null hypothesis that none of the additionally introduced variables explains anything about the risk of a job change for men must then be rejected.

The significance test of the individual coefficients once again produces the well-known picture. The variables EDU and COHO2 are not significant, and the variables prestige (M59), NOJ, LFX, and COHO3 have significant effects with the expected signs (significance level: 0.05).

It is important that $\hat{\gamma}_0$ (TDEP) is significantly different from zero and has a negative sign. In controlling the relevant covariates a monotonically falling hazard rate appears, that speaks in favor of the hypothesis of the mobility restrictive effects of job specific qualifications. If one compares the nonstandardized coefficients of job experience at the beginning of each new job activity ($\hat{\beta}_{LFX}$) with the job experience within a job ($\hat{\gamma}_0 \triangleq$ TDEP), then it becomes clear that both possess approximately the same magnitude (–0.0087 and –0.0073). With regard to job changing behavior, whether or not labor force experience was obtained in one specific job or in various jobs does not matter.

Analyzing the Residuals in the Gompertz Model

These interpretations for the Gompertz model, remain valid only under the condition that one is dealing with homogeneous subpopulations. As has been already discussed in detail in Sections 3.7.1 and 6.2.3, one can consider the cumulative hazard rate $\Lambda(t_i|\mathbf{x}_i)$ as a residual r_i given the validity of the distribution assumption, and these may be used to evaluate the model.

For the Gompertz distribution case given multiepisodes one obtains an estimation of the residuals r_{ik} as follows:

$$\hat{r}_{ik} = \hat{\Lambda}\,(v_{ik}|\mathbf{x}_{ik}) = \frac{\exp(\mathbf{x}_{ik}'\hat{\boldsymbol{\beta}})}{\hat{\gamma}_0}\,(\exp(\hat{\gamma}_0 v_{ik}) - 1).$$

From these calculated residuals, given censored data, the survivor function can once again be estimated using the product-limit method through adequate transformation and plotted against r. Under the Gompertz distribution assumption one should obtain a line with the slope of –1.

The following two program examples with the subprogram P1L from BMDP illustrate this residual test of the Gompertz model without covariates and the Gompertz model in which the λ_0 coefficient is related log-linearly to the covariate vector $\mathbf{x}$:

Program Example 6.13:

```
/INPUT UNIT IS 30.
       CODE IS DATA.
/VARIABLE NAMES ARE (63)DUR,(64)CEN,(65)EDU,(66)LFX,
                    (67)NOJ,(68)COHO2,(69)COHO3,(70)RESID1.
          ADD IS 8.
/TRANSFORM USE = (M3 EQ 1).
           DUR = M51 - M50 + 1.
           CEN = 1.
           IF (M51 EQ M47) THEN CEN = 0.
           IF (M41 EQ 1 AND M42 EQ 1) THEN EDU = 9.
           IF (M41 EQ 1 AND (M42 EQ 2 OR M42 EQ 3)) THEN EDU = 11.
           IF (M41 EQ 2 AND M42 EQ 1) THEN EDU = 10.
           IF (M41 EQ 2 AND (M42 EQ 2 OR M42 EQ 3)) THEN EDU = 12.
           IF (M41 EQ 3 AND (M42 EQ 1 OR M42 EQ 2 OR M42 EQ 3))
           THEN EDU = 13.
           IF (M42 EQ 4) THEN EDU = 17.
           IF (M42 EQ 5) THEN EDU = 19.
           COHO2 = 0.
           COHO3 = 0.
           IF(M48 GE 468 AND M48 LE 504) THEN COHO2 = 1.
           IF(M48 GE 588 AND M48 LE 624) THEN COHO3 = 1.
           LFX = M50 - M43.
           NOJ = M5 - 1.
           RESID1= (EXP(-4.080915)/(-0.008132))*(EXP((-0.008132)*DUR)-1)
/FORM TIME IS RESID1.
      STATUS IS CEN.
      RESPONSE IS 1.
/ESTIMATE METHOD IS PROD.
          PLOTS ARE LOG.
/PRINT CASES ARE 0.
/END
```

Program Example 6.14:

```
/INPUT UNIT IS 30.
       CODE IS DATA.
/VARIABLE NAMES ARE (63)DUR,(64)CEN,(65)EDU,(66)LFX,
                    (67)NOJ,(68)COHO2,(69)COHO3,(70)RESID2.
          ADD IS 8.
/TRANSFORM USE = (M3 EQ 1).
           DUR = M51 - M50 + 1.
           CEN = 1.
           IF (M51 EQ M47) THEN CEN = 0.
           IF (M41 EQ 1 AND M42 EQ 1) THEN EDU = 9.
           IF (M41 EQ 1 AND (M42 EQ 2 OR M42 EQ 3)) THEN EDU = 11.
```

```
            IF (M41 EQ 2 AND M42 EQ 1) THEN EDU = 10.
            IF (M41 EQ 2 AND (M42 EQ 2 OR M42 EQ 3)) THEN EDU = 12.
            IF (M41 EQ 3 AND (M42 EQ 1 OR M42 EQ 2 OR M42 EQ 3))
            THEN EDU = 13.
            IF (M42 EQ 4) THEN EDU = 17.
            IF (M42 EQ 5) THEN EDU = 19.
            COHO2 = 0.
            COHO3 = 0.
            IF(M48 GE 468 AND M48 LE 504) THEN COHO2 = 1.
            IF(M48 GE 588 AND M48 LE 624) THEN COHO3 = 1.
            LFX = M50 - M43.
            NOJ = M5 - 1.
            RESID2= (EXP(-3.651509+0.003205*EDU-0.004916*M59
                   +0.155978*NOJ-0.008679*LFX+0.037889*COHO2
                   +0.167833*COHO3)/(-0.007335))*(EXP((-0.007335)*DUR)-1)
/FORM TIME IS RESID2.
      STATUS IS CEN.
      RESPONSE IS 1.
/ESTIMATE METHOD IS PROD.
          PLOTS ARE LOG.
/PRINT CASES ARE 0.
/END
```

In the TRANSFORM paragraphs of the above program examples, once again the residuals (RESID1 or RESID2) are calculated based on the estimated coefficients. In the FORM paragraphs these are introduced along with the censoring variables (STATUS IS CEN) as durations (TIME IS RESID1 or TIME IS RESID2). The product-limit method (METHOD IS PROD) is used to estimate the survivor functions, and finally the logarithmic survivor functions are plotted (PLOTS ARE LOG).

The results of these runs are illustrated in Figure 6.8 (for the Gompertz model without covariates) and in Figure 6.9 (for the Gompertz model in which the λ_0 coefficient is related log-linearly with the covariate vector). In examining the Gompertz model without covariates, it appears that, given small residuals, a relatively good linear fit is obtained (see Figure 6.8). Only in the case of large residuals does one obtain large deviations in the estimation. Through inclusion of covariates, a somewhat better linear fit may be obtained (see Figure 6.9), yet even here one cannot overlook the fact that given large residuals, deviations from a linear pattern are present. If one compares this result with the residual plots of the exponential model (see Figures 6.5 and 6.6), then one may conclude that the Gompertz model with its λ_0 coefficient being log-linearly related to the covariates is the more appropriate model. However, at this point it must be noted once again that residual tests are not tests in a strict sense, since the residuals are not independent or identically distributed.

Figure 6.8: Residual Plot of the Gompertz Model Without Covariates

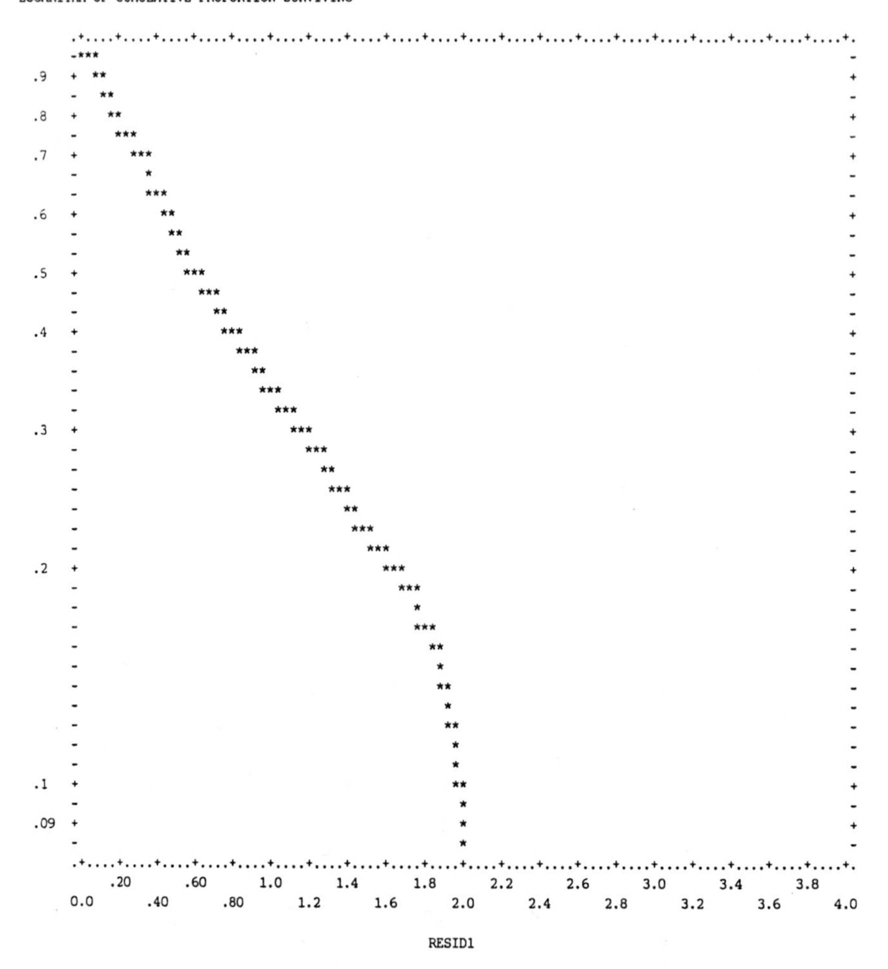

Gompertz Model With Covariates in the λ_0 and γ_0 Term

Even in the discussion of models with periodical durations, we saw that the effect of time constant covariates over the duration need not remain unchanged. Keeping the signal or filter theory in mind, the influence of the "signal" variables may even be expected to decrease with duration. The declining influence of such variables given a Gompertz distribution may be included in the model by making the coefficient γ_0 linearly dependent on the covariate vector $\mathbf{x}$: $\gamma_0(\mathbf{x}) = \mathbf{x}'\boldsymbol{\gamma}$. In so doing the following Gompertz model is obtained:

Figure 6.9: Residual Plot of the Gompertz Model With Covariates in the λ_0 Term

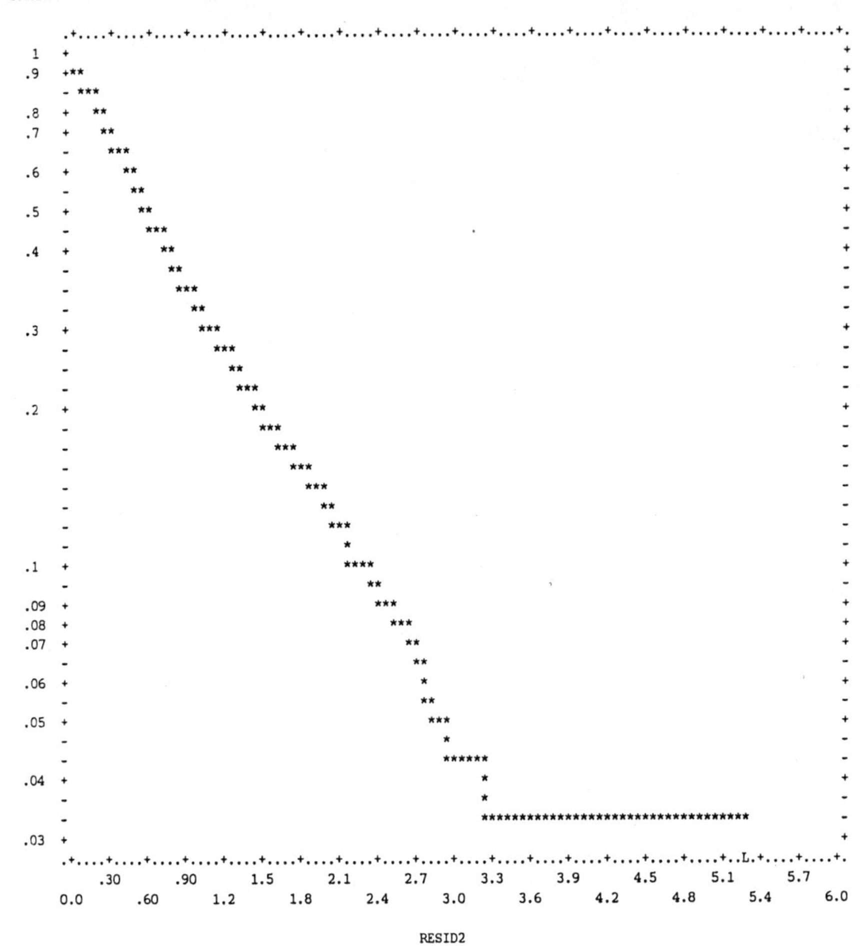

$$\lambda^k(v|\mathbf{x}_k) = \exp(\mathbf{x}_k'\boldsymbol{\beta}) \cdot \exp((\mathbf{x}_k'\boldsymbol{\gamma})v) \ , k = 1, 2, \ldots .$$

If signaling theory is correct, then the coefficients of the "filter" variables in the γ_0 term should possess exactly the opposite signs as in the λ_0 term, and their influence should become less and less important with increasing duration. To estimate this special Gompertz model, the program RATE was used. The input data file is based upon the event oriented data set of Table 6.4 (without the variable TMAR).

Program Example 6.15:

```
RUN NAME          GOMPERTZ-MODEL
N OF CASES        3516
VARIABLES         12
TS                1
TF                2
CEN               3
EDU               4
PRES              5
NOJ               6
LFX               7
COHO2             8
COHO3             9
DUR               10
JOBN              11
JOBN1             12
READ DATA
(12F5.0)
T AND S           10 3
MODEL             (4) A=0 B=1 C=-1
VECTOR            (1) 4 5 6 7 8 9 (2) 4 5 6 7 8 9
SOLVE             (2)=100
FINISH
```

In the RATE program presented above, a Gompertz-Makeham model, which is characterized by the number 4: A + B exp(Ct), is specified on the MODEL command. The letter A is set to 0. The letter B is related log-linearly (B = 1) and the letter C (C = –1) is linearly joint with the covariate vector $\mathbf{x}$, so that one obtains the following model: $\exp(\mathbf{x}'\boldsymbol{\beta}) \exp((\mathbf{x}'\boldsymbol{\gamma})v)$. In both covariate vectors, the VECTOR command includes the same variables with their RATE assigned numbers: 4 ≙ education (EDU), 5 ≙ prestige (PRES), 6 ≙ number of previously held jobs (NOJ), 7 ≙ labor force experience at the beginning of an episode (LFX), 8 ≙ cohort dummy for those born in 1939-41 (COHO2), and 9 ≙ cohort dummy for those born in 1949–51 (COHO3). The results of this estimation are presented in Table 6.12. Looking at the likelihood ratio test, it is clear that the present model, compared to the Gompertz model with covariates only in the λ_0 term is significantly better. A significant χ^2 value of

$$Lq = 2(-13843.99 - (-13854.60)) = 21.22$$

is obtained with six degrees of freedom ($\chi^2_{0.95;6} = 12.59$). Therefore, this model is somewhat better in explaining the relationship.

A comparison of the $\hat{\beta}$ coefficients in the λ_0 term (LETTER B) of the above model with the results of the first period (0 to 24 months) in the period model

Table 6.12: Results of the Gompertz Model From Program Example 6.15

DESTINATION	UNWEIGHTED FREQUENCY	WEIGHTED FREQUENCY	MAX(LOG OF L) NULL HYPOTHESIS	MAX(LOG OF L) ALTERNATIVE HYPOTHESIS	PSEUDO R-SQUARED	CHI-SQUARED	DF	PROBABILITY LEVEL
1	2586	2586.0	-1.443432D+04	-1.384399D+04	0.0409	1180.67	13	0.00D+00

DESTINATION 1 LETTER B LOG-LINEAR TIME-INDEPENDENT VECTOR

INTERNAL NUMBER	VARIABLE NUMBER	VARIABLE NAME	VECTOR 1 PARAMETER	PARAMETER STANDARD ERROR	PARAMETER F RATIO	ANTILOG OF THE PARAMETER	ANTILOG STANDARD ERROR	ANTILOG F RATIO
1		(CONSTANT)	-3.856D+00	1.704D-01	512.321	2.115D-02		
2	4	EDU	1.184D-02	1.949D-02	0.369	1.012D+00	1.972D-02	0.365
3	5	PRES	-4.360D-03	1.879D-03	5.385	9.956D-01	1.871D-03	5.409
4	6	NOJ	1.826D-01	1.603D-02	129.856	1.200D+00	1.924D-02	108.482
5	7	LFX	-9.014D-03	6.308D-04	204.174	9.910D-01	6.252D-04	206.024
6	8	COHO2	1.475D-01	6.459D-02	5.217	1.159D+00	7.486D-02	4.510
7	9	COHO3	2.874D-01	7.312D-02	15.452	1.333D+00	9.747D-02	11.672

DESTINATION 1 LETTER C LINEAR TIME-INDEPENDENT VECTOR

INTERNAL NUMBER	VARIABLE NUMBER	VARIABLE NAME	VECTOR 2 PARAMETER	PARAMETER STANDARD ERROR	PARAMETER F RATIO
8		(CONSTANT)	-2.883D-03	2.656D-03	1.178
9	4	EDU	-2.055D-04	3.036D-04	0.458
10	5	PRES	-1.266D-05	2.868D-05	0.195
11	6	NOJ	-7.644D-04	2.978D-04	6.590
12	7	LFX	9.323D-06	9.219D-06	1.023
13	8	COHO2	-2.069D-03	8.585D-04	5.811
14	9	COHO3	-3.081D-03	1.440D-03	4.580

(see Table 6.9) demonstrates, first of all, a high level of similarity with regard to the nonstandardized β coefficients. The first period in the period model thus appears to have been well chosen. In addition to the nonsignificant variable education (EDU) large differences are obtained only by the dummy variable COHO2, which in the present Gompertz model is also significant.

Of the $\hat{\gamma}$ coefficients in the γ_0 term (LETTER C), only the variables NOJ, COHO2, and COHO3 are significantly different from zero. These coefficients also possess the expected opposite sign in comparison to their complementary $\hat{\beta}$ coefficients in the λ_0 term (LETTER B). This implies that with increasing duration the absolute influence of these variables is reduced more and more.

The variables PRES and LFX have a significant effect in the λ_0 term (LETTER B), but in the γ_0 term (LETTER C) they are not significantly different from zero. Thus, the effect of these variables over the duration is constant. This was to be expected for the variable job experience. The stability of the effect of the variable prestige (PRES), however, contradicts the signal hypothesis and the results of the period model (see Section 6.4). As discussed there, the variable prestige had an effect only in the first period and then was not significantly different from zero in following periods.

Gompertz-Makeham Model With Covariates in the α_0, λ_0, and γ_0 Term

Up until now the characteristic of the Gompertz model has been disregarded, that given a negative γ_0 coefficient and an increasing duration ($v \rightarrow \infty$), the hazard rate approaches the time axis asymptotically (see Figure 3.8). For large durations, the hazard rate would then be approximately zero, and the probability of an event occurrence in this model would be quite unlikely.

In the case of job change for men this would imply, for example, that for large durations in a job the inclination to change jobs approaches zero, and thus after some specific point, practically no further job changes occur. Thus a statistical model has been chosen which does not necessarily depict the real situation. One would certainly still expect mobility processes even given a rather large duration. That the model with a Gompertz distribution is inappropriate for large durations is already evident in the residual plots of Figures 6.8 and 6.9.

As presented in Section 3.2.2, this Gompertz distribution problem may be alleviated by adding a constant α_0 ($\alpha_0 > 0$) to the Gompertz model so that one obtains the so-called Gompertz-Makeham model. For large durations the risk then no longer drops asymptotically towards zero, but rather towards the value of the constant α_0. Of course, this constant, along with the other parameters of the Gompertz distribution ($\lambda_0(\mathbf{x}) = \exp(\mathbf{x}'\boldsymbol{\beta})$; $\gamma_0(\mathbf{x}) = \mathbf{x}'\boldsymbol{\gamma}$), can be modeled as dependent upon a covariate vector $\mathbf{x}$. In order for the condition $\alpha_0 > 0$ to hold, the covariate vector $\mathbf{x}$ can be related to a α_0 parameter log-linearly: $\alpha_0(\mathbf{x}) = \exp(\mathbf{x}'\boldsymbol{\alpha})$.

If, given the example of job change for men, such a Gompertz-Makeham model is used for estimation, then a α coefficient of the α_0 term expresses how the covariates influence job change behavior in the long run. According to signaling theory, the "filter" variables such as level of education, prestige, or labor force experience accumulated before job entry in the α_0 term should not be significantly different from zero. A long term influence upon job change behavior would be more likely to occur given relatively stable personality variables. Unfortunately, in the German Life History Study (GLHS) no psychological indicators are available to classify an individual as "more stable" or "less stable." For this reason, we use a proxy variable to estimate the general tendency for an individuals job relationships to be more or less stable in nature. The number of previously held jobs (NOJ) can act as such a proxy variable in this case. The greater this number is, the more likely it is that the individual has an unstable job relationship. In order to test this hypothesis along with the signaling hypothesis, we estimate the following Gompertz model with the RATE program:

$$\lambda^k(v|\mathbf{x}_k) = \exp(\mathbf{x}_k'\boldsymbol{\alpha}) + \exp(\mathbf{x}_k'\boldsymbol{\beta}) \cdot \exp((\mathbf{x}_k'\boldsymbol{\gamma})\, v) \quad \text{with } k = 1, 2, \ldots .$$

Program Example 6.16:

```
RUN NAME        GOMPERTZ-MAKEHAM-MODEL
N OF CASES      3516
VARIABLES       12
TS              1
TF              2
```

```
CEN             3
EDU             4
PRES            5
NOJ             6
LFX             7
COHO2           8
COHO3           9
DUR             10
JOBN            11
JOBN1           12
READ DATA
(12F5.0)
T AND S         10 3
MODEL           (4) A=1 B=1 C=-1
VECTOR          (1) 4 5 6 7 8 9 (2) 4 5 6 7 8 9 (3) 4 5 6 7 8 9
SOLVE           (2)=100
FINISH
```

Table 6.13: Results of the Gompertz-Makeham Model From Program Example 6.16

DESTINATION	UNWEIGHTED FREQUENCY	WEIGHTED FREQUENCY	MAX(LOG OF L) NULL HYPOTHESIS	MAX(LOG OF L) ALTERNATIVE HYPOTHESIS	PSEUDO R-SQUARED	CHI-SQUARED	DF	PROBABILITY LEVEL
1	2586	2586.0	-1.443432D+04	-1.382968D+04	0.0419	1209.29	20	0.00D+00

DESTINATION 1 LETTER A LOG-LINEAR TIME-INDEPENDENT VECTOR

INTERNAL NUMBER	VARIABLE NUMBER	VARIABLE NAME	VECTOR 1 PARAMETER	PARAMETER STANDARD ERROR	PARAMETER F RATIO	ANTILOG OF THE PARAMETER	ANTILOG STANDARD ERROR	ANTILOG F RATIO
1		(CONSTANT)	-7.285D+00	8.060D-01	81.691	6.855D-04		
2	4	EDU	5.935D-02	6.177D-02	0.923	1.061D+00	6.555D-02	0.870
3	5	PRES	-4.507D-04	7.181D-03	0.004	9.995D-01	7.178D-03	0.004
4	6	NOJ	1.055D-01	3.880D-02	7.397	1.111D+00	4.312D-02	6.663
5	7	LFX	8.332D-04	1.768D-03	0.222	1.001D+00	1.770D-03	0.222
6	8	COHO2	-2.829D-02	2.938D-01	0.009	9.721D-01	2.856D-01	0.010
7	9	COHO3	-1.231D+01	3.672D+02	0.001	4.489D-06	1.648D-03	368073.525

DESTINATION 1 LETTER B LOG-LINEAR TIME-INDEPENDENT VECTOR

INTERNAL NUMBER	VARIABLE NUMBER	VARIABLE NAME	VECTOR 2 PARAMETER	PARAMETER STANDARD ERROR	PARAMETER F RATIO	ANTILOG OF THE PARAMETER	ANTILOG STANDARD ERROR	ANTILOG F RATIO
8		(CONSTANT)	-3.836D+00	2.060D-01	346.700	2.158D-02		
9	4	EDU	1.899D-02	2.383D-02	0.635	1.019D+00	2.429D-02	0.623
10	5	PRES	-6.956D-03	2.429D-03	8.201	9.931D-01	2.412D-03	8.258
11	6	NOJ	2.321D-01	2.173D-02	114.044	1.261D+00	2.741D-02	90.828
12	7	LFX	-1.212D-02	1.101D-03	121.153	9.880D-01	1.088D-03	122.631
13	8	COHO2	1.538D-01	8.277D-02	3.455	1.166D+00	9.653D-02	2.968
14	9	COHO3	3.841D-01	8.428D-02	20.774	1.468D+00	1.237D-01	14.323

DESTINATION 1 LETTER C LINEAR TIME-INDEPENDENT VECTOR

INTERNAL NUMBER	VARIABLE NUMBER	VARIABLE NAME	VECTOR 3 PARAMETER	PARAMETER STANDARD ERROR	PARAMETER F RATIO
15		(CONSTANT)	-1.277D-03	4.084D-03	0.098
16	4	EDU	-6.124D-04	4.543D-04	1.817
17	5	PRES	2.117D-05	4.491D-05	0.222
18	6	NOJ	-1.545D-03	5.900D-04	6.858
19	7	LFX	4.597D-06	1.724D-05	0.071
20	8	COHO2	-1.397D-03	1.313D-03	1.133
21	9	COHO3	-1.316D-03	1.751D-03	0.564

The difference between the above RATE example and Program Example 6.15, is that the MODEL command contains additionally the letter A which is related log-linearly to the covariate vector (A = 1) and that the VECTOR command introduces, for this additional covariate vector, the RATE assigned numbers of the corresponding covariates. The results of this program are presented in Table 6.13.

If one, first of all, compares this Gompertz-Makeham model with the Gompertz model with covariates in the λ_0 term and the γ_0 term, then one obtains, based upon the likelihood ratio test, a χ^2 value of

$$Lq = 2(-13829.68) - (-13843.99) = 28.62$$

with six degrees of freedom. Since this value is greater than the value $\chi^2_{0.95;6} = 12.59$, at least one of the α coefficients in the α_0 term must have a significant effect. Our assumption that the Gompertz-Makeham model is an appropriate way to model the risk of job change, given large durations, is verified. This is also evident from an examination of the residual plot in Figure 6.10.[15]

Examination of the coefficients in the α_0 term (LETTER A) shows that actually only the variable, number of previously held jobs, influences the risk of a job change. None of the other covariates are significant given large durations. This supports the signaling theory, which assumes that the influ-

---

[15]Stated below is the P1L program run to test the residuals of the Gompertz-Makeham model in which the covariates in the α_0, λ_0, and γ_0 terms are taken into consideration:

```
/INPUT UNIT IS 30.
      CODE IS DATA.
/VARIABLE NAMES ARE (63)DUR,(64)CEN,(65)EDU,(66)LFX,
                    (67)NOJ,(68)COHO2,(69)COHO3,(70)RESID,
                    (71)A,(72)B,(73)C.
         ADD IS 11.
/TRANSFORM USE = (M3 EQ 1).
          DUR = M51 - M50 + 1.
          CEN = 1.
          IF (M51 EQ M47) THEN CEN = 0.
          IF (M41 EQ 1 AND M42 EQ 1) THEN EDU = 9.
          IF (M41 EQ 1 AND (M42 EQ 2 OR M42 EQ 3)) THEN EDU = 11.
          IF (M41 EQ 2 AND M42 EQ 1) THEN EDU = 10.
          IF (M41 EQ 2 AND (M42 EQ 2 OR M42 EQ 3)) THEN EDU = 12.
          IF (M41 EQ 3 AND (M42 EQ 1 OR M42 EQ 2 OR M42 EQ 3))
          THEN EDU = 13.
          IF (M42 EQ 4) THEN EDU = 17.
          IF (M42 EQ 5) THEN EDU = 19.
          COHO2 = 0.
          COHO3 = 0.
          IF(M48 GE 468 AND M48 LE 504) THEN COHO2 = 1.
          IF(M48 GE 588 AND M48 LE 624) THEN COHO3 = 1.
          LFX = M50 - M43.
          NOJ = M5 -1.
          A = EXP(-7.285 + 0.05935*EDU - 0.0004507*M59 +
                0.1055*NOJ + 0.0008332*LFX - 0.02829 * COHO2 -
                12.31*COHO3).
          B = EXP(-3.836 + 0.01899*EDU - 0.006956*M59 +
                0.2321*NOJ - 0.01212*LFX + 0.1538*COHO2 +
                0.3841*COHO3).
```

Figure 6.10: Residual Plot of the Gompertz-Makeham Model, in Which the α_0, λ_0, and γ_0 Terms are Related to the Covariate Vector **x**

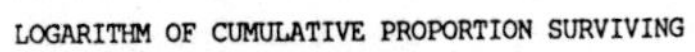

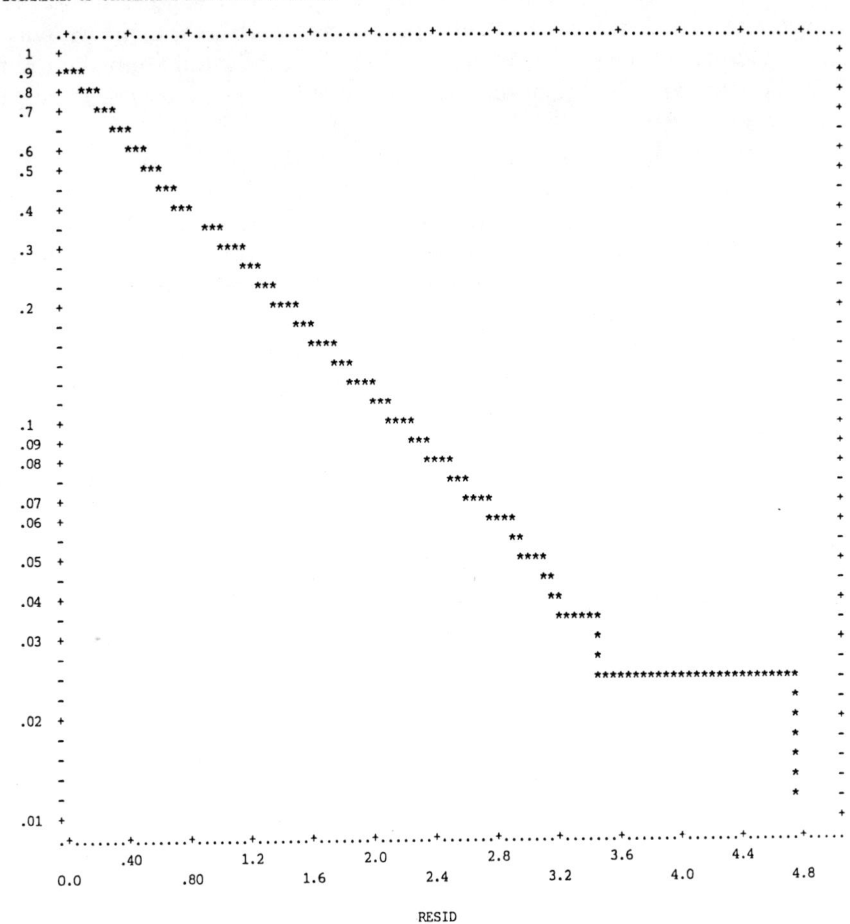

Footnote 15 continued

```
          C =    (-0.001277 - 0.0006124*EDU + 0.00002117*M59 -
                 0.001545*NOJ + 0.000004597*LFX - 0.001397*COHO2 -
                 0.001316*COHO3).
          RESID = A*DUR + B/C * (EXP(C*DUR)- 1).
/FORM TIME IS RESID.
      STATUS IS CEN.
      RESPONSE IS 1.
/ESTIMATE METHOD IS PROD.
          PLOTS ARE LOG.
/PRINT CASES ARE 0.
/END
```

ence of the "filter" variables is initially quite large for each new job and decreases slowly with increasing duration. It also supports the presumption that in the long run only relatively stable personality variables are decisive.

In the γ_0 term (LETTER C) only the variable number of previously held jobs (NOJ) has a significant effect. The $\hat{\gamma}$ coefficient has the opposite sign to the λ_0 term (LETTER B). That means, given short and middle term durations on the job, the influence of the variable, number of previously held jobs, is increasingly reduced.

Finally, in the λ_0 term (LETTER B) only prestige (PRES), labor force experience at the beginning of each new episode (LFX), and COHO3 have significant effects. For short and middle term durations their effect upon the risk of changing jobs is largely maintained. However, this does not support the filter theory which states that the influence of the variables prestige and labor force experience, prior to job entry, should diminish.

We have used this example to show that through the successive introduction of covariates in the parameters of the Gompertz-(Makeham) distribution, an increasingly complex and at the same time more realistic model of job shifts can be obtained.

In a further step, one should eliminate all nonsignificant covariates from the covariate vectors in the RATE program and estimate the model again, thus achieving a suitable model with fewer parameters. For the moment, we exclude this step here. Instead, we now turn to a special Gompertz model in which an attempt is made to separate the influence of the continuous time-dependent covariate job experience from pure duration dependency.

Gompertz Model With a Continuous Time-Dependent Covariate

Applying the Gompertz model in Program Example 6.12, we attempted to separate cumulated job experience to time t into two components: first, the general labor force experience ($x_{LFX}(t_{k-1})$) an individual has accumulated before entering a new job, and second, the accumulated job specific experience $x_{LFX}(v)$ an individual has gathered within an occupation:

$$x_{LFX}(t) = \beta_{LFX} x_{LFX}(t_{k-1}) + \gamma_0 x_{LFX}(v).$$

All effects that arise given increased job duration ($x_{DUR}(v) = t - t_{k-1}$) are then automatically set equivalent to the variable $x_{LFX}(v)$. The equation may, therefore, also be written as:

$$x_{LFX}(t) = \beta_{LFX} x_{LFX}(t_{k-1}) + \gamma_0 x_{DUR}(v).$$

Such a model is only appropriate if over the duration no effects outside of specific labor force experience within a job exist and $x_{LFX}(v)$ actually corresponds to the duration dependency $x_{DUR}(v)$. This is, however, not plausible since in addition to the job specific human capital investments further influences may be at play. For example, one could imagine that through habitual processes, the number of social networks present in a certain job environment et cetera, effects are produced, which may additionally restrict the job change behavior of employees. Alternatively given increasing duration within a job and the perception that further advancement is no longer possible in these circumstances, a rise in the risk of changing jobs is likely.

In the following model, therefore, the "pure" effect of labor force experience will be separated from the "pure" duration effect of a job. In so doing, we define labor force experience regardless of whether the knowledge collected has been accumulated within one job or over several jobs:

$$x_{LFX}(t) = x_{LFX}(t_{k-1}) + x_{LFX}(v)$$

and obtain the following Gompertz model:

$$\lambda^k(v|\mathbf{x}_k) = \exp(\mathbf{x}_k'\boldsymbol{\beta} + \beta_{LFX}x_{LFX}(t)) \exp(\gamma_0 x_{DUR}(v)) \qquad \text{with } k = 1, 2, \ldots .$$

Furthermore, if one presupposes that job experience and other influences increase as a linear function of duration v:

$$x_{LFX}(t) = x_{LFX}(t_{k-1}) + v$$

and

$$x_{DUR}(v) = v,$$

the Gompertz model can be reformulated as

$$\lambda^k(v|\mathbf{x}_k) = \exp(\mathbf{x}_k'\boldsymbol{\beta} + \beta_{LFX}(x_{LFX}(t_{k-1}) + v)) \exp(\gamma_0 v) \qquad \text{with } k = 1, 2, \ldots .$$

This can now be estimated with the special model M3 of Trond Petersen's P3RFUN program:[16]

Program Example 6.17:

```
/INPUT UNIT IS 30.
       CODE IS DATA.
/VARIABLE NAMES ARE (63)DUR,(64)CEN,(65)EDU,(66)LFXS,(67)NOJ,
                    (68)COHO2,(69)COHO3,(70)X1,(71)DP,(72)LFXF.
          ADD IS 10.
/TRANSFORM USE = (M3 EQ 1).
           DUR = M51 - M50 + 1.
           CEN = 1.
           IF (M51 EQ M47) THEN CEN = 0.
           IF (M41 EQ 1 AND M42 EQ 1) THEN EDU = 9.
           IF (M41 EQ 1 AND (M42 EQ 2 OR M42 EQ 3)) THEN EDU = 11.
           IF (M41 EQ 2 AND M42 EQ 1) THEN EDU = 10.
           IF (M41 EQ 2 AND (M42 EQ 2 OR M42 EQ 3)) THEN EDU = 12.
           IF (M41 EQ 3 AND (M42 EQ 1 OR M42 EQ 2 OR M42 EQ 3))
```

[16] β_{LFX} and γ_0 may naturally only be identified in the multiepisode case, since otherwise $x_{LFX}(t) = x_{DUR}(v) = t - t_{k-1} = v$.

```
            THEN EDU = 13.
            IF (M42 EQ 4) THEN EDU = 17.
            IF (M42 EQ 5) THEN EDU = 19.
            COHO2 = 0.
            COHO3 = 0.
            IF(M48 GE 468 AND M48 LE 504) THEN COHO2 = 1.
            IF(M48 GE 588 AND M48 LE 624) THEN COHO3 = 1.
            LFXS = M50 - M43.
            LFXF = LFXS + DUR.
            NOJ = M5 - 1.
            DP = 0.
            X1 = 0.
/REGRESS DEPENDENT IS DP.
         PARAMETERS ARE 8.
         PRINT IS 0.
         MEANSQUARE IS 1.0.
         ITERATIONS ARE 100.
         LOSS.
/PARAMETER INITIAL ARE -3.8,0.001,-0.01,0.04,-0.01,0.15,0.1,0.17.
           NAMES ARE CONST,TDEP,LFX,EDU,M59,NOJ,COHO2,COHO3.
/END.
/COMMENT'
M 3
T 70 63
D 71
C 64
I  5 65 59 67 68 69
L 66 72
 '.
```

Analogous to Program Example 6.12, the above program first processes the variables in the TRANSFORM paragraph (see Appendix 1). Since in this case labor force experience is introduced as varying over time, the initial labor force experience (LFXS), and final labor force experience (LFXF) are necessary for each episode. These variables are therefore created in the TRANSFORM paragraph.

Similar to Program Example 6.12, 8 parameters (PARAMETERS ARE 8) are estimated. For PARAMETER INITIAL ARE, however, the initial values must be rearranged since the start value of the β coefficient of the continuous time-dependent variable labor force experience LFX now must be included on the third position, following the regression constant (CONST) and the duration dependency (TDEP).

After the END statement, the parameters of the subprogram P3RFUN are entered. In the present case, the model number is 3. In comparison to Program

Example 6.12, the I command (independent) now contains only the five variables that are constant over time. The time-dependent variable labor force experience, represented by LFXS and LFXF, must be stated on the L command (labor force) specified with the respective BMDP assigned number: 66 = LFXS and 72 = LFXF. The estimation results are presented in Table 6.14.

Table 6.14: Results of the Gompertz Model With a Continuous Time-Dependent Labor Force Experience Variable From Program Example 6.17

PARAMETER	ESTIMATE	ASYMPTOTIC STANDARD DEVIATION
CONST	-3.651521	0.127242
TDEP	0.001344	0.000579
LFX	-0.008679	0.000416
EDU	0.003206	0.013920
M59	-0.004916	0.001314
NOJ	0.155966	0.009915
COHO2	0.037890	0.046693
COHO3	0.167832	0.052601

Although one can separate the effect of "pure" duration dependency from the "pure" effect of job experience in the Gompertz model presented in Table 6.14, it is not possible to test this model against the Gompertz model in Table 6.11 in which labor force experience at the beginning of each job episode was regarded as being a time constant covariate. The Gompertz model just estimated may also be formulated as follows:

$$\lambda^k(v|x_k) = \exp(x_k'\boldsymbol{\beta} + \beta_{LFX} x_{LFX}(t_{k-1})) \exp((\beta_{LFX} + \gamma_0)\, v), \qquad k = 1, 2, \ldots,$$

where labor force experience at the beginning of each job episode has the effect β_{LFX} and the duration possesses the effect $\beta_{LFX} + \gamma_0$. Thus, the model in Table 6.14 results in a value of the log likelihood function ($\triangleq$ –LOSS) equal to –13854.60, which corresponds exactly to the Gompertz model of Table 6.11. Also, the nonstandardized $\hat{\beta}$ coefficients of the time-independent covariates EDU, M59 (prestige), NOJ, COHO2, and COHO3 do not change compared to previous Gompertz model.

The advantage of the Gompertz model from Program Example 6.17 lies simply in the fact that the effect of "pure" job experience (LFX) and "pure" duration dependency (TDEP) can be differentiated. Since the coefficient of job experience $\hat{\beta}_{LFX} = -0.008679$ as well as the coefficient of duration dependency $\hat{\gamma}_0 = 0.001344$ are significantly different from zero, it follows that the mobility restricting influence of job experience within an occupation may be partially compensated by other mechanisms ($\hat{\gamma}_0 > 0$). As already discussed above, one explanation for this tendency may be that given increasing duration in one and the same job, one slowly begins to realize that an advancement may be realized only through a change of employment.

These examples of the exponential and Gompertz models give only a basic impression of the modeling and interpretative possibilities of parametric models. In the following sections we will concentrate on the Weibull, log-logistic, and lognormal distributions.

6.5.2 Weibull Model

The Weibull model is very flexible and appropriate for a wide variety of situations. Like the Gompertz distribution, the Weibull distribution can also be used to model a monotonically falling ($0 < \alpha < 1$) or monotonically increasing risk ($\alpha > 1$) (see Section 3.2.2). For the special case $\alpha = 1$ one obtains an exponential distribution, and it is, therefore, possible to test the null hypothesis of a constant event risk against the alternative $\alpha \neq 1$.

Weibull Model Without Covariates

To illustrate the interpretation of the Weibull model, we will first of all estimate a Weibull model without covariates, based upon the multiepisode case of the average risk of job change for men. The λ coefficient of the Weibull distribution is related to the regression constant β_0^* log-linearly and α is estimated directly:

$$\begin{aligned} \lambda^k(v) &= \exp(\beta_0^*)^\alpha \, \alpha \, v^{\alpha-1} \\ &= \exp(\beta_0) \, \alpha \, v^{\alpha-1} \qquad k = 1, 2, \ldots \end{aligned}$$

with $\beta_0 = \alpha\beta_0^*$.

The duration v serves in this example as a proxy variable for job specific knowledge and skills which have to be newly acquired in each occupation. If the hypothesis that the inclination towards changing jobs decreases with increasing accumulation of job specific knowledge is true, then we expect that the Weibull distribution will possess a significant $\hat{\alpha}$, lying between 0 and 1.

The estimation of this model is done with the aid of GLIM. Since it is easy to use the GLIM macros (see Roger and Peacock, 1983) which are presented in Appendix 5, only a few program lines will be presented here.

Program Example 6.18:

```
$INPUT 40 MAKROS
$UNITS 3517$
$DATA TS TF CEN EDU PRES NOJ LFX C2 C3 DUR JN JN1 TMAR$
$DINPUT 20$
$CALC U = %LOG(DUR)$
```

```
$CALC C = CEN$
$USE SETW$
$END$
$STOP$
```

First of all, in the GLIM Program Example 6.18 the GLIM macros (see Appendix 5), which are stored in an external file and are required to estimate the Weibull model, are loaded with INPUT through the input channel 40.

The input data file is once again taken from the event oriented file shown in Table 6.4 (see also Appendix 1). The variables are formated in such a way that between each column at least one blank is given, so they can be read in as free format data. With the DATA command the variable vectors are assigned a name according to their position in the input file, and the data is read in through the input channel 20 (DINPUT 20). Due to requirements of the program, the length of the GLIM vectors must be set with the command UNITS to the number of episodes +1; in the example above the number 3516 + 1 = 3517 is obtained.

The macro SETW applied to estimate the Weibull model presupposes that in the vector C the censoring information CEN has been loaded and that the logarithmic duration DUR[17] has been loaded in vector U. The command USE SETW starts the computation of the estimates. The results of this estimation are presented in Table 6.15.

Table 6.15: Results of the Weibull Model From Program Example 6.18

```
        SCALED
CYCLE  DEVIANCE      DF
   4     6237.     3515

       ESTIMATE       S.E.       PARAMETER
   1  -3.581       0.5980E-01   GM
   2  0.7866       0.1231E-01   U
```

From Table 6.15 we obtain a deviance value (SCALED DEVIANCE) of 6237. The reduction of the deviance through estimation of further parameters is asymptotically χ^2 distributed with k degrees of freedom (k $\triangleq$ number of additionally fitted parameters).

A comparison of the maximum likelihood estimates in Table 6.15 $\hat{\lambda} = \exp(-3.581/0.7866) = 0.0105$ and $\hat{\alpha} = 0.7866$ with the estimates obtained

[17]If one were to set CALC U = DUR, then the macro SETW would estimate an extreme value distribution.

from the graphic method in Section 6.1 ($\hat{\lambda} = 0.0138$ and $\hat{\alpha} = 0.8621$) demonstrates once again that these initial estimates were rather good.

If one compares the null hypothesis of a constant or monotonically increasing event risk $H_0 : \alpha \geq 1$ with the alternative hypothesis of a monotonically decreasing path of the hazard rate $H_1 : \alpha < 1$ using the following standard normal distributed test statistic (see Section 3.7.3):

$$z = \frac{\hat{\alpha} - 1}{s(\hat{\alpha})} = \frac{0.7866 - 1}{0.01231} = -17.34,$$

one finds that the null hypothesis must be rejected (significance level: 0.05). Consequently, with increasing duration a monotonically falling risk of job change exists. This supports the hypothesis that with increasing duration the inclination to change jobs decreases more and more as job specific knowledge accumulates.

If we compare, for example, a man who has been employed only one month on a job [$\lambda^k(1) = 0.0105 \cdot 0.7866 \cdot (0.0105 \cdot 1)^{-0.2134} = 0.02183$], with a man who has already worked ten years in a job [$\lambda^k(120) = 0.0105 \cdot 0.7866 \cdot (0.0105 \cdot 120)^{-0.2134} = 0.00786$], then we find that for the second man (due to his accumulation of job specific knowledge) the inclination to change jobs has decreased by approximately 64 percent [$(0.00786 - 0.02183)/0.02183 \cdot 100\% = -63.99\%$]. In the case of the Gompertz distribution we found a decrease of 62 percent (see Section 6.5.1).

Regarding the relationship (3.2.20)

$$S(t) = \exp((-\lambda t)^{\alpha})$$

one can once again estimate the median of the duration of men in a job M*, with S(M*) = 0.5:

$$S(\hat{M}^*) = \exp((-0.0105 \cdot \hat{M}^*)^{0.7866})$$
$$\hat{M}^* = 59.77 \text{ months.}$$

The median of the Weibull distribution thus lies between the median of the Gompertz distribution (about four years, see Section 6.5.1) and the median of the exponential distribution (about five and one-half years, see Section 6.2.1).

Moreover, we can calculate the likelihood that after a period of eight years a man will still work in the same job, by $\hat{S}(96) = \exp(-(0.0105 \cdot 96)^{0.7866}) = 0.3656$ or 36.56 percent. Compared to this, the Gompertz model of Section 6.5.1 has a probability of 32.43 percent and the exponential model of Section 6.2.1 has a probability of 37.56 percent.

Of course, these interpretations are only valid given the assumption that one can speak of a homogeneous population. If this is not the case, then the monotonically falling duration dependency may have resulted from aggregation across heterogeneous subgroups (see Section 3.9.1) and the explanation that the risk of changing jobs decreases as job specific knowledge is accumulated would be a false interpretation. Thus, in a further step covariates must be considered in the Weibull model.

Weibull Model With Covariates in the λ Term

In order to include covariates in the Weibull model, the parameter λ is related to the covariate vector **x** log-linearly (see Section 3.3.2): $\lambda(\mathbf{x}) = \exp(\mathbf{x}'\boldsymbol{\beta}^*)$. The Weibull model is then expressed as:

$$\lambda^k(v|\mathbf{x}_k) = \exp(\mathbf{x}_k'\boldsymbol{\beta}^*)^\alpha \alpha\, v^{\alpha-1}$$
$$= \exp(\mathbf{x}_k'\boldsymbol{\beta})\alpha\, v^{\alpha-1} \qquad , k = 1, 2, \ldots$$

with $\boldsymbol{\beta} = \alpha\boldsymbol{\beta}^*$.

The variables education (EDU), prestige (PRES), number of previously held jobs (NOJ), labor force experience before job entry (LFX), as well as the cohort dummies C2 and C3 (see Appendix 1) are once again contained in the covariate vector **x**. Job specific experience which increases given increasing duration in a job should result once again in a monotonically falling inclination towards job change ($0 < \alpha < 1$).

The parameter estimations are carried out with the program system GLIM.[18]

---

[18]If one were to calculate this Weibull model with BMDP program P3R and its integrated subprogram P3RFUN written by Trond Petersen, which assumes the rate function $\lambda^k(v|\mathbf{x}_k) = \exp(\mathbf{x}_k'\boldsymbol{\beta})v^{\alpha^*}$, then one must transform the estimated parameter $\hat{\alpha}^*$ and the constant term $\hat{\beta}_0^1$ as follows in order to obtain the parameters $\hat{\alpha}$ and $\hat{\beta}_0$ of the model stated above:

$$\hat{\alpha} = \hat{\alpha}^* + 1$$
$$\hat{\beta}_0 = \hat{\beta}_0^1 - \ln\hat{\alpha}.$$

All other regression coefficients are identical in both approaches. The BMDP program is the same as Program Example 6.12 with the exception that the model must now be specified with the number 4 (M4):

```
/INPUT UNIT IS 30.
      CODE IS DATA.
/VARIABLE NAMES ARE (63)DUR,(64)CEN,(65)EDU,(66)LFX,
                    (67)NOJ,(68)COHO2,(69)COHO3,(70)X1,(71)DP.
         ADD IS 9.
/TRANSFORM USE = (M3 EQ 1).
          DUR = M51 - M50 + 1.
          CEN = 1.
          IF (M51 EQ M47) THEN CEN = 0.
          IF (M41 EQ 1 AND M42 EQ 1) THEN EDU = 9.
          IF (M41 EQ 1 AND (M42 EQ 2 OR M42 EQ 3)) THEN EDU = 11.
          IF (M41 EQ 2 AND M42 EQ 1) THEN EDU = 10.
          IF (M41 EQ 2 AND (M42 EQ 2 OR M42 EQ 3)) THEN EDU = 12.
          IF (M41 EQ 3 AND (M42 EQ 1 OR M42 EQ 2 OR M42 EQ 3))
          THEN EDU = 13.
          IF (M42 EQ 4) THEN EDU = 17.
          IF (M42 EQ 5) THEN EDU = 19.
          COHO2 = 0.
          COHO3 = 0.
          IF(M48 GE 468 AND M48 LE 504) THEN COHO2 = 1.
          IF(M48 GE 588 AND M48 LE 624) THEN COHO3 = 1.
          LFX = M50 - M43.
          NOJ = M5 - 1.
          DP = 0.
          X1 = 0.
```

Program Example 6.19:

```
$INPUT 40 MAKROS
$UNITS 3517$
$DATA TS TF CEN EDU PRES NOJ LFX C2 C3 DUR JN JN1 TMAR$
$DINPUT 20$
$CALC EDU(%NU)=0$
$CALC PRES(%NU)=0$
$CALC NOJ(%NU)=0$
$CALC LFX(%NU)=0$
$CALC C2(%NU)=0$
```

Footnote 18 continued

```
/REGRESS DEPENDENT IS DP.
         PARAMETERS ARE 8.
         PRINT IS 0.
         MEANSQUARE IS 1.0.
         ITERATIONS ARE 100.
         LOSS.
/PARAMETER INITIAL ARE -3.8,-0.14,0.04,-0.01,0.15,-0.01,0.1,0.17.
           NAMES ARE CONST,TDEP,EDU,M59,NOJ,LFX,COHO2,COHO3.
/END.
/COMMENT'
M 4
T 70 63
D 71
C 64
I  6 65 59 67 66 68 69
'.
```

The following program example also shows the estimation of this Weibull model with the procedure LIFEREG of SAS. SAS estimates the Weibull model as a “regression model” for the log of the duration ($\log v = -\boldsymbol{\beta}^{*\prime}\mathbf{x} + \sigma\omega$), where $\sigma = \alpha^{-1}$, $\boldsymbol{\beta}^* = -\sigma\boldsymbol{\beta}$, and where ω has an extreme value distribution (see Kalbfleisch and Prentice, 1980, p. 21 and Section 3.2.2):

```
LIBNAME SAS'[MAIR.MAIR.SAS]';
DATA WEIB;
 SET SAS.SYS;
 IF M3 = 1;
 DUR = M51 - M50 + 1;
 CEN = 1;
 IF (M51 EQ M47) THEN CEN = 0;
 IF (M41 = 1 AND M42 = 1) THEN EDU = 9;
 IF (M41 = 1 AND (M42 = 2 OR M42 = 3)) THEN EDU = 11;
 IF (M41 = 2 AND M42 = 1) THEN EDU = 10;
 IF (M41 = 2 AND (M42 = 2 OR M42 = 3)) THEN EDU = 12;
 IF (M41 = 3 AND (M42 = 1 OR M42 = 2 OR M42 = 3)) THEN EDU = 13;
 IF (M42 = 4) THEN EDU = 17;
 IF (M42 =5) THEN EDU = 19;
 COHO2 = 0;
 COHO3 = 0;
 IF (M48 GE 468 AND M48 LE 504) THEN COHO2 = 1;
 IF (M48 GE 588 AND M48 LE 624) THEN COHO3 = 1;
 LFX = M50 - M43;
 NOJ = M5 - 1;
PROC LIFEREG DATA=WEIB;
 MODEL DUR*CEN(0)=EDU M59 NOJ LFX COHO2 COHO3/D=WEIBULL;
```

```
$CALC C3(%NU)=0$
$CALC U = %LOG(DUR)$
$CALC C = CEN$
$USE SETW$
$FIT+EDU+PRES+NOJ+LFX+C2+C3
$DIS E
$END$
$STOP$
```

For this program (in comparison to Program Example 6.18) the (N+1)th elements (%NU) of all variable vectors that are to be included as covariates must be set to zero. In addition, the FIT and DIS commands follow the USE SETW command. The FIT command specifies the names of the included covariates, whereas the DIS command specifies the output of the estimated values as well as their asymptotic standard errors. The results of this estimation are found in Table 6.16.[19]

From examining Table 6.16 we can see that the deviance value in comparison to the Weibull model without covariates is reduced from 6237 to 5632 through the estimation of the additional parameters. The reduction of –2logL (see Aitkin and Clayton, 1980, p. 161) corresponds to a χ^2 value of 605 (given six degrees of freedom) which implies that the model was significantly improved with the introduction of the covariates.

The estimated β coefficients are relatively consistent with those of the Gompertz model presented in Table 6.11 with regard to the direction and size of influence. The exceptions here are the cohort dummies and, once again, the

[19]In the model M4 of Trond Petersen given a value of the log likelihood function (≙ –LOSS) of –14000.8, one obtains the following result:

PARAMETER	ESTIMATE	ASYMPTOTIC STANDARD DEVIATION
CONST	-3.667665	0.136723
TDEP	-0.161433	0.016871
EDU	0.007218	0.012912
M59	-0.004928	0.001224
NOJ	0.160115	0.009639
LFX	-0.008542	0.000395
COHO2	0.124740	0.040516
COHO3	0.340973	0.049549

Disregarding the rounding off of figures, then the $\hat{\beta}$ coefficients of the covariates coincide with those of Table 6.16, and the estimates of the program P3RFUN $\hat{\alpha}^*$ (TDEP) and $\hat{\beta}_0^1$ (CONST) can be restated as GLIM estimates $\hat{\alpha}$ (U) and $\hat{\beta}_0$ (GM) as follows:

$$\hat{\alpha} = \hat{\alpha}^* + 1 = -0.161433 + 1 = 0.83856$$

$$\hat{\beta}_0 = \hat{\beta}_0^1 - \ln\hat{\alpha} = -3.6676 - \ln 0.83856 = -3.492.$$

Using the procedure LIFEREG of SAS, the following result is obtained:

Table 6.16: Results of the Weibull Model From Program Example 6.19

```
      SCALED
CYCLE  DEVIANCE      DF
  3     5632.       3509

      ESTIMATE        S.E.         PARAMETER
  1  -3.424         0.1433         GM
  2  0.8266         0.1293E-01     U
  3  0.7150E-02     0.1446E-01     EDU
  4  -0.4906E-02    0.1376E-02     PRES
  5  0.1592         0.1222E-01     NOJ
  6  -0.8552E-02    0.4506E-03     LFX
  7  0.1218         0.4648E-01     C2
  8  0.3325         0.5300E-01     C3
```

Footnote 19 continued

```
                                   SAS

                         L I F E R E G   P R O C E D U R E

DATA SET            =WORK.WEIB
DEPENDENT VARIABLE=DUR
CENSORING VARIABLE=CEN
CENSORING VALUE(S)=      0
NONCENSORED VALUES=2586  CENSORED VALUES= 930
OBSERVATIONS WITH MISSING VALUES= 330

LOGLIKELIHOOD FOR WEIBULL        -5402.07

VARIABLE   DF   ESTIMATE STD ERR CHISQUARE  PR>CHI LABEL/VALUE

INTERCPT    1     4.14095 0.154863  714.999  0.0001 INTERCEPT
EDU         1 -.00848996 .0175949 0.232828  0.6294
M59         1 0.00591752 .0016823  12.3727  0.0004
NOJ         1  -0.192875 .0149262  166.976  0.0001
LFX         1  0.0103681  5.7E-04  328.082  0.0001
COHO2       1  -0.147338 0.056381  6.82914  0.0090
COHO3       1   -0.40244 0.063316  40.3994  0.0001
SCALE       1    1.20974 .0189593                   EXTREME VALUE SCALE PARAMETER
```

Given the parametrization in SAS, the estimated coefficients $\hat{\boldsymbol{\beta}}^*$ and the "SCALE parameter" $\hat{\sigma}$ must be transformed in order to obtain the parameters $\hat{\alpha}$ and $\hat{\boldsymbol{\beta}}$ in Table 6.16:

$$\hat{\alpha} = \frac{1}{\hat{\sigma}},$$

$$\hat{\boldsymbol{\beta}} = -\frac{1}{\hat{\sigma}}\hat{\boldsymbol{\beta}}^*.$$

For example:

$$\hat{\alpha} = \frac{1}{1.20974} = 0.8266,$$

$$\hat{\beta}_0 = -\frac{4.14095}{1.20974} = -3.423.$$

nonsignificant variable education (EDU). Independent of the distribution (Gompertz or Weibull), we have come to more or less the same substantive conclusions regarding the influence of the covariates.

A test of the null hypothesis $H_0 : \alpha \geq 1$ against the alternative hypothesis $H_1 : \alpha < 1$ demonstrates that even controlling for the covariates a monotonically decreasing inclination to change jobs remains:

$$z = \frac{\hat{\alpha} - 1}{s(\hat{\alpha})} = \frac{0.8266 - 1}{0.01293} = -13.41.$$

Thus, consequently one obtains the same conclusions as in the case of the Gompertz model in Table 6.11.

Examining Residuals in the Weibull Model

As previously discussed in detail in Sections 3.7.1 and 6.2.3, one can generally regard the cumulative hazard rates $\Lambda(t_i|\mathbf{x}_i)$ as residuals r_i and use these in evaluating the model. In the case of a Weibull distribution one obtains an estimation of the residuals r_{ik} as follows:

$$\hat{r}_{ik} = \hat{\Lambda}(v_{ik}|\mathbf{x}_{ik}) = \exp(\mathbf{x}_{ik}'\hat{\boldsymbol{\beta}})v_{ik}^{\hat{\alpha}}.$$

Using the residuals calculated in this way, one can, given censored data, estimate the survivor function with the help of the product-limit method and then transform and plot this function against r. If the assumption of a Weibull distributed hazard rate is correct, then a line with a slope of –1 is obtained.

The following two program runs with the BMDP subprogram P1L illustrate this residual test for the Weibull model without covariates and the Weibull model in which the λ coefficient is related in a log-linear fashion to the covariate vector:

Program Example 6.20:

```
/INPUT UNIT IS 30.
       CODE IS DATA.
/VARIABLE NAMES ARE (63)DUR,(64)CEN,(65)EDU,(66)LFX,
                    (67)NOJ,(68)COHO2,(69)COHO3,
                    (70)RESID1,(71)ALPHA,(72)LAMBDA.
          ADD IS 10.
/TRANSFORM USE = (M3 EQ 1).
           DUR = M51 - M50 + 1.
           CEN = 1.
           IF (M51 EQ M47) THEN CEN = 0.
           IF (M41 EQ 1 AND M42 EQ 1) THEN EDU = 9.
           IF (M41 EQ 1 AND (M42 EQ 2 OR M42 EQ 3)) THEN EDU = 11.
           IF (M41 EQ 2 AND M42 EQ 1) THEN EDU = 10.
           IF (M41 EQ 2 AND (M42 EQ 2 OR M42 EQ 3)) THEN EDU = 12.
           IF (M41 EQ 3 AND (M42 EQ 1 OR M42 EQ 2 OR M42 EQ 3))
           THEN EDU = 13.
```

```
           IF (M42 EQ 4) THEN EDU = 17.
           IF (M42 EQ 5) THEN EDU = 19.
           COHO2 = 0.
           COHO3 = 0.
           IF(M48 GE 468 AND M48 LE 504) THEN COHO2 = 1.
           IF(M48 GE 588 AND M48 LE 624) THEN COHO3 = 1.
           LFX = M50 - M43.
           NOJ = M5 -1.
           ALPHA = 0.7866.
           LAMBDA = EXP(-3.581).
           RESID1 = LAMBDA * DUR ** ALPHA.
/FORM TIME IS RESID1.
      STATUS IS CEN.
      RESPONSE IS 1.
/ESTIMATE METHOD IS PROD.
          PLOTS ARE LOG.
/PRINT CASES ARE 0.
/END
```

Program Example 6.21:

```
/INPUT UNIT IS 30.
       CODE IS DATA.
/VARIABLE NAMES ARE (63)DUR,(64)CEN,(65)EDU,(66)LFX,
                    (67)NOJ,(68)COHO2,(69)COHO3,
                    (70)RESID2,(71)ALPHA,(72)LAMBDA.
          ADD IS 10.
/TRANSFORM USE = (M3 EQ 1).
           DUR = M51 - M50 + 1.
           CEN = 1.
           IF (M51 EQ M47) THEN CEN = 0.
           IF (M41 EQ 1 AND M42 EQ 1) THEN EDU = 9.
           IF (M41 EQ 1 AND (M42 EQ 2 OR M42 EQ 3)) THEN EDU = 11.
           IF (M41 EQ 2 AND M42 EQ 1) THEN EDU = 10.
           IF (M41 EQ 2 AND (M42 EQ 2 OR M42 EQ 3)) THEN EDU = 12.
           IF (M41 EQ 3 AND (M42 EQ 1 OR M42 EQ 2 OR M42 EQ 3))
           THEN EDU = 13.
           IF (M42 EQ 4) THEN EDU = 17.
           IF (M42 EQ 5) THEN EDU = 19.
           COHO2 = 0.
           COHO3 = 0.
           IF(M48 GE 468 AND M48 LE 504) THEN COHO2 = 1.
           IF(M48 GE 588 AND M48 LE 624) THEN COHO3 = 1.
           LFX = M50 - M43.
           NOJ = M5 -1.
           ALPHA = 0.8266.
           LAMBDA = EXP(-3.424 + 0.007150*EDU - 0.004906*M59 +
```

```
                  0.1592*NOJ - 0.008552*LFX + 0.1218*COHO2 +
                  0.3325*COHO3).
         RESID2 = LAMBDA * DUR ** ALPHA.
/FORM TIME IS RESID2.
      STATUS IS CEN.
      RESPONSE IS 1.
/ESTIMATE METHOD IS PROD.
          PLOTS ARE LOG.
/PRINT CASES ARE 0.
/END
```

Figure 6.11: Residual Plot for the Weibull Model Without Covariates

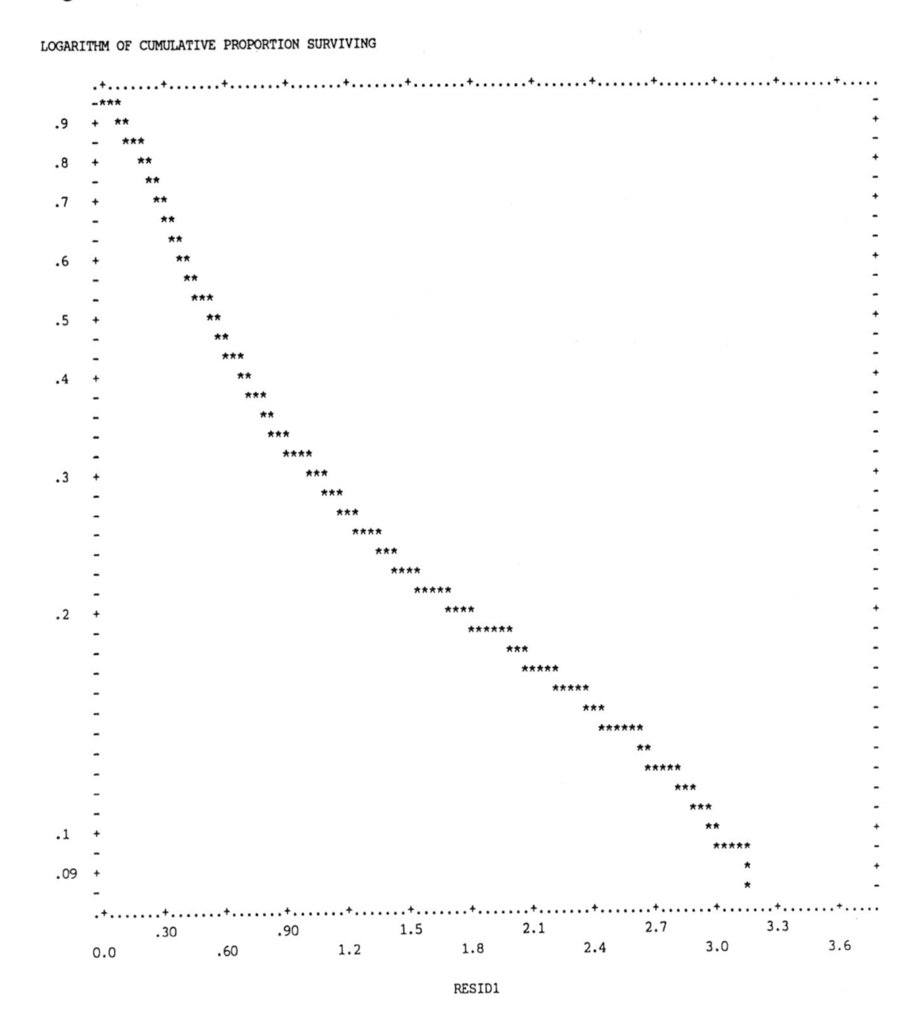

Figure 6.12: Residual Plot for the Weibull Model With Covariates in the λ Term

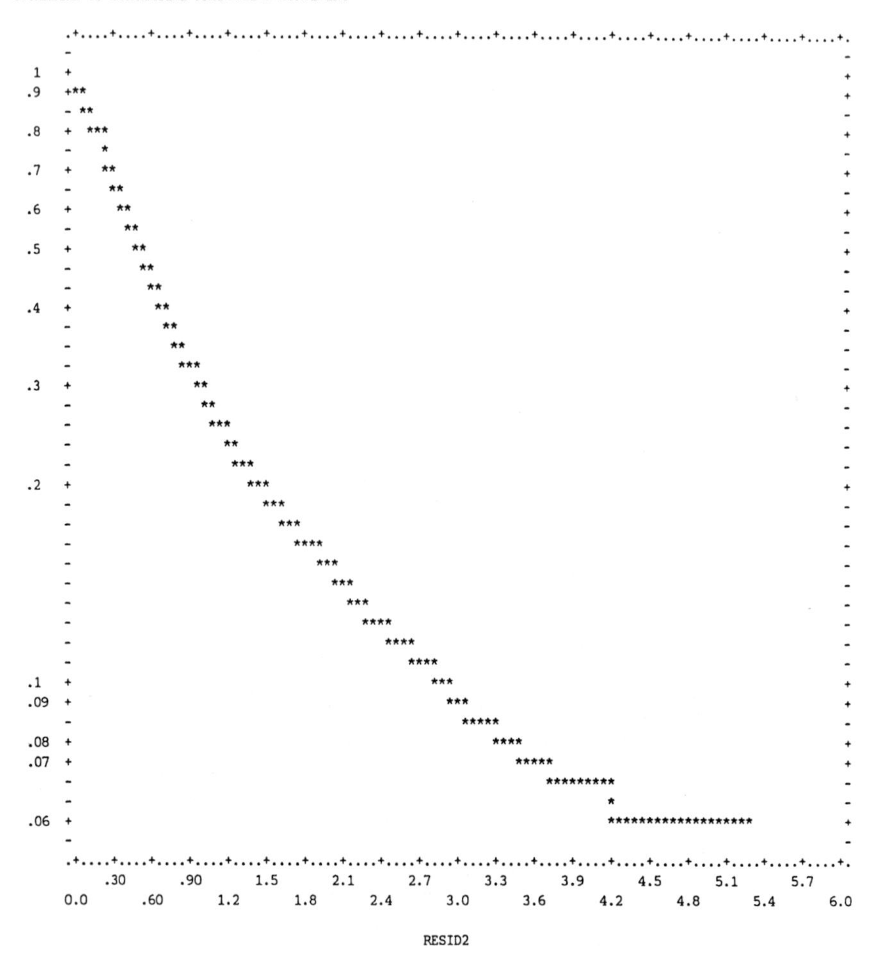

In the TRANSFORM paragraphs of the above program examples the residuals (RESID1 or RESID2) are calculated once again based on the coefficient estimates in Tables 6.15 and 6.16. In the FORM paragraphs, these are then implemented next to the censoring variables (STATUS IS CEN) as durations (TIME IS RESID1 or TIME IS RESID2). To estimate the survivor function the product-limit method (METHOD IS PROD) is applied each time, and finally the logarithmic survivor functions (PLOTS ARE LOG) are plotted.

The results of these runs are presented in Figures 6.11 (for the Weibull model without covariates) and 6.12 (for the Weibull model with covariates). Both plots clearly deviate from a line with a slope of –1, so that in this case the Weibull distribution does not characterize the path of job change risk as well as the Gompertz distribution.

6.5.3 Log-Logistic Model

The log-logistic distribution is, like the Weibull distribution, extremely flexible and may be applied in different situations. For $\alpha \leq 1$ one obtains a monotonically falling hazard rate, and for $\alpha > 1$ the risk rises monotonically at first till some specific point of the duration to a maximum ($v_{max} = (\alpha - 1)^{1/\alpha}/\lambda$) and then falls monotonically (see Figure 3.7). In the literature, the log-logistic model, along with the lognormal distribution, are the most commonly recommended distributions, if an initial increasing and then a decreasing risk is presumed to exist.[20]

Log-Logistic Model Without Covariates

To illustrate the use of the log-logistic model a model without covariates is estimated, once again applying the multiepisode case of job change risk for men. The λ coefficient of the log-logistic model is related to the regression constant β_0^* log-linearly and the parameter α is estimated directly:

$$\lambda^k(v) = \frac{\exp(\beta_0^*)^\alpha \ \alpha \ v^{\alpha-1}}{1 + \exp(\beta_0^*)^\alpha \ v^\alpha}$$

$$= \frac{\exp(\beta_0)\alpha \ v^{\alpha-1}}{1 + \exp(\beta_0)v^\alpha} \qquad , k = 1, 2, \ldots$$

with $\beta_0 = \alpha\beta_0^*$.

From the viewpoint of our example, this model can now be used to test the hypothesis of a monotonically decreasing job change risk ($\alpha \leq 1$), compared to the alternative hypothesis of an initially increasing and then decreasing job change risk ($\alpha > 1$). The nonmonotonic path of the job change risk is plausible if we take into account the usual adjustment processes that occur with each new job appointment, including the increased risk of job change at this time. However, as the expectations of the employer and employee adjust to one another, the desire to change jobs, given the increased accumulation of job specific knowledge, diminishes.

[20]For modeling such nonmonotonic paths in economics and the social sciences, Diekmann und Mitter (1983, 1984) have also proposed the use of the sickle distribution.

The estimation of this model is again carried out with the program system GLIM.

Program Example 6.22:

```
$INPUT 40 MAKROS
$UNITS 3517$
$DATA TS TF CEN EDU PRES NOJ LFX C2 C3 DUR JN JN1 TMAR$
$DINPUT 20$
$CALC U = %LOG(DUR)$
$CALC C = CEN$
$USE SETL$
$END$
$STOP$
```

In the above program example, the command USE SETW, used in Program Example 6.18, is replaced by the command USE SETL.[21] The results of the estimation are presented in Table 6.17.

Table 6.17: Results of the Log-Logistic Model From Program Example 6.22

```
        SCALED
CYCLE   DEVIANCE        DF
   4     3881.        3515
    ESTIMATE        S.E.        PARAMETER
1  -4.464       0.7530E-01   GM
2   1.144       0.1856E-01   U
```

First of all, given a value of the deviance function equal to 3881, we obtain a value $\hat{\lambda} = \exp(-4.464/1.144) = 0.0202$ and an α estimation of 1.144. Both of these estimates are once again very similar to the estimates achieved by the graphic method of Section 6.1 ($\hat{\lambda} = 0.024$ and $\hat{\alpha} = 1.1313$).

A comparative analysis of the null hypothesis of a monotonically falling job change risk ($H_0 : \alpha \leq 1$) and the alternative of an initially rising and then decreasing job change risk ($H_1 : \alpha > 1$) shows that the null hypothesis must be rejected (level of significance: 0.05):

$$z = \frac{1.144 - 1}{0.01856} = 7.759.$$

The risk of job change thus rises initially after the recruitment of a new employee due to the evolving adjustment processes, reaches a climax after

[21]If one were to set CALC U = DUR, then the macro SETL would estimate a logistic distribution.

about nine months [$v_{max} = (1.144 - 1)^{1/1.144}/0.0202) = 9.10$] and then decreases monotonically with increasing job specific knowledge.

Applying the relationship (3.2.29)

$$S(t) = \frac{1}{1 + (\lambda t)^{\alpha}}$$

the median of the duration of men in a job M*, with S(M*) = 0.5, can be estimated:

$$S(\hat{M}^*) = \frac{1}{1 + (0.0202\ \hat{M}^*)^{1.144}}$$

$$\hat{M}^* = 49.50 \text{ months.}$$

According to the log-logistic model the median is thus somewhat more than four years and is similar to the level of the Gompertz distribution (49.90 months, see Section 6.5.1). This median, however, lies under the level of the Weibull distribution (approximately five years, see Section 6.5.2), and under the level of the exponential distribution (approximately five and one-half years, see Section 6.2.1).

In addition to this, one can calculate the likelihood that a man, after a period of eight years, still works in the same occupation. A value of $\hat{S}(96) = (1 + (0.0202 \cdot 96)^{1.144})^{-1} = 0.3192$ or 31.92 percent is obtained. In comparison, the Gompertz model (see Section 6.5.1) yields a probability of 32.56 percent, the Weibull model (see Section 6.5.2) a probability of 36.56 percent, and the exponential model (see Section 6.2.1) a probability of 37.56 percent.

These interpretations are of course only valid given the condition that for men one can speak of a homogeneous population and that an average risk of changing jobs actually exists. In a further step we can include the known covariates in the model (see Appendix 1), and test whether or not the form of the duration dependency is affected.

Log-Logistic Model With Covariates in the λ Term

Once again, covariates are introduced into the log-logistic model by a log-linear relationship of the λ parameter and the covariate vector **x** (see Section 3.3.2): $\lambda(\mathbf{x}) = \exp(\mathbf{x}'\boldsymbol{\beta}^*)$. The log-logistic model may then be stated as:

$$\lambda^k(v|\mathbf{x}_k) = \frac{\exp(\mathbf{x}_k'\boldsymbol{\beta}^*)^{\alpha}\ \alpha\ v^{\alpha-1}}{1 + \exp(\mathbf{x}_k'\boldsymbol{\beta}^*)^{\alpha}\ v^{\alpha}}$$

$$= \frac{\exp(\mathbf{x}_k'\boldsymbol{\beta})\ \alpha\ v^{\alpha-1}}{1 + \exp(\mathbf{x}_k'\boldsymbol{\beta})\ v^{\alpha}} \qquad k = 1, 2, \ldots$$

with $\boldsymbol{\beta} = \alpha\boldsymbol{\beta}^*$.

Again, the covariate vector consists of the variables education (EDU), prestige (PRES), number of previously held jobs (NOJ), labor force experience before job entrance (LFX), as well as the cohort dummies C2 and C3 (see Appendix 1). If, for each new job appointment, adjustment processes that increase the job change risk actually do occur, then the nonmonotonic path of job change risk ($\alpha > 1$) should still be found after controlling for these covariates.

The program system GLIM is used to estimate the parameters.[22]

---

[22]If one were to calculate this log-logistic model with the BMDP program P3R and its integrated subprogram P3RFUN written by Trond Petersen which postulates the following rate function

$$\lambda^k(v|\mathbf{x}_k) = \frac{\exp(\mathbf{x}_k'\boldsymbol{\beta})\,(\alpha^* + 1)\,v^{\alpha^*}}{1 + \exp(\mathbf{x}_k'\boldsymbol{\beta})\,v^{\alpha^*+1}},$$

then the estimated $\hat{\alpha}^*$ coefficient must be transformed as follows in order to obtain the parameter $\hat{\alpha}$ of the above model:

$$\hat{\alpha} = \hat{\alpha}^* + 1.$$

The $\hat{\beta}$ coefficients (including the constant term $\hat{\beta}_0$) are identical for both approaches. The BMDP program required is the same as Program Example 6.12 with the exception that the model is specified with the number 5 (M5):

```
/INPUT UNIT IS 30.
       CODE IS DATA.
/VARIABLE NAMES ARE (63)DUR,(64)CEN,(65)EDU,(66)LFX,
                    (67)NOJ,(68)COHO2,(69)COHO3,(70)X1,(71)DP.
          ADD IS 9.
/TRANSFORM USE = (M3 EQ 1).
           DUR = M51 - M50 + 1.
           CEN = 1.
           IF (M51 EQ M47) THEN CEN = 0.
           IF (M41 EQ 1 AND M42 EQ 1) THEN EDU = 9.
           IF (M41 EQ 1 AND (M42 EQ 2 OR M42 EQ 3)) THEN EDU = 11.
           IF (M41 EQ 2 AND M42 EQ 1) THEN EDU = 10.
           IF (M41 EQ 2 AND (M42 EQ 2 OR M42 EQ 3)) THEN EDU = 12.
           IF (M41 EQ 3 AND (M42 EQ 1 OR M42 EQ 2 OR M42 EQ 3))
           THEN EDU = 13.
           IF (M42 EQ 4) THEN EDU = 17.
           IF (M42 EQ 5) THEN EDU = 19.
           COHO2 = 0.
           COHO3 = 0.
           IF(M48 GE 468 AND M48 LE 504) THEN COHO2 = 1.
           IF(M48 GE 588 AND M48 LE 624) THEN COHO3 = 1.
           LFX = M50 - M43.
           NOJ = M5 - 1.
           DP = 0.
           X1 = 0.
/REGRESS DEPENDENT IS DP.
         PARAMETERS ARE 8.
         PRINT IS 0.
         MEANSQUARE IS 1.0.
         ITERATIONS ARE 100.
         LOSS.
/PARAMETER INITIAL ARE -3.8,1.14,0.04,-0.01,0.15,-0.01,0.1,0.17.
           NAMES ARE CONST,TDEP,EDU,M59,NOJ,LFX,COHO2,COHO3.
/END.
/COMMENT'
M 5
T 70 63
D 71
C 64
I  6 65 59 67 66 68 69
 '.
```

Program Example 6.23:

```
$INPUT 40 MAKROS
$UNITS 3517$
$DATA TS TF CEN EDU PRES NOJ LFX C2 C3 DUR JN JN1 TMAR$
$DINPUT 20$
$CALC EDU(%NU)=0$
$CALC PRES(%NU)=0$
$CALC NOJ(%NU)=0$
$CALC LFX(%NU)=0$
$CALC C2(%NU)=0$
$CALC C3(%NU)=0$
$CALC U = %LOG(DUR)$
$CALC C = CEN$
$USE SETL$
$FIT+EDU+PRES+NOJ+LFX+C2+C3
$DIS E
$END$
$STOP$
```

In the program example above (in comparison to Program Example 6.22) it is necessary to first set the N+1th elements (%NU) of all variable vectors, which are to be included as covariates, equal to zero. Thereafter, the USE

Footnote 22 continued

The following program example shows the estimation of this log-logistic model with the procedure LIFEREG of SAS. SAS estimates the log-logistic model as a "regression model" for the log of the duration ($\log v = -\boldsymbol{\beta}^{*\prime}\mathbf{x} + \sigma\omega$), where $\sigma = \alpha^{-1}$, $\boldsymbol{\beta}^* = -\sigma\boldsymbol{\beta}$, and where ω has a logistic density (see Kalbfleisch and Prentice, 1980, p. 27 and Section 3.2.2):

```
LIBNAME SAS'[MAIR.MAIR.SAS]';
DATA LLOG;
 SET SAS.SYS;
 IF M3 = 1;
 DUR = M51 - M50 + 1;
 CEN = 1;
 IF (M51 EQ M47) THEN CEN = 0;
 IF (M41 = 1 AND M42 = 1) THEN EDU = 9;
 IF (M41 = 1 AND (M42 = 2 OR M42 = 3)) THEN EDU = 11;
 IF (M41 = 2 AND M42 = 1) THEN EDU = 10;
 IF (M41 = 2 AND (M42 = 2 OR M42 = 3)) THEN EDU = 12;
 IF (M41 = 3 AND (M42 = 1 OR M42 = 2 OR M42 = 3)) THEN EDU = 13;
 IF (M42 = 4) THEN EDU = 17;
 IF (M42 = 5) THEN EDU = 19;
 COHO2 = 0;
 COHO3 = 0;
 IF (M48 GE 468 AND M48 LE 504) THEN COHO2 = 1;
 IF (M48 GE 588 AND M48 LE 624) THEN COHO3 = 1;
 LFX = M50 - M43;
 NOJ = M5 - 1;
PROC LIFEREG DATA=LLOG;
 MODEL DUR*CEN(0)=EDU M59 NOJ LFX COHO2 COHO3/D=LLOGISTIC;
```

SETL command is followed by the additional commands FIT and DIS. The FIT command specifies the names of the relevant covariates (each one is marked with a positive sign), whereas the DIS command produces an output of the estimates and their asymptotic standard errors. The results of this estimation are presented in Table 6.18.[23]

Table 6.18: Results of the Log-Logistic Model From Program Example 6.23

```
        SCALED
CYCLE  DEVIANCE        DF
   2     3294.        3509

        ESTIMATE         S.E.         PARAMETER
   1  -4.333          0.1992          GM
   2    1.234         0.1954E-01  U
   3  0.1038E-01  0.2088E-01  EDU
   4 -0.7245E-02  0.1963E-02  PRES
   5  0.2661          0.1990E-01  NOJ
   6 -0.1264E-01  0.6258E-03  LFX
   7  0.1244          0.7263E-01  C2
   8  0.3244          0.8027E-01  C3
```

First of all, looking at Table 6.18 it is evident that compared to the log-logistic model without covariates, the deviance value has been reduced from 3881 to 3294 through the estimation of the additional parameters. This reduction of –2logL corresponds to a χ^2 value of 587 (given six degrees of freedom) and implies that the model was significantly improved with the introduction of the covariates.

---

[23]Using Trond Petersen's model M5 and given a value of the log likelihood function (≙ -LOSS) of –13830.3, the following result is obtained:

PARAMETER	ESTIMATE	ASYMPTOTIC STANDARD DEVIATION
CONST	-4.338965	0.200337
TDEP	0.236721	0.021759
EDU	0.010408	0.020212
M59	-0.007254	0.001841
NOJ	0.268493	0.017634
LFX	-0.012778	0.000553
COHO2	0.124088	0.071432
COHO3	0.324118	0.082421

Disregarding the rounding off of figures once again, it is observed that the $\hat{\beta}$ coefficients of the covariates (including the regression coefficient $\hat{\beta}_0$) are in accordance with those in Table 6.18, and the estimations of the coefficient α may be transformed as follows:

$$\hat{\alpha} = \hat{\alpha}^* + 1 = 0.236721 + 1 = 1.236721.$$

Using the procedure LIFEREG of SAS, the following result is obtained:

Although the nonstandardized β coefficients of the log-logistic model in Table 6.18 clearly differ from the β coefficients of the Gompertz (see Section 6.5.1) and the Weibull models (see Section 6.5.2), the direction of influence of the covariates and their significance remains unchanged.

Again, an examination of the null hypothesis $H_0 : \alpha \leq 1$ in comparison to the alternative hypothesis $H_1 : \alpha > 1$ shows that even controlling for the covariates, a nonmonotonic risk of job change is maintained:

$$z = \frac{\hat{\alpha} - 1}{s(\hat{\alpha})} = \frac{1.234 - 1}{0.01954} = 11.98.$$

Again, this supports the hypothesis that the initial increased risk of job change diminishes after a phase of adjustment to the new job has taken place and job specific knowledge has been accumulated.

Footnote 23 continued

```
                                   SAS

                      L I F E R E G   P R O C E D U R E

DATA SET          =WORK.LLOG
DEPENDENT VARIABLE=DUR
CENSORING VARIABLE=CEN
CENSORING VALUE(S)=     0
NONCENSORED VALUES=2586  CENSORED VALUES= 930
OBSERVATIONS WITH MISSING VALUES= 330

LOGLIKELIHOOD FOR LLOGISTC       -5232.06

VARIABLE   DF   ESTIMATE STD ERR CHISQUARE  PR>CHI LABEL/VALUE

INTERCPT    1     3.50845 0.151579    535.74  0.0001 INTERCEPT
EDU         1 -.00841777 0.017145 0.241058  0.6234
M59         1 0.00586536 .0016193  13.1196  0.0003
NOJ         1   -0.217112 .0162976  177.468  0.0001
LFX         1    0.010333  5.3E-04  386.223  0.0001
COHO2       1   -0.100343 .0594566  2.84824  0.0915
COHO3       1   -0.262084 .0652341  16.1411  0.0001
SCALE       1    0.808588 .0131956                  LOG LOGISTIC SCALE PARAMETER
```

Given the parametrization in SAS, the estimated coefficients $\hat{\boldsymbol{\beta}}^*$ and the "SCALE parameter" $\hat{\sigma}$ must be transformed as follows in order to obtain the parameters $\hat{\alpha}$ and $\hat{\boldsymbol{\beta}}$ in Table 6.18:

$$\hat{\alpha} = \frac{1}{\hat{\sigma}},$$

$$\hat{\boldsymbol{\beta}} = -\frac{1}{\hat{\sigma}}\hat{\boldsymbol{\beta}}^*.$$

For example:

$$\hat{\alpha} = \frac{1}{0.808588} = 1.2367,$$

$$\hat{\beta}_0 = -\frac{3.5085}{0.808588} = -4.33898.$$

Examination of the Residuals in the Log-Logistic Model

As has been extensively discussed in Sections 3.7.1 and 6.2.3, in the log-logisitic model the cumulative hazard rates $\Lambda(t_i|x_i)$ can be regarded as residuals r_i, and consequently used to evaluate the model. The residual estimates $\hat{r}_{ik}$ are obtained as follows:

$$\hat{r}_{ik} = \hat{\Lambda}(v_{ik}|\mathbf{x}_{ik}) = \ln(1 + \exp(\mathbf{x}'_{ik}\hat{\boldsymbol{\beta}})v_{ik}^{\hat{\alpha}}).$$

From the residuals calculated in this way, given censored data, the survivor function can be estimated with the aid of the product-limit method and transformed and plotted against r. If the assumption of a log-logistically distributed hazard rate is correct, then a line with the slope of –1 should result.

The following two program runs, with the subprogram P1L of BMDP, demonstrate this residual test for the log-logistic model without covariates and the log-logistic model in which the λ coefficient is related in a log-linear fashion to the covariate vector.

Program Example 6.24:

```
/INPUT UNIT IS 30.
       CODE IS DATA.
/VARIABLE NAMES ARE (63)DUR,(64)CEN,(65)EDU,(66)LFX,
                    (67)NOJ,(68)COHO2,(69)COHO3,(70)RESID1.
          ADD IS 8.
/TRANSFORM USE = (M3 EQ 1).
           DUR = M51 - M50 + 1.
           CEN = 1.
           IF (M51 EQ M47) THEN CEN = 0.
           IF (M41 EQ 1 AND M42 EQ 1) THEN EDU = 9.
           IF (M41 EQ 1 AND (M42 EQ 2 OR M42 EQ 3)) THEN EDU = 11.
           IF (M41 EQ 2 AND M42 EQ 1) THEN EDU = 10.
           IF (M41 EQ 2 AND (M42 EQ 2 OR M42 EQ 3)) THEN EDU = 12.
           IF (M41 EQ 3 AND (M42 EQ 1 OR M42 EQ 2 OR M42 EQ 3))
           THEN EDU = 13.
           IF (M42 EQ 4) THEN EDU = 17.
           IF (M42 EQ 5) THEN EDU = 19.
           COHO2 = 0.
           COHO3 = 0.
           IF(M48 GE 468 AND M48 LE 504) THEN COHO2 = 1.
           IF(M48 GE 588 AND M48 LE 624) THEN COHO3 = 1.
           LFX = M50 - M43.
           NOJ = M5 -1.
           RESID1 = LN(1+EXP(-4.464)*DUR**1.144).
/FORM TIME IS RESID1.
      STATUS IS CEN.
      RESPONSE IS 1.
```

```
/ESTIMATE METHOD IS PROD.
          PLOTS ARE LOG.
/PRINT CASES ARE 0.
/END
```

Program Example 6.25:

```
/INPUT UNIT IS 30.
       CODE IS DATA.
/VARIABLE NAMES ARE (63)DUR,(64)CEN,(65)EDU,(66)LFX,
                    (67)NOJ,(68)COHO2,(69)COHO3,
                    (70)RESID2,(71)ALPHA,(72)LAMBDA.
          ADD IS 10.
/TRANSFORM USE = (M3 EQ 1).
           DUR = M51 - M50 + 1.
           CEN = 1.
           IF (M51 EQ M47) THEN CEN = 0.
           IF (M41 EQ 1 AND M42 EQ 1) THEN EDU = 9.
           IF (M41 EQ 1 AND (M42 EQ 2 OR M42 EQ 3)) THEN EDU = 11.
           IF (M41 EQ 2 AND M42 EQ 1) THEN EDU = 10.
           IF (M41 EQ 2 AND (M42 EQ 2 OR M42 EQ 3)) THEN EDU = 12.
           IF (M41 EQ 3 AND (M42 EQ 1 OR M42 EQ 2 OR M42 EQ 3))
           THEN EDU = 13.
           IF (M42 EQ 4) THEN EDU = 17.
           IF (M42 EQ 5) THEN EDU = 19.
           COHO2 = 0.
           COHO3 = 0.
           IF(M48 GE 468 AND M48 LE 504) THEN COHO2 = 1.
           IF(M48 GE 588 AND M48 LE 624) THEN COHO3 = 1.
           LFX = M50 - M43.
           NOJ = M5 -1.
           ALPHA = 1.234.
           LAMBDA = EXP(-4.333 + 0.01038*EDU -0.007245*M59 +
                    0.2661*NOJ - 0.01264*LFX +0.1244*COHO2 +
                    0.3244*COHO3).
           RESID2 = LN(1 + LAMBDA * DUR ** ALPHA).
/FORM TIME IS RESID2.
      STATUS IS CEN.
      RESPONSE IS 1.
/ESTIMATE METHOD IS PROD.
          PLOTS ARE LOG.
          NO PRINT.
/PRINT CASES ARE 0.
/END
```

Figure 6.13: Residual Plot for the Log-Logistic Model Without Covariates

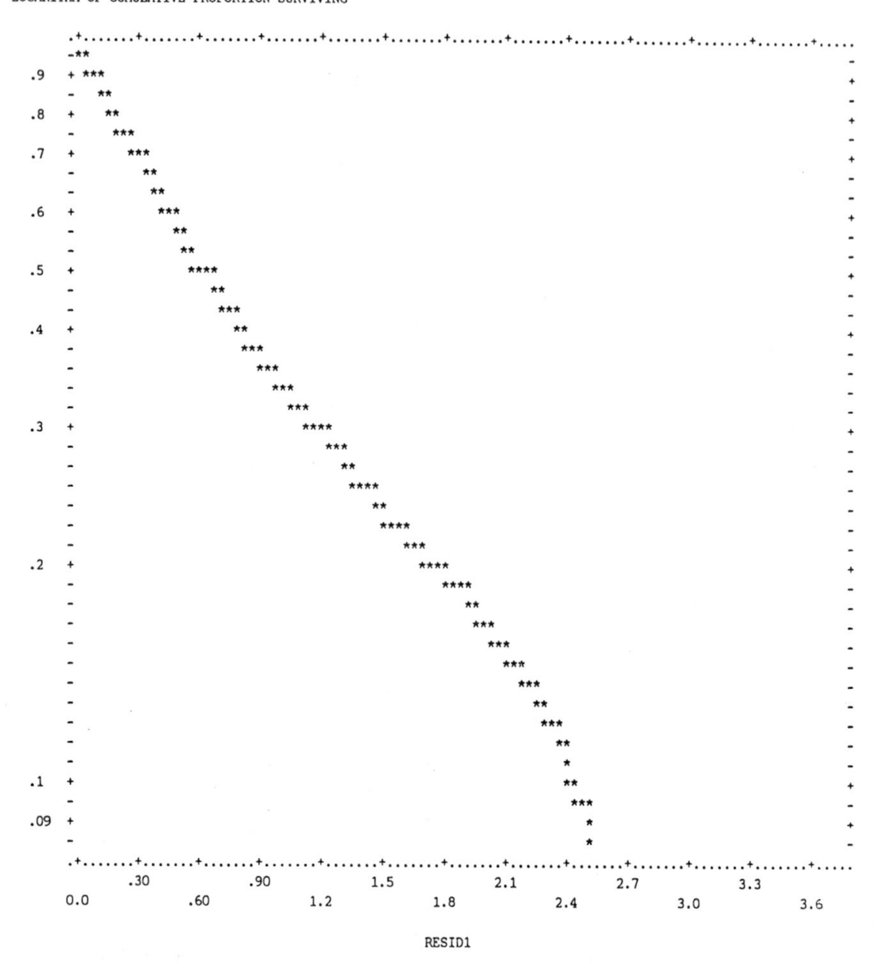

In the TRANSFORM paragraphs of the above program runs, first the residuals (RESID1 or RESID2) are calculated, based upon the coefficient estimates from Tables 6.17 and 6.18. The FORM paragraphs then insert these as durations (TIME IS RESID1 or TIME IS RESID2) along with the censoring variables (STATUS IS CEN). The survivor functions are then estimated using the product-limit method (METHOD IS PROD), and finally logarithmic survivor functions (PLOTS ARE LOG) are plotted.

The results of these tests are shown in Figures 6.13 (for the log-logistic model without covariates) and 6.14 (for the log-logistic model with covari-

Figure 6.14: Residual Plot for the Log-Logistic Model With Covariates in the λ Term

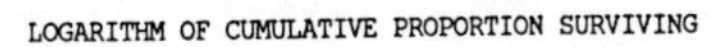

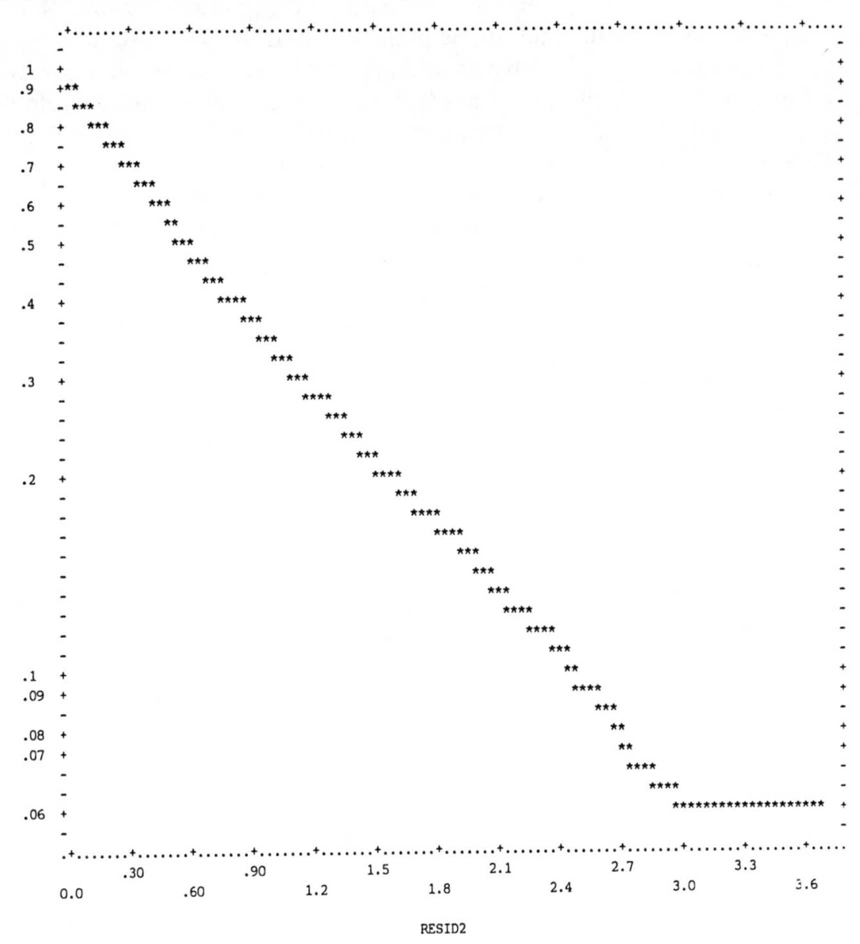

ates). Both plots demonstrate paths that coincide relatively well with a line possessing the slope of –1, and the inclusion of covariates even improves this fit. If these residual plots are compared with the residual plots of a Gompertz distribution, we observe that both distributions describe the process of job shift process given small and middle size residuals relatively well. However, in the area of large residuals clear deviations exist. This implies that, especially given long durations, both models are less than completely appropriate.

Analysis of the Marriage Process With the Aid of a Log-Logistic Model

An instructive example of the use of the log-logistic distribution is seen in some research by Papastefanou (1988) who, based upon the GLHS, examined the process of marriage for men and women after the legal marriage age. Since marriage rate has a nonmonotonic path, which at first increases and then decreases, the log-logistic distribution is applicable. This nonmonotonic marriage rate path is not only historically (see Papastefanou, 1988) and interculturally (see Coale, 1971) stable, but can also be explained theoretically (see Keely, 1979; Hernes, 1972; Sørensen and Sørensen, 1984).

The aim of Papastefanou's study was to show the extent to which the marriage process is determined by social structural attributes (such as educational level, occupational position, labor force experience, etc.), in which women and men exhibit large differences, and also to look at effects due to remaining sex specific differences. According to Coleman (1984a), general marriage rate differences between men and women can be traced to processes of psychobiological maturation, which exert more pressure on women to marry than on men. This then is the reason why marriage chances for women based upon attractiveness and fertility, and independent of social structural attributes such as level of education or occupational position, are clearly more pronounced in a specific segment of the lifetime of women as opposed to men.

Figure 6.15: Average Marriage Rate of Men and Women Given the Control of Relevant Social Structural Characteristics (Estimated With a Log-Logistic Model)

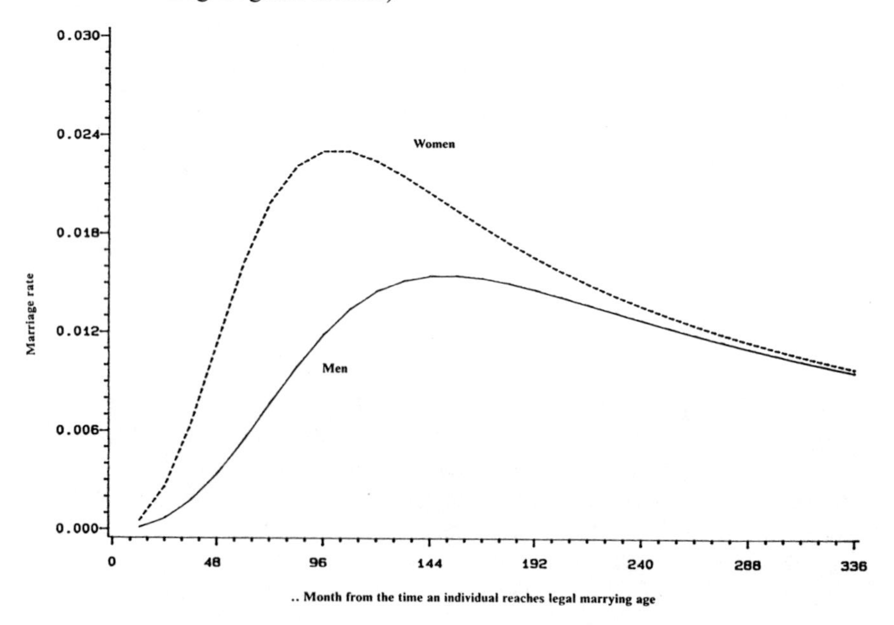

The beginning of the duration or spell in this example is defined by the legally prescribed minimum marriage age. Duration is then determined as the number of months from the time an individual reaches legal marrying age to the first marriage. We are thus dealing here with a one-episode case with two states (not married, married).

The estimation of the log-logistic model shows first of all that the difference between the marriage rate of men and women is highly significant and exists even after controlling for relevant social structural variables (see Papastefanou, 1988). The partial effect of gender upon the marriage rate, however, may not be directly interpreted in the log-logistic model, since the log-logistic rate function is dependent upon duration and varies in magnitude according to the duration. This is evident in Figure 6.15, which illustrates the log-logistic model estimations for an "average" marriage rate path for women and men, in which the relevant social structural variables have been controlled using the respective means. The difference between men and women at the beginning of the process increases, reaching its maximum some eight years after legal marriage age and decreases steadily with time. Thus, the inclination for women to get married is especially large in the first phase of the process in comparison to men. This can be interpreted as some support for the Coleman (1984a) hypothesis that the propensity for women to get married is clearly more concentrated within a specific lifetime segment due to the psychobiological conditions.

6.5.4 Lognormal Distribution

The lognormal distribution, like the log-logistic model, is also a widely used model of time dependence that implies a nonmonotonic relationship between hazard rate and the duration: The hazard rate initially increases to a maximum and then decreases. The lognormal distribution is especially simple to use if there is no censoring, but with censoring the computations quickly become formidable (see Kalbfleisch and Prentice, 1980). Like the exponential (see Footnote 4), the Weibull (see Footnote 18), and the log-logistic model (see Footnote 22), the lognormal model may be estimated with the procedure LIFEREG of SAS. SAS estimates the lognormal model as a "regression model" for the log of the duration

$$\log v = -\boldsymbol{\beta}^*\mathbf{x} + \sigma\omega,$$

where ω has a standard normal distribution. The density function for v can be written (see Kalbfleisch and Prentice, 1980, p. 24)

$$f(v) = (2\pi)^{-1/2}\,\alpha v^{-1} \exp\left(\frac{-\alpha^2 (\log\lambda v)^2}{2}\right),$$

where $\sigma = \alpha^{-1}$, $\lambda = \exp(-\boldsymbol{\beta}^*\mathbf{x})$.

The estimation of this model is shown in Program Example 6.26.

Program Example 6.26:

```
LIBNAME SAS'[MAIR.MAIR.SAS]';
DATA LNORM;
 SET SAS.SYS;
 IF M3 = 1;
 DUR = M51 - M50 + 1;
 CEN = 1;
 IF (M51 EQ M47) THEN CEN = 0;
 IF (M41 = 1 AND M42 = 1) THEN EDU = 9;
 IF (M41 = 1 AND (M42 = 2 OR M42 = 3)) THEN EDU = 11;
 IF (M41 = 2 AND M42 = 1) THEN EDU = 10;
 IF (M41 = 2 AND (M42 = 2 OR M42 = 3)) THEN EDU = 12;
 IF (M41 = 3 AND (M42 = 1 OR M42 = 2 OR M42 = 3)) THEN EDU = 13;
 IF (M42 = 4) THEN EDU = 17;
 IF (M42 = 5) THEN EDU = 19;
 COHO2 = 0;
 COHO3 = 0;
 IF (M48 GE 468 AND M48 LE 504) THEN COHO2 = 1;
 IF (M48 GE 588 AND M48 LE 624) THEN COHO3 = 1;
 LFX = M50 - M43;
 NOJ = M5 - 1;
PROC LIFEREG DATA=LNORM;
 MODEL DUR*CEN(0)=EDU M59 NOJ LFX COHO2 COHO3/D=LNORMAL;
```

After transformation, the variables are used in the same way shown in the previous sections. The PROC LIFEREG statement invokes the procedure. The input SAS data set LLOG is used. The MODEL statement specifies what variables are to be used in the regression. Again, the covariate vector consists of the variables education (EDU), prestige (PRES), number of previously held jobs (NOJ), labor force experience (LFX), as well as the cohort dummies COHO2 and COHO3. It also indicates that the duration is contained in a variable named DUR and that if the variable CEN takes on the value 0 the observation is right censored. D = LNORMAL tells SAS to use a lognormal model.

The results of this estimation are presented in Table 6.19. The lognormal model has a log likelihood value of –5225.4 and fits therefore a little better than the log-logistic model (log likelihood = –5232.4). But the lognormal model leads to similar β coefficients (see Footnote 23) with similar interpretations. Given the parametrization in SAS, the "SCALE parameter" $\hat{\sigma}$ must be transformed in order to obtain $\hat{\alpha}$: $\hat{\alpha} = \frac{1}{\hat{\sigma}} = \frac{1}{1.39748} = 0.7155$. This means that we observe at first a steeply rising risk of the job change, which decreases after an initial phase of adjustment in each new job.

Table 6.19: Results of the Lognormal Model From Program Example 6.26

```
                                        SAS

                            L I F E R E G   P R O C E D U R E

DATA SET            =WORK.LNORM
DEPENDENT VARIABLE=DUR
CENSORING VARIABLE=CEN
CENSORING VALUE(S)=      0
NONCENSORED VALUES=2586  CENSORED VALUES= 930
OBSERVATIONS WITH MISSING VALUES= 330

LOGLIKELIHOOD FOR LNORMAL          -5225.4

VARIABLE   DF    ESTIMATE STD ERR CHISQUARE  PR>CHI LABEL/VALUE

INTERCPT    1     3.46408 0.152254  517.653  0.0001 INTERCEPT
EDU         1 -.00195817 0.017053 .0131855  0.9086
M59         1 0.00577093 .0015849  13.2591  0.0003
NOJ         1   -0.197323 .0154865  162.348  0.0001
LFX         1 0.00938467  4.9E-04  372.965  0.0001
COHO2       1 -0.0718115 .0593385  1.46458  0.2262
COHO3       1   -0.249695 .0662887  14.1887  0.0002
SCALE       1     1.39748 .0202814                  LOG NORMAL SCALE PARAMETER
```

6.6 Models With Unobserved Heterogeneity

In the parametric model examples, we presumed that all of the relevant influences had been measured and included in the model. The hazard rate was completely determined by the independent variables, including duration dependency which was used as a proxy variable. Differences in the hazard rate could consequently only be a result of variation in these independent characteristics.

Such an assumption is naturally restrictive, and may seldom be fulfilled. In addition to the covariates included in the model, further characteristics that were not recorded or not known influence the hazard rate. If in estimating the hazard rate one aggregates these unobserved differences, then, as discussed in Section 3.9.1, an apparent duration dependency occurs. At the level of the hazard rate to be analyzed, it is no longer possible to differentiate whether the hazard rate falls with increasing duration for each individual or if this is simply a methodological artifact due to neglected differences between individuals. In the examples concerned with the risk of a job change for men we could therefore also argue that the negative duration dependency was not founded on job specific human capital investment increases, but rather was the consequence of unobserved heterogeneity.

In the job change example in this section we will examine the negative duration dependency from this methodological viewpoint and allow for unobserved heterogeneity in the form of an error term. In so doing, we will first use Tuma's (1978) relatively simple model in which individual hazard rates are postulated to be constant over time. As such, no duration dependency is permitted at the individual level. Individuals may, however, depending upon their unobserved heterogeneity, show different rates. The distribution over these rates is considered to be characterized by a gamma distribution. As already mentioned in Section 3.9.2, unobserved heterogeneity results in a negative duration dependency of the estimated hazard rate.

Model Without Observed Heterogeneity

In order to illustrate the interpretation of this model, we first compute a model for the job change risk of men based upon the multiepisode case, given unobserved heterogeneity. As in the previous Section 3.9.2, we analyze a model of the form

$$\lambda^k(v|\epsilon) = \lambda\epsilon \qquad , k = 1, 2, \ldots .$$

Consequently, the individual level hazard rates are presumed to be time independent, and the heterogeneity component is characterized by a gamma distribution. In Section 3.9.2 we saw that due to $E(\epsilon) = 1$ the gamma distribution has only one freely varying parameter α (see relation (3.9.8)), which is set in RATE to $\alpha = \exp(-\gamma_0)$. Furthermore, in RATE we estimate $\lambda = \exp(\beta_0)$.

Program Example 6.27:

```
RUN NAME        GAMMA-MODEL
N OF CASES      3516
VARIABLES       12
TS              1
TF              2
CEN             3
EDU             4
PRES            5
NOJ             6
LFX             7
COHO2           8
COHO3           9
DUR             10
JOBN            11
JOBN1           12
```

```
READ DATA
(12F5.0)
T AND S          10 3
MODEL            (2) A=1 B=1
VECTOR           (1) (2)
SOLVE
FINISH
```

In the above RATE program, the model is specified on the MODEL statement by the number (2), such that the letters A and B are each modeled as related in a log-linear fashion with the covariate vector $\mathbf{x}$ (A = 1, B = 1) A = $\exp(\mathbf{x}'\boldsymbol{\beta})$, B = $\exp(\mathbf{x}'\boldsymbol{\gamma})$. Since the VECTOR command does not include covariates on either the first (1) or the second (2) vector: only the constants β_0 and γ_0 are estimated. The results are presented below in Table 6.20.

Table 6.20: Results From Program Example 6.27 With Unobserved Heterogeneity

DESTINATION	UNWEIGHTED FREQUENCY	WEIGHTED FREQUENCY	MAX(LOG OF L) NULL HYPOTHESIS	MAX(LOG OF L) ALTERNATIVE HYPOTHESIS	PSEUDO R-SQUARED	CHI-SQUARED	DF	PROBABILITY LEVEL
1	2586	2586.0	-1.443432D+04	-1.415316D+04	0.0195	562.34	1	0.00D+00

DESTINATION 1 LETTER A LOG-LINEAR TIME-INDEPENDENT VECTOR

INTERNAL NUMBER	VARIABLE NUMBER	VARIABLE NAME	VECTOR 1 PARAMETER	PARAMETER STANDARD ERROR	PARAMETER F RATIO	ANTILOG OF THE PARAMETER	ANTILOG STANDARD ERROR	ANTILOG F RATIO
1		(CONSTANT)	-3.974D+00	3.481D-02	13029.359	1.880D-02		

DESTINATION 1 LETTER B LOG-LINEAR TIME-INDEPENDENT VECTOR

INTERNAL NUMBER	VARIABLE NUMBER	VARIABLE NAME	VECTOR 2 PARAMETER	PARAMETER STANDARD ERROR	PARAMETER F RATIO	ANTILOG OF THE PARAMETER	ANTILOG STANDARD ERROR	ANTILOG F RATIO
2		(CONSTANT)	-1.244D-01	5.622D-02	4.893	8.831D-01		

From Table 6.20 we obtain an estimation for β_0 of –3.974 and for γ_0 a value of –0.1244. Thus, the estimated error variance is

$$\hat{\mathrm{Var}}(\epsilon) = \frac{1}{\hat{\alpha}} = \exp(\hat{\gamma}_0) = 0.833.$$

Using $\phi(\mathbf{x}'\boldsymbol{\beta}) = \exp(\beta_0)$ and $\alpha = \exp(-\gamma_0)$ in (3.9.12), one obtains for the hazard rate

$$\lambda(\mathrm{v}) = \frac{\exp(\beta_0 - \gamma_0)}{\exp(\beta_0)\,\mathrm{v} + \exp(-\gamma_0)} = \frac{\exp(-\gamma_0)}{\exp(-\beta_0 - \gamma_0) + \mathrm{v}}$$

$$= \frac{\exp(0.1244)}{\exp(3.974 + 0.1244) + \mathrm{v}} = \frac{1.1325}{60.2438 + \mathrm{v}}.$$

Although the individual hazard rates were assumed to be time-independent, a monotonically falling path with increased duration is obtained for the estimated hazard rate.

Regression Model With Observed and Unobserved Heterogeneity

After having calculated a measure of the total heterogeneity present using the error variance $Var(\epsilon)$, it is possible in a second step to determine the magnitude of the total amount of variance which can be explained by the covariates included in the model (observed heterogeneity) as well as the amount of unobserved heterogeneity that remains. As was the case in the exponential model the covariate vector **x** is related log-linearly to λ ($\lambda(\mathbf{x}) = \exp(\mathbf{x}'\boldsymbol{\beta})$) and the variance of the error term ϵ is related log-linearly to the parameter γ_0 ($Var(\epsilon) = \exp(\gamma_0)$), so that the following rate equation may be formulated

$$\lambda^k(v|\mathbf{x},\epsilon) = \exp(\mathbf{x}'\boldsymbol{\beta})\epsilon \qquad \text{with } k = 1, 2, \ldots .$$

Program Example 6.28:

```
RUN NAME          GAMMA-MODEL
N OF CASES        3516
VARIABLES         12
TS                1
TF                2
CEN               3
EDU               4
PRES              5
NOJ               6
LFX               7
COHO2             8
COHO3             9
DUR               10
JOBN              11
JOBN1             12
READ DATA
(12F5.0)
T AND S           10 3
MODEL             (2) A=1 B=1
VECTOR            (1) 4 5 6 7 8 9 (2)
SOLVE
FINISH
```

Compared to Program Example 6.27, covariates in the above RATE program run are specified with their RATE assigned numbers (see Appendix 1) in the first vector on the VECTOR command. The results are presented in Table 6.21.

After inclusion of the covariates, an error variance of $Var(\epsilon) = \exp(-0.4523) = 0.6362$ remains. With the aid of a PRE measure, with which the error variance of the model without covariates $Var(\epsilon_0)$ is related to the error

Table 6.21: Results With Observed and Unobserved Heterogeneity for Program Example 6.28

DESTINATION	UNWEIGHTED FREQUENCY	WEIGHTED FREQUENCY	MAX(LOG OF L) NULL HYPOTHESIS	MAX(LOG OF L) ALTERNATIVE HYPOTHESIS	PSEUDO R-SQUARED	CHI-SQUARED	DF	PROBABILITY LEVEL
1	2586	2586.0	-1.443432D+04	-1.387753D+04	0.0386	1113.59	7	0.00D+00

DESTINATION 1 LETTER A LOG-LINEAR TIME-INDEPENDENT VECTOR

INTERNAL NUMBER	VARIABLE NUMBER	VARIABLE NAME	VECTOR 1 PARAMETER	PARAMETER STANDARD ERROR	PARAMETER F RATIO	ANTILOG OF THE PARAMETER	ANTILOG STANDARD ERROR	ANTILOG F RATIO
1		(CONSTANT)	-3.617D+00	1.677D-01	465.263	2.686D-02		
2	4	EDU	6.965D-03	1.882D-02	0.137	1.007D+00	1.895D-02	0.136
3	5	PRES	-6.173D-03	1.779D-03	12.043	9.938D-01	1.768D-03	12.118
4	6	NOJ	2.193D-01	1.751D-02	156.814	1.245D+00	2.180D-02	126.442
5	7	LFX	-1.103D-02	5.797D-04	361.990	9.890D-01	5.733D-04	366.008
6	8	COHO2	9.830D-02	6.409D-02	2.352	1.103D+00	7.071D-02	2.134
7	9	COHO3	2.650D-01	7.074D-02	14.031	1.303D+00	9.221D-02	10.828

DESTINATION 1 LETTER B LOG-LINEAR TIME-INDEPENDENT VECTOR

INTERNAL NUMBER	VARIABLE NUMBER	VARIABLE NAME	VECTOR 2 PARAMETER	PARAMETER STANDARD ERROR	PARAMETER F RATIO	ANTILOG OF THE PARAMETER	ANTILOG STANDARD ERROR	ANTILOG F RATIO
8		(CONSTANT)	-4.523D-01	6.511D-02	48.265	6.361D-01		

variance of the model presented above $\text{Var}(\epsilon)$, we find that with the inclusion of the covariates the total measured error variance can be reduced by only 23.65 percent and 76.35 percent remains unexplained:

$$\text{PRE} = \frac{\text{Var}(\epsilon_0) - \text{Var}(\epsilon)}{\text{Var}(\epsilon_0)} = \frac{0.833 - 0.636}{0.833} = 23.65.$$

This result is naturally not surprising, since we obtained a significant effect of the duration dependency for the Gompertz, Weibull, log-logistic, and lognormal distributions.

A similar model with unobserved heterogeneity can be estimated with SAS (see also Hannan and Freeman, 1988). SAS introduces a gamma-distributed disturbance into the regression model for the log of the duration:

$$\log v = -\boldsymbol{\beta}^{*\prime}\mathbf{x} + \omega,$$

where the density of an error term, ω, equals:

$$f(\omega) = \frac{\exp(k\omega - e^{\omega})}{\Gamma(k)}.$$

The rate is a monotonic increasing function of duration if $k > 1$ and a decreasing function of duration if $k < 1$. This model has as a special case the exponential ($k = 1$), which means that there is neglible unobserved heterogeneity (see Hannan and Freeman, 1988). This model may therefore be used to construct a likelihood ratio test of the gamma model against the exponential. The estimation of this model is shown in Program Example 6.29. The MODEL statement indicates that the duration is contained in variable DUR, that the variable CEN takes on the value 0, if the observation is censored, and that the variables EDU, PRES, NOJ, LFX, COHO2, and COHO3 are used as covariates. D = GAMMA NOSCALE tells SAS to use a gamma model.

Program Example 6.29:

```
LIBNAME SAS'[MAIR.MAIR.SAS]';
DATA GAM;
 SET SAS.SYS;
 IF M3 = 1;
 DUR = M51 - M50 + 1;
 CEN = 1;
 IF (M51 EQ M47) THEN CEN = 0;
 IF (M41 = 1 AND M42 = 1) THEN EDU = 9;
 IF (M41 = 1 AND (M42 = 2 OR M42 = 3)) THEN EDU = 11;
 IF (M41 = 2 AND M42 = 1) THEN EDU = 10;
 IF (M41 = 2 AND (M42 = 2 OR M42 = 3)) THEN EDU = 12;
 IF (M41 = 3 AND (M42 = 1 OR M42 = 2 OR M42 = 3)) THEN EDU = 13;
 IF (M42 = 4) THEN EDU = 17;
 IF (M42 = 5) THEN EDU = 19;
 COHO2 = 0;
 COHO3 = 0;
 IF (M48 GE 468 AND M48 LE 504) THEN COHO2 = 1;
 IF (M48 GE 588 AND M48 LE 624) THEN COHO3 = 1;
 LFX = M50 - M43;
 NOJ = M5 - 1;
PROC LIFEREG DATA=GAM;
 MODEL DUR*CEN(0)=EDU M59 NOJ LFX COHO2 COHO3/D=GAMMA NOSCALE;
```

The results of this estimation are presented in Table 6.22. Because the GAMMA parameter is significant and not equal to 1, there is unobserved heterogeneity that can not be neglected.

In SAS it is also possible to estimate a generalized gamma model. With this model we can test, whether the generalized model including the gamma distributed disturbance is a significant improvement over the Weibull with covariates and an exponential with covariates (see Hannan and Freeman, 1988). This model is also specified in terms of the regression model of the log of duration:

$$\log v = -\boldsymbol{\beta}^{*\prime}\mathbf{x} + \sigma\omega,$$

where $\sigma = \alpha^{-1}$ and ω has the density

$$f(\omega) = \frac{\exp(k\omega - e^{\omega})}{\Gamma(k)}.$$

Table 6.22: Results of the Gamma Model From Program Example 6.29

```
                                    SAS

                         L I F E R E G  P R O C E D U R E

DATA SET            =WORK.GAM
DEPENDENT VARIABLE=DUR
CENSORING VARIABLE=CEN
CENSORING VALUE(S)=     0
NONCENSORED VALUES=2586  CENSORED VALUES= 930
OBSERVATIONS WITH MISSING VALUES= 330

LOGLIKELIHOOD FOR GAMMA         -5471.79

VARIABLE   DF   ESTIMATE STD ERR CHISQUARE  PR>CHI LABEL/VALUE

INTERCPT    1     4.17548 0.127421   1073.82  0.0001 INTERCEPT
EDU         1 -0.0125919 .0139095 0.819518  0.3653
M59         1 0.00525305 0.001326  15.6937  0.0001
NOJ         1   -0.174064 .0119789  211.146  0.0001
LFX         1 0.00870388  4.4E-04  395.465  0.0001
COHO2       1   -0.168383 .0453477  13.7875  0.0002
COHO3       1   -0.454416 .0510537  79.2231  0.0001
SCALE       0          1        0                  GAMMA SCALE PARAMETER
GAMMA       1   0.749133 .0525093                  GAMMA SHAPE PARAMETER
LAGRANGE MULTIPLIER CHI-SQUARE FOR SCALE     767.3138 PR>CHI .0001
```

This model has as special cases, the exponential model ($\sigma = 1$, $k = 1$), the Weibull model ($k = 1$, $\sigma \neq 1$), and the gamma model ($\sigma = 1$, $k \neq 1$). The estimation of this generalized gamma model is shown in Program Example 6.30. D = GAMMA tells SAS to use a generalized gamma model.

Program Example 6.30:

```
LIBNAME SAS'[MAIR.MAIR.SAS]';
DATA GGAM;
 SET SAS.SYS;
 IF M3 = 1;
 DUR = M51 - M50 + 1;
 CEN = 1;
 IF (M51 EQ M47) THEN CEN = 0;
 IF (M41 = 1 AND M42 = 1) THEN EDU = 9;
 IF (M41 = 1 AND (M42 = 2 OR M42 = 3)) THEN EDU = 11;
 IF (M41 = 2 AND M42 = 1) THEN EDU = 10;
 IF (M41 = 2 AND (M42 = 2 OR M42 = 3)) THEN EDU = 12;
```

```
 IF (M41 = 3 AND (M42 = 1 OR M42 = 2 OR M42 = 3)) THEN EDU = 13;
 IF (M42 = 4) THEN EDU = 17;
 IF (M42 = 5) THEN EDU = 19;
 COHO2 = 0;
 COHO3 = 0;
 IF (M48 GE 468 AND M48 LE 504) THEN COHO2 = 1;
 IF (M48 GE 588 AND M48 LE 624) THEN COHO3 = 1;
 LFX = M50 - M43;
 NOJ = M5 - 1;
PROC LIFEREG DATA=GGAM;
 MODEL DUR*CEN(0)=EDU M59 NOJ LFX COHO2 COHO3/D=GAMMA;
```

The results of this estimation are presented in Table 6.23. The GAMMA and the SCALE parameter are significant ($p = 0.05$) and do not equal 1. This means, we have a Weibull model with a decreasing hazard rate ($\hat{\alpha} = \frac{1}{\hat{\sigma}} = \frac{1}{1.41319} = 0.708$) and there is unobserved heterogeneity, left in the model (GAMMA : $-0.13771 \neq 1$).

Table 6.23: Results of the Generalized Gamma Model From Program Example 6.30

```
                                        SAS

                           L I F E R E G   P R O C E D U R E

DATA SET           =WORK.GGAM
DEPENDENT VARIABLE=DUR
CENSORING VARIABLE=CEN
CENSORING VALUE(S)=      0
NONCENSORED VALUES=2586  CENSORED VALUES= 930
OBSERVATIONS WITH MISSING VALUES= 330

LOGLIKELIHOOD FOR GAMMA           -5222.81
```

VARIABLE	DF	ESTIMATE	STD ERR	CHISQUARE	PR>CHI	LABEL/VALUE
INTERCPT	1	3.37135	0.15707	460.706	0.0001	INTERCEPT
EDU	1	-.00062143	.0169325	.0013469	0.9707	
M59	1	0.00570094	.0015649	13.2713	0.0003	
NOJ	1	-0.195245	.0154824	159.031	0.0001	
LFX	1	0.00911273	4.9E-04	348.321	0.0001	
COHO2	1	-0.0591734	.0597826	0.979724	0.3223	
COHO3	1	-0.23319	.0668574	12.1652	0.0005	
SCALE	1	1.41319	.0213357			GAMMA SCALE PARAMETER
GAMMA	1	-0.13771	0.060107			GAMMA SHAPE PARAMETER

Appendices

Appendix 1: List of Variable Names Used in Examples

Variable Name	Definition
NOJ	Number of previously held jobs
LFX, ILFX[2]	Labor force experience in months
LFXS	Labor force experience in months, measured at the beginning of each episode
LFXF	Labor force experience in months, measured at the end of each episode
EDU, ED[1]	Education, measured in years: 9 Years ≙ Lower secondary school qualification (completion of compulsory education) 10 Years ≙ Middle School qualification (certificate from *Realschule*) 11 Years ≙ Lower secondary school qualification with additional vocational training (apprenticeship or certificate from specialized vocational school) 12 Years ≙ Middle school qualification with additional vocational training degree (apprenticeship or certificate from specialized vocational school) 13 Years ≙ *Abitur* (included in this category are certificates from a *Gymnasium, Kolleg* or *Wirtschaftsgymnasium;* also certificates from a secondary technical school, the *Fachoberschule* or the *Höhere Berufsfachschule*) 17 Years ≙ Professional college qualification (certificate from a higher technical college or a professional college, the *Fachhochschule, Ingenieurschule* or *Höhere Fachschule*) 19 Years ≙ University degree (from all institutions of higher education)

Appendix 1 continued

Variable Name	Definition
DP	Variable that serves as a **dep**endent variable in P3RFUN (Trond Petersen)
DUR, IDUR[2]	Job duration in months
SEX	Sex: 1 ≙ men 2 ≙ women
MAR	Marital status: 0 ≙ not married 1 ≙ married
COHO	Birth cohort: 1 ≙ cohort 1929–31 2 ≙ cohort 1939–41 3 ≙ cohort 1949–51
COHO2, C2[1], COH2[4]	Dummy variable for cohort 1939–41: 1 ≙ cohort 1939–41 0 ≙ other
COHO3, C3[1], COH3[4]	Dummy variable for cohort 1949–51: 1 ≙ cohort 1949–51 0 ≙ other
CONST	Intercept in P3RFUN outputs
M3	Dummy variable for men: 1 ≙ men 0 ≙ other
M5	Number of the job
M41	Highest general educational grade: 1 ≙ Lower secondary school qualification 2 ≙ Middle school qualification 3 ≙ Abitur
M42	Highest vocational training degree: 1 ≙ no vocational training 2 ≙ vocational training degree 3 ≙ master craftsman or technician degree 4 ≙ professional college degree 5 ≙ university degree
M43	Time of entry into the labor market, measured in months from the beginning of the century
M47	Time of interview, measured in months from the beginning of the century
M48	Time of birth, measured in months from the beginning of the century
M50, TS[3]	Starting time of the job, measured in months from the beginning of the century
M51, TF[3]	Ending time of the job, measured in months from the beginning of the century

Appendix 1 continued

Variable Name	Definition
M59, PRES[3]	Wegener's (1985) prestige score for job i
M61, JOBN[3], JN[1]	Occupational group for job i (Table 4.4)
M62, JOBN1[3], JN1[1]	Occupational group for job i + 1 (Table 4.4)
RESID, RESID1[5], RESID2[5]	Residual
M69, TMAR[3], TMA[4]	Time of marriage, measured in months from the beginning of the century
U	Logarithmic duration for GLIM
TDEP	Time dependence in parametric models
X1	Starting time of the episodes for P3RFUN
CEN, ICEN[2], C[1]	Censoring variable: 1 ≙ event 0 ≙ censored
Z2	Interaction variable sex * duration

[1]Names for GLIM. This program distinguishes only the first three characters.

[2]Names for episode splitting.

[3]Names for SPSS runs to prepare data for RATE.

[4]Names for episode splitting.

[5]Differentiates residuals.

Appendix 2: Listing of the FORTRAN Program P3RFUN Written by Trond Petersen

```
C       WRITTEN BY TROND PETERSEN, DAVID DICKENS AND NANCY WILLIAMSON.
      SUBROUTINE P3RFUN( F,     DF,    P,     X,      N,
     *                  KASE, NVAR, NPAR, IPASS, XLOSS,
     *                  IDEP )
C
C     P3RFUN - FUNCTION SUBROUTINE
C                                                                  **
C    F      = FUNCTION VALUE (OUTPUT)
C    DF     = ARRAY OF DERIVATIVES WITH RESPECT TO P (OUTPUT)
C    P      = CURRENT VALUE OF PARAMETERS (INPUT)
C    X      = CURRENT CASE (INPUT)
C    N      = CODE NUMBER FOR FUNCTION (INPUT)
C    KASE   = CURRENT CASE NUMBER (INPUT)
C    NVAR   = NUMBER OF VARIABLES (INPUT)
C    NPAR   = NUMBER OF PARAMETERS (INPUT)
C    IPASS  = INDEX OF PASS (INPUT, SEE MANUAL)
C    XLOSS = LOSS VALUE (OUTPUT, SEE MANUAL) NOT USED HERE.
C         INCLUDED ONLY SO USER SUPPLIED FUN CAN USE IT.
C    IDEP   = INDEX OF THE DEPENDENT VARIABLE (INPUT)
C *********************************************************************
      IMPLICIT REAL*8 (A-H,O-Z)
      DIMENSION INDEP(100),A(6),H(9),DF(NPAR),P(NPAR),X(NVAR)
C
C
      LOGICAL ERR,ALPHA(26),FIRST,FDPROG,FDMODL
        DATA ERR,ALPHA,FIRST/28*.FALSE./
        DATA FDPROG,FDMODL/2*.FALSE./
      CHARACTER * 256 CONTRL

C     VALMAX IS THE MAXIMUM VALUE THAT SHOULD BE USED AS AN
C     ARGUMENT TO THE EXP FUNCTION TO AVOID OVER AND UNDERFLOWS
C     IN SUBROUTINE LSTSQ. ( 2*VALMAX ) SHOULD NOT OVERFLOW.

      DATA VALMAX / 40. /
      IF (.NOT. FIRST) THEN
               FIRST = .TRUE.

997        DO 9971 I = 1,256
9971         CONTRL(I:I) = ' '
                     READ (1,1000,END=980,ERR=8888)CONTRL
1000         FORMAT (A256)
8888                WRITE (40,2222) CONTRL
2222      FORMAT (A256)
          REWIND 40
1818      DO 999 I = 1,256
               IF (CONTRL(I:I) .NE. ' ') GO TO 998
999        CONTINUE

998        CONTINUE


        IF (CONTRL(I:I) .EQ. '''') GO TO 990
```

```
      IF (.NOT. FDPROG) THEN
            FDPROG = .TRUE.
            IF (.NOT.(CONTRL(I:I+1) .EQ. '/C')) THEN
                  WRITE(35,1001) CONTRL(1:60)
1001              FORMAT (' FIRST BMDP CONTROL CARD READ SHOULD
     * BE /C OR /C AND',
     *         ' INSTEAD HAVE READ'/5X,A60)
                  ERR = .TRUE.
                  GO TO 980
            END IF
            GO TO 997
      END IF

      IF (.NOT.FDMODL) THEN
            FDMODL = .TRUE.
            IF (CONTRL(I:I) .EQ.'M' .OR. CONTRL(I:I) .EQ. 'M')THEN
                  ALPHA(13) = .TRUE.
                  READ (40,2223) IMOD
2223   FORMAT (1X,I3)
       REWIND 40
                  WRITE (35,10021) IMOD
10021              FORMAT (/' MODEL CARD SAYS RUNNING MODEL',I5)
                  GO TO 997
            ELSE
                  WRITE(35,1002) CONTRL(1:60)
1002              FORMAT (' FIRST P3RFUN CARD SHOULD BE MODEL CARD,
     *      BUT INSTEAD IS '/5X,A60)
                  ERR = .TRUE.
                  GO TO 980
            END IF
      END IF

      IF (CONTRL(I:I) .EQ. 'T' .OR. CONTRL(I:I) .EQ. 'T')THEN
            ALPHA(20) = .TRUE.
            READ (40,2224) ITIME1,ITIME2
2224        FORMAT(1X,2I3)
            REWIND 40
            WRITE (35,1003) ITIME1,ITIME2
1003      FORMAT (/' TIME CARD SHOWS BEGIN AND END TIME VARIABLES
     *        AS',2I5)

      ELSE IF (CONTRL(I:I) .EQ. 'C' .OR. CONTRL(I:I) .EQ. 'C')THEN
            ALPHA(3) = .TRUE.
            READ (40,2225) ICNVAR
2225        FORMAT(1X,I3)
            REWIND 40
            WRITE (35,1004)  ICNVAR
1004      FORMAT (/' CENSOR CARD SHOWS CENSOR VARIABLE AS',I5)


      ELSE IF (CONTRL(I:I) .EQ. 'D' .OR. CONTRL(I:I) .EQ. 'D')THEN
            ALPHA(4) = .TRUE.
            READ(40,2226) IDEPVR
```

```
2226          FORMAT(1X,I3)
              REWIND 40
              WRITE (35,1005) IDEPVR

1005        FORMAT (/' DEPENDENT VARIABLE CARD SHOWS DEPVAR AS',I5)
        ELSE IF (CONTRL(I:I) .EQ. 'I' .OR.CONTRL(I:I) .EQ. 'I')THEN
              READ (40,2227)MDNPAR
2227          FORMAT (1X,I3)
              REWIND 40

              IF (MDNPAR .EQ. 0) THEN
                    ALPHA(9) = .TRUE.
                    WRITE(35,1009)
1009                    FORMAT(/' INDEPENDENT VARIABLE CARD SHOWS NO',
     *                    ' COVARIATES')
                    GO TO 997
              END IF
                    ALPHA(9) = .TRUE.
           READ (40,2228)(INDEP(J),J=1,MDNPAR)
2228    FORMAT (4X,40I3)
        REWIND 40
                    WRITE (35,1006) MDNPAR,(INDEP(J),J=1,MDNPAR)
1006              FORMAT (/' INDEPENDENT VARIABLE CARD SHOWS',I3,
     *    ' COVARIATES TO BE',
     *             /2X,20I3/2X,20I3)
       REWIND 40

        ELSE IF (CONTRL(I:I) .EQ. 'L' .OR. CONTRL(I:I) .EQ. 'L') THEN
              READ (40,2229)LFX1,LFX2
              ALPHA(12) = .TRUE.
              WRITE (35,1007) LFX1,LFX2
2229   FORMAT(1X,2I3)
1007              FORMAT (/' LABOR FORCE EXPERIENCE (OR SIMILAR
     *     VARIALE) AT',
     *              ' BEGINNING AND END'/5X,'OF RECORD SHOWN AS
     *             VARIABLES',2I5)
                  REWIND 40
        ELSE
              WRITE (35,1008) CONTRL(1:60)
1008              FORMAT(' READ A CONTROL CARD TYPE NOT
     *   EXPECTING'/11X,A60)
                   ERR = .TRUE.
        END IF

        GO TO 997


980      WRITE (35,981)
981        FORMAT (//'***  RUN STOPPED BECAUSE OF FATAL ERRORS ***')
             STOP
990       CONTINUE
        IF (.NOT.( ALPHA(13) .AND. ALPHA(20) .AND. ALPHA(3) .AND.
     *      ALPHA(9) .AND. ALPHA(4))) THEN
             WRITE (35,991)
```

```
991          FORMAT (//' DID NOT FIND ALL REQUIRED (M,T,C,I,AND D)',
     *        ' CONTROLS CARDS')
              ERR = .TRUE.
        END IF
        IF (IMOD .EQ. 3 .AND. .NOT. ALPHA(12)) THEN
              WRITE (35,992)
992              FORMAT (//' SPECIFIED MODEL 3 BUT NO L CONTROL CARD')
              ERR = .TRUE.
        END IF

        IF (ERR) GO TO 980

        WRITE (35,995)
995          FORMAT(//' END OF READING P3RFUN COMMAND INFORMATION',
     *    ' FROM TITLE CARD OF BMDP COMMAND FILE.')
      END IF


C       THERE ARE ALTOGETHER FIVE MODELS.
C       THESE ARE:
C       MODEL 1: THE EXPONENTIAL DISTRIBUTION
C       MODEL 2: THE GOMPERTZ DISTRIBUTION
C       MODEL 3: THE GOMPERTZ DISTRIBUTION WITH AGE OR
C            LABOR FORCE EXPERIENCE IN ADDITION TO AGE
C            AS A CONTINUOUSLY VARYING VARIABLE WITHIN
C            THE STATE.
C       MODEL 4: THE WEIBULL DISTRIBUTION
C       MODEL 5: THE LOG-LOGISTIC DISTRIBUTION.

C       ALL MODELS ALLOW FOR THE INCLUSION OF TIME-DEPENDENT
C       COVARIATES.

C       THE PARAMETERS TO BE ESTIMATED ALWAYS COME IN THIS ORDER.

C       COMMON TO ALL MODELS:
C       P(1): THE CONSTANT TERM

C       IN MODEL 1:
C       P(2),P(3) ETC.: THE EFFECTS OF VARIABLES WHICH
C       FOLLOW STEP-FUNCTIONS OVER TIME.

C       IN MODEL 2-5:
C       P(2): THE DURATION DEPENDENCE

C       IN MODEL 2,4,5:
C       P(3),P(4) ETC.: THE EFFECTS OF VARIABLES WHICH FOLLOW
C       STEP FUNCTIONS OVER TIME.

C       IN MODEL 3:
C       P(3): THE EFFECT OF AGE OR LABOUR FORCE EXPERIENCE
C       OR SOME SIMILAR VARIABLE AS A CONTINUOUSLY VARYING
C       VARIABLE WITHIN A STATE IN ADDITION TO DURATION.
C       THIS VARIABLE MUST BE A DETERMINISTIC AND LINEAR
```

```
C       FUNCTION OF SOME INITIAL VALUE AT THE BEGINNING OF
C       A STATE AND THE TIME SPENT IN THE STATE.

C       P(4),P(5) ETC.: THE EFFECTS OF VARIABLES WHICH
C       FOLLOW STEP-FUNCTIONS OVER TIME.

C  ESTABLISH PARAMETERS FOR SPECIFIC MODELS, FOLLOWED BY
C  CALCULATION OF CV BEFORE BRANCHING BASED ON MODELS.

      IF (IMOD .EQ. 2 .OR. IMOD .EQ. 4 .OR. IMOD .EQ. 5)      THEN
        IBEGIN = 3
      ELSE IF (IMOD .EQ. 1)THEN
        IBEGIN = 2
      ELSE IF (IMOD .EQ. 3) THEN
        IBEGIN = 4
      END IF


C       WE START WITH DEFINING THE CRUCIAL VARIABLES
C       ON DURATION, CENSORING AND THE COVARIATES.

      CV=0.0
      IF (NPAR .GE. IBEGIN) THEN
        DO 11 I=IBEGIN,NPAR
11        CV=CV+P(I)*X(INDEP(I-(IBEGIN-1)))
      END IF

      IF (IMOD .EQ. 1) THEN

C       THEN COMES A BUNCH OF FUNCTIONS WHICH WE NEED IN
C       WRITING DOWN THE LOG-LIKELIHOOD AS WELL AS THE GRADIENT
C       VECTOR USED IN THE MODIFIED SCORING ALGORITHM.

        H(1)=P(1)+CV
        DUR=X(ITIME2)-X(ITIME1)
        A(1) = DUR*DEXP(H(1))

C       HERE COMES THE LOGLIKELIHOOD.

        F= X(ICNVAR)*H(1) - A(1)


        DF(1) = X(ICNVAR) - A(1)


      ELSE IF (IMOD .EQ. 2) THEN

C       NOW COMES THE MODEL FOR THE GOMPERTZ DISTRIBUTION. THIS
C       IS THE MODEL WHERE WE DO NOT TAKE ACCOUNT OR AGE OR
C       LABOR FORCE EXPERIENCE AS A CONTINUOUSLY VARYING VARIABLE
C       WITHIN A SPELL.
```

```
      H(1)=P(1)+P(2)*X(ITIME2)+CV
      H(2)=DEXP(H(1))
      H(3)=DEXP(P(1)+P(2)*X(ITIME1)+CV)
      A(1)=(1.0/P(2))*(H(2)-H(3))
      A(2)=(1.0/(P(2)*P(2)))*(H(2)-H(3))
      A(3)=(1.0/P(2))*(X(ITIME2)*H(2)-X(ITIME1)*H(3))

C     THEN COMES THE LOGLIKELIHOOD.

      F=X(ICNVAR)*H(1)-A(1)

C     THEN COMES THE GRADIENT VECTOR.

      DF(1)=X(ICNVAR) - A(1)
      DF(2)=X(ICNVAR)*X(ITIME2)+A(2)-A(3)



     ELSE IF (IMOD .EQ. 3) THEN

C     THIS IS THE SET-UP FOR DOING THE GUMPERTZ DISTRIBUTION
C     WITH TIME-DEPENDENT COVARIATES, AND IN PARTICULAR WITH
C     AGE OR LABOR FORCE EXPERIENCE IN ADDITION TO DURATION AS
C     A CONTINUOUSLY VARYING VARIABLE WITHIN THE STATE.


C     NOW, COMES SOME AUXILLIARY FUNCTIONS.
C     THESE ARE USED IN WRITING DOWN THE LOG-LIKELIHOOD
C     AND THE GRADIENT VECTOR.

      H(1)=P(1)+P(2)*X(ITIME2)+P(3)*X(LFX2)+CV
      H(2) = DEXP(H(1))
      H(3) = DEXP(P(1) + P(2)*X(ITIME1) + P(3)*X(LFX1) + CV)
      A(1) = (1.0/(P(2)+P(3)))*(H(2)-H(3))
      A(2) = (1.0/((P(2)+P(3))*(P(2)+P(3))))*(H(2)-H(3))
      A(3) = (1.0/(P(2)+P(3)))*(X(ITIME2)*H(2)-X(ITIME1)*H(3))
      A(4) = (1.0/(P(2)+P(3)))*(X(LFX2)*H(2)-X(LFX1)*H(3))

C     THEN COMES THE LOGLIKELIHOOD FUNCTION.

      F = X(ICNVAR)*H(1) - A(1)

C     THEN COMES THE GRADIENT VECTOR.

      DF(1) = X(ICNVAR) - A(1)
      DF(2) = X(ICNVAR)*X(ITIME2) + A(2) - A(3)
      DF(3) = X(ICNVAR)*X(LFX2) + A(2) - A(4)


     ELSE IF (IMOD .EQ. 4) THEN

C     NOW COMES THE MODEL FOR THE WEIBULL DISTRIBUTION. THIS
C     IS THE MODEL WHERE WE DO NOT TAKE ACCOUNT OR AGE OR
C     LABOR FORCE EXPERIENCE AS A CONTINUOUSLY VARYING VARIABLE
```

```
C       WITHIN A SPELL.
C       THE HAZARD IS OF THE FORM:
C       H(T) = EXP(BX + A*LN(T))
C       WHEN A IS LESS THAT ZERO THERE IS NEGATIVE DURATION
C       DEPENDENCE, OTHERWISE IT IS POSITIVE. THE A COEFFICIENT
C       SHOULD BE GREATER THAN -1.0.

C       THIS PARAMETRIZATION IS SOMEWHAT DIFFERENT FROM THE
C       ONE WE USUALLY FIND IN THE STATISTICAL LITERATURE. IN THE
C       USUAL PARAMETRIZATION THE DURATION DEPENDENCE PARAMETER
C       IS ALWAYS GREATER THAN 0.0, AND WE HAVE NEGATIVE DURATION
C       DEPENDENCE WHEN THE PARAMETER IS LESS THAN 1.0.
C       THE USUAL PARAMETER CAN BE OBTAINED FROM THE MODEL
C       ESTIMATED HERE BY JUST ADDING 1.0 TO THE PRESENT ESTIMATE
C       OF THE DURATION DEPENDENCE PARAMETER.



        H(1)=P(1)+CV
        H(2)=DEXP(H(1))
        H(3) = DLOG(X(ITIME2))
        H(4) = P(2) + 1.0
        H(5) = (1/H(4))
        H(6) = X(ITIME2)**H(4)

        IF (X(ITIME1) .EQ. 0.0) THEN
              H(7) = 0.0
        ELSE
              H(7) = X(ITIME1)**H(4)
        END IF

        H(8) = H(6) - H(7)

        IF (X(ITIME1) .LE. 1.0) THEN
              H(9) = H(6)*H(3)
        ELSE
               H(9) = H(6)*H(3) - H(7)*DLOG(X(ITIME1))
        END IF

        A(1) = H(5)*H(2)*H(8)
        A(2) = X(ICNVAR) - A(1)
        A(3) = H(5)*H(2)*(H(5)*H(8) - H(9))

C       THEN COMES THE LOG-LIKELIHOOD.

        F = X(ICNVAR)*(H(1) + P(2)*H(3)) - A(1)

C       THEN COMES THE GRADIENT VECTOR.

        DF(1) = A(2)
        DF(2) = X(ICNVAR)*H(3) + A(3)


      ELSE IF (IMOD .EQ. 5) THEN
```

Appendix 2 continued

```
C       NOW COMES THE MODEL FOR THE LOG-LOGISTIC DISTRIBUTION. THIS
C       IS THE MODEL WHERE WE DO NOT TAKE ACCOUNT OR AGE OR
C       LABOR FORCE EXPERIENCE AS A CONTINUOUSLY VARYING VARIABLE
C       WITHIN A SPELL.

C       THE HAZARD IS:
C       R(T)= (P+1)*(EXP(BX + P*LN(T)))/(1 + EXP(BX + P*LN(T)))
C       HERE, P SHOULD BE GREATER THAN -1.0.
C       THERE IS MONOTONE NEGATIVE DURATION DEPENDENCE WHEN
C       P IS LESS THAN ZERO. FOR P LARGER THAN ZERO THE DURATION
C       DEPENDENCE IS FIRST POSITIVE AND AFTER SOME TIME IT
C       BECOMES NEGATIVE. THE SWITCH POINT DEPENDS BOTH ON P
C       AND ON BX. IT SHOULD BE CALCULATED FROM THE ESTIMATES.

C       NOTE THAT THIS PARAMETRIZATION IS SOMEWHAT DIFFERENT
C       FROM THE ONE WE USUALLY FIND IN THE LITERATURE.
C       IN PARTICULAR: THE TRADITIONAL DURATION DEPENDENCE PARAMETER
C       SHOULD ALWAYS BE LARGER THAN 0.0, WHICH GIVES MONOTONE NEGATIVE
C       DURATION DEPENDENCE FOR ALL VALUES OF THE PARAMETER EQUAL
C       TO OR LESS THAN 1.0 . THE TRADIONAL DURATION DEPENDENCE
C       CAN BE OBTAINED FROM THE PRESENT BY ADDING 1.0 TO THE
C       DURATION DEPENDENCE PARAMETER ESTIMATED HERE.
C       I FIND THE PRESENT PARAMETRIZATION PREFERABLE BECAUSE IT
C       MAXIMIZES COMPARABILITY OF THE DURATION DEPENDENCE ACROSS THE
C       FIVE MODELS.

            H(1)=P(1)+CV
            H(2)=DEXP(H(1))
            H(3) = P(2) + 1.0
            H(4) = X(ITIME2)**H(3)

            IF (X(ITIME1) .EQ. 0.0) THEN
                  H(5) = 0.0
            ELSE
                  H(5) = X(ITIME1)**H(3)
            END IF

            H(6) = DLOG(X(ITIME2))

            IF (X(ITIME1) .LE. 1.0) THEN
                  H(7) = 0.0
            ELSE
                  H(7) = DLOG(X(ITIME1))
            END IF

            H(8) = H(2)*H(5)
            H(9) = H(2)*H(4)
            A(1) = DLOG(1 + H(8))
            A(2) = DLOG(1 + H(9))
            A(3) = (1./(1+H(8)))* H(8)
            A(4) = (1./(1+H(9)))* H(9)
            A(5) = A(3)*H(7)
            A(6) = A(4)*H(6)
```

```
C       THEN COMES THE LOG-LIKELIHOOD.

              IF (X(ICNVAR) .EQ. 1.0) THEN
                    F = H(1) +DLOG(H(3)) + P(2)*H(6) + A(1) - 2.*A(2)
              ELSE
                    F = A(1) - A(2)
              END IF

C       THEN COMES THE GRADIENT VECTOR.

              IF (X(ICNVAR) .EQ. 1.0) THEN
                    DF(1) = 1. + A(3) - 2.*A(4)
                    DF(2) = (1./H(3)) + H(6) + A(5) - 2.*A(6)
              ELSE
                    DF(1) = A(3) - A(4)
                    DF(2) = A(5) - A(6)
              END IF
      END IF

C  NOW THE CALCULATIONS THAT ARE DONE IN COMMON FOR ALL MODELS.


      DO 21 I=IBEGIN,NPAR
21      DF(I)=X(INDEP(I-(IBEGIN-1)))*DF(1)

C       THIS IS THE TRICK TO REDEFINE THE DEPENDENT VARIABLE
C       IN THE SCORING ALGORITHM USED FOR NON-LINEAR LEAST SQUARES
C       PROBLEMS OR FOR MAXIMUM LIKELIHOOD PROBLEMS IN MODELS
C       FALLING WITHIN AN EXPONENTIAL FAMILY.

      X(IDEPVR)= F+1.0

C       FINALLY COMES THE LOG-LIKELIHOOD CONVERGENCE CRITERION.

      XLOSS=-F
          RETURN
       END
```

Appendix 3: Listing of the FORTRAN Program for Episode Splitting Given Discrete Time-Dependent Covariates

```
      PROGRAM TRANSF
      INTEGER TS,TF,CEN,EDU,PRES,NOJ,LFX
      INTEGER COHO2,COHO3,DUR,JOBN,JOBN1,TMAR
      INTEGER MAR
      N = 0
      M = 0
      MDUR = 0
      NDUR = 0
1     READ(20,1001,END=999) TS,TF,CEN,EDU,PRES,NOJ,LFX,
     *                      COHO2,COHO3,DUR,JOBN,JOBN1,TMAR
      M = M + 1
      MDUR = MDUR + DUR
      IF(TMAR .EQ. 0) TMAR = 10000
      IF(TMAR .GE. TF) THEN
        MAR = 0
        IDUR = DUR
        ITS = TS
        ITF = TF
        ICEN = CEN
        WRITE(30,1002) ITS,ITF,ICEN,EDU,PRES,NOJ,LFX,
     *                 COHO2,COHO3,IDUR,JOBN,JOBN1,TMAR,MAR
        N = N + 1
        NDUR = NDUR + IDUR
      ELSE IF(TMAR .LE. TS) THEN
        MAR = 1
        IDUR = DUR
        ITS = TS
        ITF = TF
        ICEN = CEN
        WRITE(30,1002) ITS,ITF,ICEN,EDU,PRES,NOJ,LFX,
     *                 COHO2,COHO3,IDUR,JOBN,JOBN1,TMAR,MAR
        N = N + 1
        NDUR = NDUR + IDUR
      ELSE
        MAR = 0
        IDUR = TMAR - TS
        ITS = TS
        ITF = TMAR
        ICEN = 0
        WRITE(30,1002) ITS,ITF,ICEN,EDU,PRES,NOJ,LFX,
     *                 COHO2,COHO3,IDUR,JOBN,JOBN1,TMAR,MAR
        N = N + 1
        NDUR = NDUR + IDUR
        MAR = 1
        IDUR = TF - TMAR
        ITS = TF - IDUR
        ITF = TF
        ICEN = CEN
        WRITE(30,1002) ITS,ITF,ICEN,EDU,PRES,NOJ,LFX,
     *                 COHO2,COHO3,IDUR,JOBN,JOBN1,TMAR,MAR
```

```
          N = N + 1
          NDUR = NDUR + IDUR
        END IF
        GOTO 1
999     WRITE (2,2000) M,N,MDUR,NDUR
2000    FORMAT(I5,2X,I6,2X,I10,2X,I15)
1001    FORMAT(13I5)
1002    FORMAT(14I5)
        END
```

Appendix 4: Listing of the FORTRAN Program for Episode Splitting Given Continuous Time-Dependent Covariates

```
      PROGRAM TRANSF
      INTEGER TS,TF,CEN,EDU,PRES,NOJ,LFX
      INTEGER COHO2,COHO3,DUR,JOBN,JOBN1,TMAR
      INTEGER INTERVAL/60/,UPPERLIMIT
      N = 0
      M = 0
      MDUR = 0
      NDUR = 0
1     READ(20,1000,END=999) TS,TF,CEN,EDU,PRES,NOJ,LFX,
     *                      COHO2,COHO3,DUR,JOBN,JOBN1,TMAR
      M = M + 1
      MDUR = MDUR + DUR
      UPPERLIMIT = INTERVAL
      ILFXS = LFX
      ITS = TS
2     IF (DUR .LE. INTERVAL) THEN
         ILFX = ILFXS
         ICEN = CEN
         IDUR = DUR
         ITS = ITS
         ITF = ITS + INTERVAL
         IF (ITF .GT. TF) ITF = TF
      ELSE
         ILFX = ILFXS
         ICEN = 0
         IDUR = INTERVAL
         ITS = ITS
         ITF = ITS + INTERVAL
         IF (ITF .GT. TF) ITF = TF
      END IF
      WRITE(30,1000) ITS,ITF,ICEN,EDU,PRES,NOJ,ILFX,
     *               COHO2,COHO3,IDUR,JOBN,JOBN1,TMAR
      N = N + 1
      NDUR = NDUR + IDUR
      UPPERLIMIT = UPPERLIMIT + INTERVAL
      ILFXS = ILFXS + INTERVAL
      DUR = DUR - INTERVAL
      ITS = ITS + INTERVAL
      IF (DUR .LE. 0) GOTO 1
      GOTO 2
999   WRITE (2,2000) M,N,MDUR,NDUR
2000  FORMAT(I5,2X,I6,2X,I10,2X,I15)
1000  FORMAT(13I5)
      END
```

Appendix 5: Listing of the GLIM Macros to Estimate the Weibull and Log-Logistic Models of Roger and Peacock

```
$SUBFILE MAKROS
$M ML1
$CA %LP=%IF(%LT(%LP,78),%LP,78)
$CA %LP=%IF(%GT(%LP,-78),%LP,-78)
$CA %FV=N/(1+%EXP(-%LP))
$CA %FV(%NU)=%W/2
$END
$C
$C
$C
$M ML2
$CA %DR=N/(%FV*(N-%FV))
$CA %DR(%NU)=%LP(%NU)/%FV(%NU)
$END
$C
$C
$C
$M ML3
$CA %VA=%FV*(N-%FV)/N
$CA %VA(%NU)=%W/4
$END
$C
$C
$C
$M ML4
$CA %DI=2*((R-N)*%LOG(N-%FV)-R*%LOG(%FV))
$CA %DI(%NU) = -2*%W*%LOG(%LP(%NU))
$END
$C
$C
$C
$M SETL
$CA GM=1
$CA GM(%NU)=0
$CA U(%NU)=1
$CA C(%NU)=0
$CA N=%EQ(C,1)
$CA %W=%CU(N)
$CA N=N+1
$CA R=%NE(C,0)
$CA R(%NU)=N(%NU)=%W
$YVAR R
$OWN ML1 ML2 ML3 ML4
$SCALE 1$
$CA %LP=%LOG(R+0.1)/(N-R+0.1)
$CA %LP(%NU)=0.5
$FIT GM - %GM + U
$DIS E $
$END
$C
$C
```

Appendix 5 continued

```
$C
$MAC MW1 $
$CA %LP=%IF(%LT(%LP,78),%LP,78)
$CA %LP=%IF(%GT(%LP,-78),%LP,-78)
$CA %FV=%EXP(%LP)
$CA %FV(%NU)=%W/2
$END
$C
$C
$C
$MAC MW2 $
$CA %DR=1/%FV
$CA %DR(%NU)=%LP(%NU)/%FV(%NU)
$END
$C
$C
$C
$MAC MW3 $
$CA %VA=%FV
$CA %VA(%NU)=%W/4
$END
$C
$C
$C
$MAC MW4 $
$CA %DI=2*(%FV-C*(%LP+1))
$CA %DI(%NU)=-2*%W*%LOG(%LP(%NU))
$END
$C
$C
$C
$MAC SETW $
$CA GM=1
$CA GM(%NU)=0
$CA U(%NU)=1
$CA C(%NU)=0
$CA %W=%CU(C)
$CA C(%NU)=%W
$YVAR C
$OWN MW1 MW2 MW3 MW4
$SCALE 1 $
$CA %LP=%LOG(C*0.8+0.1)
$CA %LP(%NU)=0.5
$FIT GM-%GM+U
$DIS E$
$END
$C
$C
$C
$RETURN
$FINISH
```

References

Allison, P. D. (1982). Discrete time methods for the analysis of event histories. In S. Leinhardt (Ed.), *Sociological methodology* (pp. 61–98). San Francisco: Jossey Bass.

Allison, P. D. (1984). *Event history analysis. Regression for longitudinal event data.* Beverly Hills, CA: Sage.

Amemiya, T. (1985). *Advanced econometrics.* Cambridge, MA: Basil Blackwell.

Andersen, P. K. (1982). Testing goodness of fit of Cox's regression and life model. *Biometrics, 38,* 67–77.

Andersen, P. K., & GILL, R. D. (1982). Cox's regression model for counting processes: A large sample study. *Annals of Statistics, 10,* 1100–1120.

Andress, H. J. (1985). *Multivariate Analyse von Verlaufsdaten* (ZUMA-Methodentexte, Vol. 1). Mannheim.

Arminger, G. (1984a). Modelltheoretische und methodische Probleme bei der Analyse von Paneldaten mit qualitativen Variablen. *Vierteljahreshefte des Deutschen Instituts für Wirtschaftsforschung, 4,* 470–480.

Arminger, G. (1984b). EM estimation for compound generalized linear models. Manuscript, Wuppertal.

Arminger, G. (1986). Testing against misspecification in parametric rate models. International Conference on Applications of Life History Analysis in Life Course Research, Max Planck Institute for Human Development and Education, Berlin, June.

Arrow, K. (1973). Higher education as a filter. *Journal of Public Economics, 2,* 193–216.

Bailey, K. R. (1983). The asymptotic joint distribution of regression and survival parameter estimates in the Cox regression model. *Annals of Statistics, 11,* 39–48.

Bajenescu, T. I. (1985). *Zuverlässigkeit elektronischer Komponenten.* Berlin and Offenbach: VDE-Verlag.

Baker, R. J., & Nelder, J. A. (1978). *The GLIM system.* Oxford: NAG.

Birg, H., et al. (1985). Arbeitsmarktdynamik, Familienentwicklung und generatives Verhalten—Eine biographietheoretische Konzeption für Untersuchungen demographisch relevanter Verhaltensweisen. In J. Schmid & I. Schwarz (Eds.), *Politische und prognostische Tragweite von Forschungen zum generativen Verhalten* (pp. 209–223). Berlin: Deutsche Gesellschaft für Bevölkerungsforschung.

Blossfeld, H.-P. (1985a). Berufseintritt und Berufsverlauf. Eine Kohortenanalyse über die Bedeutung des ersten Berufs in der Erwerbsbiographie. *Mitteilungen aus der Arbeitsmarkt- und Berufsforschung, 2,* 177–197.

Blossfeld, H.-P. (1985b). *Bildungsexpansion und Berufschancen.* Frankfurt a. M. and New York: Campus.

Blossfeld, H.-P. (1986). Career opportunities in the Federal Republic of Germany: A dynamic approach to the study of life course, cohort, and period effects. *European Sociological Review, 2,* 208–225.

Blossfeld, H.-P. (1987). Zur Repräsentativität der SfB-3-Lebensverlaufsstudie: Ein Vergleich mit Daten aus der amtlichen Statistik. *Allgemeines Statistisches Archiv, 71,* 126–144.

Blossfeld, H.-P. (1987a). Labor market entry and the sexual segregation of careers in the Federal Republic of Germany. *American Journal of Sociology, 93,* 89–118.

Blossfeld, H.-P. (1987b). Entry into the labor market and occupational career in the Federal Republic—A comparison with American studies. *International Journal of Sociology, 17,* 86–115.
Blossfeld, H.-P., & Hamerle, A. (1988a). *Using Cox models to study multiepisode processes: An illustration from a German career mobility survey.* Preprint, Berlin.
Blossfeld, H.-P., & Hamerle, A. (1988b). Unobserved heterogeneity in hazard rate models: A test and an illustration from a study of career mobility. In K. U. Mayer & N. B. Tuma (Eds.), *Event histories analysis in life course research.* Madison, WI: University of Wisconsin Press.
Blossfeld, H.-P., & Mayer, K. U. (1988). Labor market segmentation in the Federal Republic of Germany: An empirical study of segmentation theories from a life course perspective. *European Sociological Review, 4,* 1–18.
Brännäs, K. (1986). On heterogeneity in econometric duration models. *Sankhya: The Indian Journal of Statistics, 48,* 284–295.
Braun, H., & Hoem, J. (1978). Modelling cohabitational birth intervals in the current Danish population. A progress report. Working Paper No. 24, Copenhagen.
Breslow, N. E. (1970). A generalized Kruskal-Wallis test for comparing K samples subject to unequal patterns of censorship. *Biometrika, 57,* 579–594.
Breslow, N. E. (1974). Covariance analysis of censored survival data. *Biometrics, 30,* 89–100.
Brückner, E., et al. (1984). Methodenbericht "Lebensverläufe" (ZUMA-Arbeitsbericht No. T 84/08), Mannheim.
Buckley, J., & James, I. (1979). Linear regression with censored data. *Biometrika, 66,* 429–436.
Burdett, K., Kiefer, N. M., & Sharma, S. (1985). Layoffs and duration dependence in a model of turnover. *Journal of Econometrics, 28,* 51-69.
Carr-Hill, R. A., & Macdonald, K. I. (1973). Problems in the analysis of life histories. *Sociological Review Monograph, 19,* 57–95.
Carroll, G. R. (1984). Organizational ecology. *Annual Review of Sociology, 10,* 71–93.
Carroll, G. R., & Delacroix, J. (1982). Organizational mortality in the newspaper industries of Argentina and Ireland: An ecological approach. *Administrative Science Quarterly, 27,* 169–198.
Carroll, G. R., & Huo, Y. P. (1985). Organizational task and institutional environments in ecological perspective: Findings from the local newspaper industry. American Sociological Association Meetings. Washington, D.C.
Carroll, G. R., & Huo, Y. P. (1986). Losing by winning: The paradox of electoral success by organized labor parties in the Knights of Labor era. Technical Report No. OBIR-6. Center for Research in Management, University of California, Berkeley, CA.
Carroll, G. R., & Mayer, K. U. (1986). Job-shift patterns in the Federal Republic of Germany: The effects of social class, industrial sector and organizational size. *American Sociological Review, 51,* 323–341.
Chamberlain, G. (1980). Analysis of covariance with qualitative data. *Review of Economic Studies, 47,* 225–238.
Clayton, D. G., & Cuzick, J. (1985). Multivariate generalizations of the proportional hazards model. *Journal of the Royal Statistical Society A, 148,* 82–117.
Coale, A. J. (1971). Age patterns of marriage. *Population Studies, 25,* 193–214.
Coleman, J. S. (1981). *Longitudinal data analysis.* New York: Basic Books.
Coleman, J. S. (1984a). Stochastic models of market structures. In A. Diekmann & P. Mitter (Eds.), *Stochastic modelling of social processes* (pp. 189–213). New York: Academic Press.
Coleman, J. S. (1984b). Interdependence among qualitative attributes. *Journal of Mathematical Sociology, 10,* 29–50.
Courgeau, D. (1984). Relations entre cycle de vie et migrations. *Population, 3,* 483–514.
Courgeau, D. (1985). Interrelation between spatial mobility, family, and career life-cycle: A French survey. *European Sociological Review, 1,* 139–163.
Crowley, J., & Storer, B. E. (1983). Comment. *Journal of the American Statistical Association, 78,* 277–281.
Cox, D. R. (1972). Regression models and life-tables (with discussion). *Journal of the Royal Statistical Society B, 34,* 187–220.
Cox, D. R. (1975). Partial likelihood. *Biometrika, 62,* 269–276.
Cox, D. R., & Hinkley, D. V. (1974). *Theoretical statistics.* London: Chapman & Hall.
Cox, D. R., & Oakes, D. (1974). *Analysis of survival data.* London: Chapman & Hall.

Cox, D. R., & Snell, E. J. (1968). A general definition of residuals (with discussion). *Journal of the Royal Statistical Society B, 30,* 248–275.
David, H. A., & Moeschberger, M. L. (1978). *The theory of competing risks.* London: Griffin.
Dempster, A. P., Laird, N. M., & Rubin, D. B. (1977). Maximum likelihood from incomplete data via the EM algorithm (with discussion). *Journal of the Royal Statistical Society B, 39,* 1–38.
Diekmann, A. (1980). *Dynamische Modelle sozialer Prozesse.* München: Oldenburg.
Diekmann, A. (1987). Determinanten des Heiratsalters und Scheidungsrisikos. Eine Analyse sozio-demographischer Umfragedaten mit Modellen und statistischen Schätzmethoden. Habilitationthesis, Ludwig-Maximilians-Universität, München.
Diekmann, A., & Mitter, P. (1983). The "Sickle Hypothesis." *Journal of Mathematical Sociology, 9,* 85–101.
Diekmann, A., & Mitter, P. (1984). A comparison of the "Sickle Function" with alternative stochastic models of divorce rates for Austrian and U.S. marriage cohorts. In A. Diekmann & P. Mitter (Eds.), *Stochastic modelling of social processes.* New York.
Dixon, W. J., et al. (1983). *BMDP statistical software.* Berkeley, Los Angeles, and London: University of California Press.
Elbers, C., & Ridder, G. (1982). True and spurious duration dependence: The identifiability of the proportional hazards model. *Review of Economic Studies, 49,* 403–410.
Fahrmeir, L., & Hamerle, A. (1984). *Multivariate statistische Verfahren.* Berlin: DeGruyter.
Featherman, D. I. (1979–1980). Retrospective longitudinal research: Methodological considerations. *Journal of Economics and Business, 32,* 152–169.
Featherman, D. I., & Sørensen, A. B. (1983). Societal transformation in Norway and change in the life course transition into adulthood. *Acta Sociologica, 26,* 105–126.
Felmlee, D., & Eder, D. (1983). Contextual effects in the classroom: The impact of ability groups on student attention. *Sociology of Education, 56,* 77–87.
Flinn, Ch. J., & Heckman, J. J. (1982). Models for the analysis of labor force dynamics. In R. Basmann & G. Rhodes (Eds.), *Advances in econometrics* (Vol. I, pp. 35–95). Greenwich, CT.
Flinn, Ch. J., & Heckman, J. J. (1983). Are unemployment and out of the labor force behaviorally distinct labor force states? *Journal of Labor Economics, 1,* 28–42.
Freeman, J., Carroll, G. R., & Hannan, M. T. (1983). The liability of newness: Age dependence in organizational death rates. *American Sociological Review, 48,* 692–710.
Gail, M. H. (1975). A review and critique of some models used in competing risks analysis. *Biometrics, 31,* 209–222.
Galler, H.-P. (1988). Ratenmodelle mit stochastisch abhängigen konkurrierenden Risiken. Unpublished manuscript, University of Bielefeld.
Gehan, E. A. (1965). A generalized Wilcoxon test for comparing arbitrarily single-censored samples. *Biometrika, 52,* 203–223.
Gross, A. J., & Clark, V. A. (1975). *Survival distributions: Reliability applications in the Biomedical Sciences.* New York.
Hamerle, A. (1984). Zur statistischen Analyse von Zeitverläufen. Working paper No. 180, University of Regensburg.
Hamerle, A. (1985a). Zählprozeß-Modelle zur statistischen Analyse von Ereignisdaten mit Kovariablen bei konkurrierenden Risiken und mehreren Episoden. Working paper No. 90/s, University of Konstanz.
Hamerle, A. (1985b). Regressionsmodelle für gruppierte Verweildauern und Lebenszeiten. *Journal for Operations Research, Serie B: Praxis, 29,* 243–260.
Hamerle, A. (1986). Regression analysis for discrete event history or failure time data. *Statistical Papers, 27,* 207–225.
Hamerle, A. (1988a). In the incorporation of left censored observations in analysis of survival or duration data. Preprint, University of Konstanz.
Hamerle, A. (1988b). Multiple spell regression for duration data. *Applied Statistics, 37* (forthcoming).
Hamerle, A., & Pape, H. (1977). Über einen stochastischen Ansatz zur Lösung von Klassifikationsproblemen. *Statistische Hefte, 18,* 142–146.
Hamerle, A., Kemény, P., & Tutz, G. (1984). Kategoriale Regression. In L. Fahrmeir & A. Hamerle (Eds.), *Multivariate statistische Verfahren* (Chap. 6). Berlin: Springer.
Hamerle, A., & Tutz, G. (1986). Diskrete Modelle zur Analyse von Verweildauern. Manuscript, Konstanz and Regensburg.

Hamerle, A., & Tutz, G. (1988). Diskrete Modelle der Survival-Analyse. Lecture Notes in Medizinischer Informatik und Statistik. Heidelberg and Berlin: Springer.

Hanefeld, U. (1984). Das Sozio-ökonomische Panel—Eine Längsschnittstudie für die Bundesrepublik Deutschland. *Vierteljahreshefte zur Wirtschaftsforschung, 4,* 391–406.

Hanefeld, U. (1987). *Das Sozio-ökonomische Panel. Grundlagen und Konzeption.* Frankfurt a. M. and New York: Campus.

Handl, J., Mayer, K. U., & Müller, W. (1977). *Klassenlagen und Sozialstruktur.* Frankfurt a. M. and New York: Campus.

Hannan, M. T., & Freeman, J. (1977). The population ecology of organizations. *American Journal of Sociology, 82,* 929–964.

Hannan, M. T., Blossfeld, H.-P., & Schömann, K. (1988). Education, labor market segment, and growth in wages: Experiences of West German men in the Post-war period. Unpublished Manuscript.

Hannan, M. T., & Freeman, J. (1988). *Organizational ecology.* Cambridge, MA: Harvard University Press (forthcoming).

Heckman, J. J., & Borjas, G. (1980). Does unemployment cause future unemployment? Definitions, questions and answers from a continuous time model of heterogeneity and state dependence. *Econometrica, 47,* 247–283.

Heckman, J. J., & Singer, B. (1980). The identification problem in econometric models for duration data. In W. Hildenbrand (Ed.), *Advances in econometrics: Proceedings of world meetings of the Econometric Society* (pp. 39–77). Cambridge, MA: Cambridge University Press.

Heckman, J. J., & Singer, B. (1984a). Econometric duration analysis. *Journal of Econometrics, 24,* 63–132.

Heckman, J. J., & Singer, B. (1984b). A method for minimizing the impact of distributional assumptions in econometric models for duration data. *Econometrica, 52,* 271–320.

Helberger, C. (1980). Veränderungen der bildungsspezifischen Einkommensunterschiede zwischen 1969/71 und 1978. Working paper No. 51 of the SfB 3, Frankfurt a. M. and Mannheim.

Hernes, G. (1972). The process of entry into first marriage. *American Sociological Review, 37,* 173–182.

Holt, J. D. (1978). Competing risks analysis with special reference to matched pair experiments. *Biometrika, 65,* 159–166.

Hougaard, Ph. (1984). Life table methods for heterogeneous populations: Distributions describing the heterogeneity. *Biometrika, 71,* 75–83.

Huinink, J. (1987). Soziale Herkunft, Bildung und das Alter bei der Geburt des ersten Kindes. *Zeitschrift für Soziologie, 16* (5), 367–384.

Hujer, R., & Schneider, H. (1986). Ökonomische Ansätze zur Analyse von Paneldaten: Schätzung und Vergleich von Übergangsratenmodellen. Manuscript, Frankfurt a. M.

Hujer, R., & Schneider, H. (1988). Unemployment duration as a function of individual characteristics and economic trends. In K. U. Mayer & N. B. Tuma (Eds.), *Event histories in life course research.* Madison, WI: University of Wisconsin Press.

Hull, C. H., & Nie, N. H. (Eds.) (1981). *SPSS-Update 7–9.* New York: McGraw-Hill.

James, I., & Buckley, J. (1982). Note on a Monte Carlo comparison of a distribution-free method and MLE for linear regression and censored data. Working paper, University of Western Australia.

Johansen, S. (1978). The product limit estimator as maximum likelihood estimator. *Scandinavian Journal of Statistics, 5,* 195–199.

Kalbfleisch, J. D., & McIntosh, A. A. (1977). Efficiency in survival distributions with time-dependent covariables. *Biometrika, 64,* 47–50.

Kalbfleisch, J. D., & Prentice, R. L. (1980). *The statistical analysis of failure time data.* New York: Wiley.

Kaplan, E. L., & Meier, P. (1958). Nonparametric estimation from incomplete observations. *Journal of the American Statistical Association, 53,* 457–481.

Kay, R. (1977). Proportional hazard regression models and the analysis of censored survival data. *Applied Statistics, 26,* 227–237.

Keeley, M. C. (1979). An analysis of the age pattern of first marriage. *International Economic Review, 20,* 527–544.

Kemény, P., Rothmeier, F., & Hamerle, A. (1986). Explorative Variablenselektion und Anpassungstests bei Regressionsmodellen zur Analyse der stationären Aufenthaltsdauer nach Unfallverletzungen im Schulsport. *EDV in Medizin und Biologie, 17,* 61–71.
Kiefer, J., & Wolfowitz, J. (1956). Consistency of the maximum likelihood estimator in the presence of infinitely many incidental parameters. *Annals of Mathematical Statistics, 27,* 887–906.
Kiefer, N. (1984). A simple test for heterogeneity in exponential models of duration. *Journal of Labour Economics, 2,* 539–549.
Koul, H., Susarla, V., & van Ryzin, J. (1981). Regression analysis with randomly right censored data. *Annals of Statistics, 9,* 1276–1288.
Krupp, H.-J. (1985). Das Sozio-ökonomische Panel. Bericht über die Forschungstätigkeit 1983–1985. Antrag auf Förderung der Forschungsphase 1986–1988, Frankfurt a. M. and Berlin.
Krupp, H.-J., & Hanefeld, U. (Eds.) (1987). *Lebenslagen im Wandel: Analysen 1987.* Frankfurt a. M. and New York: Campus.
Lagakos, S. W. (1979). General right censoring and its impact on the analysis of survival data. *Biometrics, 35,* 139–156.
Lagakos, S. W. (1981). The graphical evaluation of explanatory variables in proportional hazard regression models. *Biometrika, 68,* 93–98.
Lancaster, T. (1985). Generalized residuals and heterogeneous duration models with applications to the Weibull model. *Journal of Econometrics, 23,* 155–169.
Lawless, J. F. (1982). *Statistical models and methods for life-time data.* New York: Wiley.
Lee, E., & Desu, M. (1972). A computer program for comparing K samples with right-censored data. *Computer Programs in Biomedicine, 2,* 315–321.
Lee, E., Desu, M., & Gehan, E. H. (1975). A Monte Carlo study of the power of some two-sample tests. *Biometrika, 62,* 425–432.
Lindsay, B. G. (1983a). The geometry of mixture likelihoods: A general theory. *Annals of Statistics, 11,* 86–94.
Lindsay, B. G. (1983b). The geometry of mixture likelihoods, part II: The exponential family. *Annals of Statistics, 11,* 783–792.
Luenberger, D. G. (1973). *Introduction to linear and nonlinear programming.* Reading, PA: Maddison/Wesley.
Mantel, N. (1966). Evaluation of survival data and two new rank order statistics arising in its consideration. *Cancer Chemotherapy Reports, 50,* 163–170.
Mantel, N., & Myers, M. (1971). Problems of convergence of maximum likelihood iterative procedures in multiparameter situations. *Journal of the American Statistical Association, 66,* 484–491.
Maritz, J. S. (1971). *Empirical Bayes methods.* London: Chapman Hall.
Matras, J. (1983). On schooling and employment in the transition of Israeli males to adulthood. Manuscript, Brookdale Institute, Jerusalem.
Mayer, K. U. (1984a). Lebensverläufe und Wohlfahrtsentwicklung. Bericht über die Forschungstätigkeit in der zweiten Forschungsphase 1982–1984 (pp. 119–142), Frankfurt a. M. and Mannheim.
Mayer, K. U. (1984b). Lebensverläufe und Wohlfahrtsentwicklung. Antrag auf Förderung für die dritte Forschungsphase 1985–1987 (pp. 131–171), Frankfurt a. M. and Mannheim.
Mayer, K. U. (1986). Lebensverlaufsforschung. In W. Voges (Ed.), *Soziologie der Lebensalter. Methoden der Biographie und Lebenslaufforschung.* Opladen: Leske & Budrich.
Mayer, K. U., & Wagner, M. (1986). Wann verlassen die Kinder ihr Elternhaus? Untersuchungen zu den Geburtsjahrgängen 1929–31, 1939–41, 1949–51. IBS-Materialien, Institut für Bevölkerungsforschung und Sozialpolitik, University of Bielefeld.
Mayer, K. U., & Tuma, N. B. (Eds.) (1987). Application of event history analysis in life course research. Materialien aus der Bildungsforschung No. 30, Max Planck Institute for Human Development and Education, Berlin.
Mayer, K. U., & Tuma, N. B. (Eds.) (1988). *Event history in life course research.* Madison, WI: University of Wisconsin Press.
Mayer, K. U., et al. (1987). Lebensverläufe und Wohlfahrtsentwicklung. Materialien zur Konzeption, Design und Methodik der Hauptuntersuchung 1981/82. Max Planck Institute for Human Development and Education, Berlin.
McCullagh, P., & Nelder, J. A. (1983). *Generalized linear models.* London: Chapman Hall.

Meulemann, H., et al. (1984). Bildung und Lebenslauf. Antrag auf Gewährung einer Sachbeihilfe (Neuantrag). Manuscript.
Michael, R. T., & Tuma, N. B. (1985). Entry into marriage and parenthood by young men and women: The influence of family background. *Demography, 22,* 515–544.
Miller, R. G. (1976). Least squares regression with censored data. *Biometrika, 63,* 449–464.
Miller, R. G. (1981). *Survival analysis.* New York: Wiley.
Müller, W. (1978). Klassenlage und Lebenslauf. Habilitationthesis, Mannheim.
Naes, T. (1982). The asymptotic distribution of the estimator for the regression parameter in Cox's regression model. *Scandinavian Journal of Statistics, 9,* 107–115.
Namboodiri, K., & Suchindran, C. M. (1987). *Life table techniques and their applications.* Orlando. FL: Academic Press.
Newman, J. L., & McCulloch, C. E. (1984). A hazard rate approach to the timing of births. *Econometrica, 52,* 939–961.
Papastefanou, G. (1980). Zur Güte von retrospektiven Daten—Eine Anwendung gedächtnispsychologischer Theorie und Ergebnisse einer Nachbefragung. Working paper No. 29 of the SfB 3 "Mikroanalytische Grundlagen der Gesellschaftspolitik," Frankfurt a. M. and Mannheim.
Papastefanou, G. (1986). Veränderungen der Familienbildung in der Bundesrepublik seit dem zweiten Weltkrieg. Manuscript, Berlin.
Papastefanou, G. (1987). Gender differences in family formation: Modelling the life course specificity of social differentiation. In K. U. Mayer & N. B. Tuma (Eds.), *Applications of event history analysis in life course research.* Materialien aus der Bildungsforschung No. 30, Max Planck Institute for Human Development and Education, Berlin.
Papastefanou, G. (1988). Familiengründung und Lebensverlauf. Eine empirische Analyse sozialstruktureller Bedingungen der Familiengründung bei den Kohorten 1929–31, 1939–41 und 1949–51. Dissertationthesis, Free University of Berlin.
Petersen, T. (1985). Incorporating time-dependent covariates in models for analysis of duration data. CDE Working paper.
Petersen, T. (1988). Specification and estimation of continuous state space hazard rate models. In C. C. Clogg (Ed.), *Sociological methodology.* San Francisco (forthcoming).
Prentice, R. L. (1975). Discrimination among some parametric models. *Biometrika, 62,* 607–614.
Prentice, R. L., & Breslow, N. E. (1978). Retrospective studies and failure time models. *Biometrika, 65,* 153–158.
Prentice, R. L., & Self, S. G. (1983). Asymptotic distribution theory for Cox-type regression models with general relative risk form. *Annals of Statistics, 11,* 804–813.
Prentice, R. L., et al. (1978). The analysis of failure time in the presence of competing risks. *Biometrics, 34,* 541–554.
Rao, C. R. (1973). *Lineare statistische Methoden und ihre Anwendungen.* Berlin: Academic-Verlag.
Robinson, B. N., et al. (1980). *SIR, scientific information retrieval, user's* manual, version 2. Evanston, IL: SIR Inc.
Roger, J. H., & Peacock, S. D. (1983). Fitting the scale as a GLIM parameter for Weibull, extreme value, logistic and log-logistic regression models with censored data. *GLIM-Newsletter, 6,* 30–37.
Sandefuhr, G. D., & Scott, W. J. (1981). A dynamic analysis of migration: An assessment of the effects of age, family, and career variables. *Demography, 18,* 355–368.
Schaich, E., & Hamerle, A. (1984). *Verteilungsfreie statistische Prüfverfahren.* Heidelberg and Berlin: Springer.
Schneider, H., & Weissfeld, L. (1986). Estimation in linear models with censored data. *Biometrika, 73,* 741–745.
Schoenfeld, D. (1980). Goodness-of-fit tests for the proportional hazards regression model. *Biometrika, 67,* 145–154.
Schoenfeld, D. (1982). Partial residuals for the proportional hazards regression model. *Biometrika, 69,* 239–241.
Schulz, M., & Strohmeier, K. P. (1985). Familienkarriere und Berufskarriere. In H.-W. Franz (Ed.), *22. Deutscher Soziologentag 1984, Beiträge der Sektions- und Ad-hoc-Gruppen* (pp. 167–171). Opladen: Westdeutscher Verlag.
Seal, H. L. (1977). Studies in the history of probability and statistics. Multiple decrements or competing risks. *Biometrika, 64,* 429–439.

Sørensen, A., & Sørensen, A. B. (1986). An event history analysis of the process of entry into first marriage. In D. I. Kertzer, *Family relations in life course perspective.* Greenwich, CT: JAI Press.
Sørensen, A. B. (1979). A model and a metric for the analysis of the intragenerational status attainment process. *American Journal of Sociology, 85,* 361–384.
Sørensen, A. B. (1984). Interpreting time dependency in career processes. In A. Diekmann & P. Mitter (Eds.), *Stochastic modelling of social processes* (pp. 89–122). New York: Academic Press.
Sørensen, A. B. (1988). Employment sector and unemployment processes. In K. U. Mayer & N. B. Tuma (Eds), *Event histories in life course research.* Madison, WI: University of Wisconsin Press.
Sørensen, A. B., & Sørensen, A. (1983). Modeling interdependence of life course events with event history data. Paper prepared for the meeting of the American Sociological Association, Detroit, September.
Sørensen, A. B., & Sørensen, A. (1986). An event history analysis of the process of entry into first marriage. *Current Perspectives on Aging and Life Cycle, 2,* 53–71.
Sørensen, A. B., & Tuma, N. B. (1981). Labor market structures and job mobility. *Research in Social Stratification and Mobility, 1,* 67–94.
Spence, A. M. (1973). Job market signaling. *Quarterly Journal of Economics, 87,* 355–374.
Spence, A. M. (1974). *Market signaling.* Cambridge, MA: Harward University Press.
Stoer, J. (1976). *Einführung in die numerische Mathematik.* 2nd ed., Berlin: Springer.
Strohmeier, K. P., Schultz, M., & Kaufmann, F.-X. (1985). Modellierung und Mikrosimulation von Prozessen der Familienentwicklung. Bericht aus dem Projekt "Generatives Verhalten in Nordrhein-Westfalen." In J. Schmid & K. Schwarz (Eds.), *Politische und prognostische Tragweite von Forschungen zum generativen Verhalten.* Berlin: Deutsche Gesellschaft für Bevölkerungsforschung.
Tarone, R. E., & Ware, J. (1977). On distribution-free tests for equality of survival distributions. *Biometrika, 64,* 156–160.
Tölke, A. (1980). Zuverlässigkeit retrospektiver Verlaufsdaten—Qualitative Ergebnisse einer Nachbefragung. Working paper No. 30 of the SfB 3 "Mikroanalytische Grundlagen der Gesellschaftspolitik," Frankfurt a. M. and Mannheim.
Treiman, D. J. (1977). *Occupational prestige in comparative perspective.* New York: Academic Press.
Trussel, J., & Richards, T. (1985). Correcting for unmeasured heterogeneity in hazard models using the Heckman-Singer procedure. In N. B. Tuma (Ed.), *Sociological methodology* (pp. 242–276). San Francisco: Jossey Bass.
Tsiatis, A. A. (1981). A large sample study of Cox's regression model. *Annals of Statistics, 9,* 93–108.
Tuma, N. B. (1985). Effects of labor market structure on job-shift patterns. In J. J. Heckman & B. Singer (Eds.), *Longitudinal analysis of labor market data.* Cambridge, MA: Cambridge University Press.
Tuma, N. B., & Hannan, M. T. (1984). *Social dynamics: Models and methods.* New York: Academic Press.
Tuma, N. B., Hannan, M. T., & Groeneveld, L. P. (1979). Dynamic analysis of event histories. *American Journal of Sociology, 84,* 820–854.
Vaupel, J. W., Manton, K. G., & Stallard, E. (1979). The impact of heterogeneity in individual frailty on the dynamics of morality. *Demography, 16,* 439–454.
Wagner, M. (1987a). Räumliche Mobilität im Lebensverlauf. Eine empirische Untersuchung sozialer Bedingungen der Migration. Dissertationthesis, Free University of Berlin.
Wagner, M. (1987b). Bildung und Migration. *Raumforschung und Raumordnung, 45* (3), 97–106.
Wagner, M., & Mayer, K. U. (1988). Wann verlassen Kinder ihr Elternhaus? Untersuchungen zu den Geburtsjahrgängen 1929–31, 1939–41, 1949–51. In A. Herlth & K. P. Strohmeier (Eds.), *Lebenslauf und Familienentwicklung.* Leverkusen: Leske & Budrich.
Wegener, B. (1985). Gibt es Sozialprestige? *Zeitschrift für Soziologie, 14,* 209–235.
Witting, H., & Nölle, G. (1970). *Angewandte mathematische Statistik.* Stuttgart: Teubner.
Wu, L. (1988). Simple graphical goodness-of-fit tests for hazard rate models. In K. U. Mayer & N. B. Tuma (Eds.), *Event histories in life course research.* Madison, WI: University of Wisconsin Press.

Index